An Introduction to
ANIMAL PHYSIOLOGY

FIG. 7.18. Toads with their chromatophores in different states. The two had been under similar conditions for an hour before they were photographed, except that the dark one was on a black background and the light one on a white background. They were placed together on a white background to be photographed. Photograph by P. A. Davenport.

An Introduction to ANIMAL PHYSIOLOGY

BY W. B. YAPP

THIRD EDITION

OXFORD
AT THE CLARENDON PRESS
1970

Oxford University Press, Ely House, London W. 1
GLASGOW NEW YORK TORONTO MELBOURNE WELLINGTON
CAPE TOWN SALISBURY IBADAN NAIROBI DAR ES SALAAM LUSAKA ADDIS ABABA
BOMBAY CALCUTTA MADRAS KARACHI LAHORE DACCA
KUALA LUMPUR SINGAPORE HONG KONG TOKYO

FIRST EDITION 1939
SECOND EDITION 1960
THIRD EDITION 1970

'Come, tell me how you live,' I cried,
'And what it is you do.'

The White Knight

PRINTED IN GREAT BRITAIN

Preface to Third Edition

ALTHOUGH interests in animal physiology have changed, and our knowledge of it has greatly increased, in the thirty years since the first edition of this book was written, the ways in which the animal body works have not. I have therefore rewritten almost the whole of the text, but have retained the basic pattern. Few paragraphs remain from the first, or even from the second, edition. I have added an introductory chapter on the material background, and to make room for this have reluctantly removed the chapter on behaviour. Research in this field has tended to become less physiological, and other good texts are available.

These changes, and some alterations which, on our present knowledge, seem logical, have meant that the paragraphs have had to be renumbered. In doing this, I have preceded each number by a letter which indicates the nature of the matter discussed. 'G' shows that it is of general application to all animals, or is comparative; 'M', that it deals with mammals; 'V', that it deals with vertebrates other than mammals, and 'J' that it deals with invertebrates. While I hope that zoologists will still find it worth while to read the book straight through, this coding should enable more specialized students to pick out what they want.

Thirty years ago I thought that it was possible for one man to read almost all that was published, at least in English, on what was then called experimental zoology. If it was true then, it is certainly no longer so, and I have had to rely much more than I like on secondary sources. There is no lack of these, both reviews and symposia. Unfortunately, while the words they use are usually English, the sentences in which the words are arranged often are not. This makes it difficult for an outsider to know what is meant, and I apologize to my readers (but not to those authors who do not trouble to express themselves clearly) if I have misinterpreted anything. The creation of jargon is a useful part of the development of any science, but it can be overdone, and some recent physiological constructions, such as *pheromone*, are painful hearing to anyone with a sense of language. I have avoided such barbarous words wherever there is a suitable alternative.

All statements of physical quantities are in accordance with Système International d'Unités (SI). This means the loss of some old friends, such as the calorie, but as the system will in a few years be the only one accepted in most scientific journals, students cannot begin to become familiar with it too early. A note on some unfamiliar units is given below.

As before, many friends have given me advice and help, and I thank them all. Once again, Professor W. H. Thorpe gave me much assistance with biochemistry; Mrs. M. I. Smith read the typescript and gave me valuable advice from her experience as a teacher. Acknowledgement to those who have helped with illustrations is given in the legends. I apologize if I have inadvertently omitted any.

Finally, I must apologize to my readers that, through no fault of mine, the new edition has been so long in coming.

W. B. Y.

Church End,
October 1969

Preface to Second Edition

IT is twenty-one years since this book was written, but in spite of the great advances that have been made in our knowledge of some aspects of physiology, it has worn well and has sold steadily throughout its life. I began work on a revision two years ago, and I apologize to my readers for the delay in the preparation of a new edition. Many parts have been completely rewritten, and the whole has been brought as nearly up-to-date as I have found possible. No doubt the book contains errors, but I hope that there is little in it that the conscientious student will later have to unlearn. In spite of some arguments to the contrary, I have left the arrangement of the matter much as it was before, with only a little renumbering of the sections and some subdivision. I hope the second edition will bring me as many friends as did the first.

My colleague Mr. H. Asher gave me much help with the section on sense-organs, and Dr. R. W. Murray and Dr. B. Schofield patiently answered my inquiries on particular points. Dr. W. V. Thorpe read the biochemical portions in manuscript. To all of these I am grateful.

W. B. Y.

Glenridding
April 1958

From the Preface to the First Edition

I AM well aware that in attempting to cover in one elementary volume almost the whole range of animal physiology, I am treading on ground where no English angel has been before me. I wish that it were otherwise, for the scope of physiology is so vast that few men can claim personal acquaintance with more than one or two small parts of it, and the risk of error is great. But in the country of the blind the one-eyed man is king, and I have written this book with the object of giving students a fuller and more accurate account of animal physiology than is at present available.

The book was written not for examination requirements but because I believe that young zoologists should know more physiology than they usually do. Nevertheless, it will, I hope, be useful for the First M.B., Higher School Certificate,[1] and College Scholarship examinations, for all of which an increasing knowledge of physiology is demanded. In addition it should serve for junior university students.

No author could have been more fortunate in his helpers than I have been. To all of them I offer my thanks, but especially to Dr. G. R. de Beer and Mr. P. F. Haggart, both of whom read the whole manuscript and left it better than they found it; to Mr. C. W. Carter for reading the book in proof; to Mr. J. Z. Young for reading section 5, and to Professor C. M. Yonge for reading most of section 1; to Professor J. H. Orton, Professor A. D. Ritchie, Mr. H. Lob, Mr. D. M. Hall, and Mr. E. G. MacGregor for advice on particular points; and to Mr. William Holmes for advice and for assistance with the Index.

Lastly, I must record my thanks for the great help I have received from the officials of the Clarendon Press.

W. B. Y.

Manchester
December 1938

[1] Now (1969) replaced by the General Certificate of Education Advanced level.

Contents

1. THE MATERIAL BACKGROUND

G 1.1. PROTOPLASM AND CELLS 1
G 1.11. Chemical analysis 1
G 1.111. Proteins 1
G 1.112. Fats 4
G 1.113. Carbohydrates 5
G 1.12. Physical structure 5
G 1.13. Organization as cells 7

G 1.2. FINE STRUCTURE 8
G 1.21. Electron micrographs and their interpretation 8
G 1.22. Robertson membranes 9
G 1.23. The nucleus 11
G 1.24. Inclusions 12

G 1.3. CHEMICAL PROCESSES 13
G 1.31. Energetics 13
G 1.32. Enzymes and catalysts 15
G 1.33. Movement 18
G 1.331. Physical movement 18
G 1.332. Chemical movement; assisted transport 21
G 1.333. Pinocytosis 22

2. NUTRITION

G 2.1. THE FOODSTUFFS: THEIR NATURE AND SOURCE 24
G 2.11. Proteins 25
G 2.12. Carbohydrates 27
G 2.13. Fats 28
G 2.14. Other organic compounds 29
M 2.141. The B-vitamins in mammals 30
M 2.142. Other vitamins in mammals 34
V 2.14. Vitamins in other vertebrates 37
J 2.14. Vitamins in invertebrates 37
G 2.15. Inorganic substances 39

G 2.2. FEEDING AND FEEDING MECHANISMS 40
G 2.21. Microphagous feeders 40
G 2.22. Macrophagous feeders 43
G 2.23. Detritus feeders 46

G 2.24. Fluid feeders 46
G 2.25. Symbiosis 46

G 2.3. DIGESTION 47
M 2.31. Digestion in the mammal, especially man 48
M 2.311. Digestion of proteins 49
M 2.312. Digestion of carbohydrates 52
M 2.313. Digestion of fats 55
M 2.314. Absorption 55
M 2.315. The movement of the food and control of digestion 58
V 2.31. Digestion in other vertebrates 60
J 2.31. Digestion in invertebrates 62

3. RESPIRATION AND TRANSPORT

G 3.1. THE RESPIRATORY QUOTIENT 72

G 3.2. RESPIRATORY ORGANS 73
M 3.21. The lungs and breathing of mammals 74
V 3.21. Breathing of other vertebrates: lungs 75
V 3.22. Breathing of other vertebrates: gills 76

J 3.2. BREATHING AND RESPIRATORY ORGANS OF INVERTEBRATES 78

G 3.3. COMPARISON OF AQUATIC AND AERIAL RESPIRATION 81

G 3.4. TRANSPORT OF OXYGEN AND CARBON DIOXIDE BY A CIRCULATORY SYSTEM 82
M 3.41. The haemoglobin of mammals 85
V 3.41. The blood of other vertebrates 89
J 3.41. The blood of invertebrates 90
MV 3.42. Transport of carbon dioxide in mammals and other vertebrates 93
J 3.42. Transport of carbon dioxide in invertebrates 94

G 3.5. OTHER FUNCTIONS AND PROPERTIES OF A CIRCULATORY SYSTEM 94

G 3.6. CONTROL OF THE CIRCULATORY SYSTEM 96

J 3.7. TRANSPORT BY TRACHEAE 100

G 3.8. LIFE IN LOW CONCENTRATIONS OF OXYGEN 101

4. METABOLISM AND CELLULAR PHYSIOLOGY

G 4.0. Energy and Synthesis 106

M 4.1. Metabolism in Mammals 107
M 4.11. Metabolism of proteins and nitrogen 107
M 4.12. Metabolism of carbohydrates 114
M 4.13. Metabolism of fats 125
M 4.14. Interrelationships of carbohydrates, fats, and proteins 128

V 4.1. Metabolism in other Vertebrates 129

J 4.1. Metabolism in Invertebrates 132

5. EXCRETION

G 5.01. Substances excreted 135
G 5.02. Methods of excretion 136

MV 5.1. The Vertebrate Kidney 137

MV 5.2. Other Vertebrate Excretory Organs 144

J 5.3. Excretion in Invertebrates 145
J 5.31. Excretion in Protozoa 146
J 5.32. Excretion in Coelenterata 147
J 5.33. Excretion in unsegmented worms 148
J 5.34. Excretion in Annelida 148
J 5.35. Excretion in Arthropoda 150
J 5.36. Excretion in Mollusca 153
J 5.37. Excretion in Echinodermata 154
J 5.38. Excretion in Hemichordata 154

6. COORDINATION OF FUNCTION

G 6.0. Introduction 155

G 6.1. Hormones 156

MV 6.1. Hormones in Vertebrates 159
MV 6.11. The gut and pancreas 159
MV 6.12. The pharynx 159
MV 6.121. The thyroid 159
MV 6.122. The parathyroids 162
MV 6.123. Other pharyngeal derivatives 162
MV 6.13. The pituitary 163
MV 6.131. The adenohypophysis 163
M 6.132. The neurohypophysis of mammals 166
V 6.132. The neurohypophysis of other vertebrates 167
V 6.133. The urophysis 168

MV 6.14. The suprarenals or adrenals 168
M 6.141. The adrenal medulla of mammals 168
V 6.141. Chromaffin tissue of other vertebrates 169
M 6.142. The adrenal cortex of mammals 170
V 6.142. The adrenal cortex of other vertebrates 170
MV 6.15. The pineal 170
MV 6.16. The gonads and related structures 171

J 6.1. HORMONES IN INVERTEBRATES 171
J 6.11. Hormones in annelids and molluscs 171
J 6.12. Hormones in crustacea 172
J 6.13. Hormones in insects 173

G 6.2. ECTOHORMONES 175

G 6.3. THE NERVOUS SYSTEM 176
G 6.30. General properties of a nervous system 176
MV 6.31. Propagation of the nervous impulse in vertebrates 178
J 6.31. Propagation of the nervous impulse in invertebrates 183
MV 6.32. The synapse in vertebrates 185
J 6.32. The synapse in invertebrates 188
G 6.33. The central nervous system 189
G 6.34. The nerve net 190

6.4. SENSE ORGANS 192
G 6.40. General properties of sense organs 192
MV 6.41. Proprioceptors in vertebrates 197
J 6.41. Proprioceptors in invertebrates 199
MV 6.42. The senses of the skin: touch, temperature, and pain in vertebrates 199
J 6.42. Touch, temperature, and pain in invertebrates 202
MV 6.43. Chemoreception in vertebrates 203
MV 6.431. Taste in vertebrates 204
MV 6.432. Smell in vertebrates 205
MV 6.433. The common chemical sense in vertebrates 207
MV 6.434. Osmoreception and internal chemoreception in vertebrates 207
J 6.43. Chemoreception in invertebrates 208
MV 6.44. Receptors for balance and movement in vertebrates 211
J 6.44. Receptors for balance and movement in invertebrates 217
M 6.45. Hearing in mammals 218
V 6.45. Hearing in other vertebrates 221
J 6.45. Hearing in invertebrates 222
G 6.46. Photoreception 224

M 6.46. Vision in mammals 225
V 6.46. Vision in other vertebrates 231
J 6.46. Vision in invertebrates 233
V 6.47. The electrical sense 236

7. EFFECTORS

MVJ 7.1. Cilia and Flagella 238

MVJ 7.2. Pseudopodia 244

7.3. Muscle 247
MV 7.31. Vertebrate skeletal muscle 247
MV 7.32. Vertebrate smooth muscle 255
MV 7.33. Vertebrate cardiac muscle 258

J 7.3. Invertebrate Muscle 260

G 7.4. Glands 266

V 7.5. Electric Organs 267

VJ 7.6. Luminescent Organs 269

J 7.7. Nematocysts 271

G 7.8. Chromatophores 273

V 7.8. Chromatophores in Vertebrates 274

J 7.8. Chromatophores in Invertebrates 278

8. REPRODUCTION

G 8.1. Asexual Reproduction 280
G 8.11. Fission 280
G 8.12. Fragmentation 281
G 8.13. Gemmation or budding 283

8.2. Sexual Reproduction 284
G 8.21. Gametes and fertilization 284
G 8.22. Parthenogenesis and gynogenesis 287
M 8.23. Sexual reproduction in mammals 289
V 8.23. Sexual reproduction in other vertebrates 299
J 8.23. Sexual reproduction in invertebrates 302
G 8.24. Hermaphroditism 305
G 8.25. Viviparity; birth rate 307

9. REGULATION

9.1. CHEMICAL REGULATION: THE COMPOSITION OF BODY-FLUIDS 309
G 9.11. Water 309
G 9.12. Salts 310
G 9.121. Osmotic regulation 310
G 9.122. Ionic regulation 314
G 9.13. Hydrogen-ion concentration 315

9.2. TEMPERATURE 317

M 9.2. TEMPERATURE CONTROL IN MAMMALS 319

V 9.2. TEMPERATURE CONTROL IN BIRDS 323

J 9.2. TEMPERATURE CONTROL IN INVERTEBRATES 325

G 9.3. BEHAVIOURAL ADAPTATIONS 325
MV 9.31. Hibernation 326
J 9.32. Diapause 328
G 9.33. Migration 328
G 9.34. Daily rhythm 331

INDEX 333

Some units in SI

Physical quantity	Name of unit	Symbol	Definition	Approximate equivalent
force	newton	N	kg m s^{-2}	0·1 kilogramme-force (conveniently, 'the weight of an apple')
energy	joule	J	kg m^2 s^{-2} (= N m)	0·24 calories
power	watt	W	kg m^2 s^{-3} (= J s^{-1})	0·0013 horse-power (= 0·9 kcal per hour)
pressure	newtons per square metre		N m^{-2}	0·0075 millimetres of mercury
luminous intensity	candela	cd	basic unit	one candle power
luminance (brightness)	candela per square metre		cd m^{-2}	0·3 millilambert
frequency	hertz	Hz	cycles per second	

1. The Material Background

G 1.1. PROTOPLASM AND CELLS

G 1.11. Chemical Analysis

We have long known that protoplasm, the active material of which the living body is made, is mostly water. Suspended in this, or in true or colloidal solution, are other substances that together make up about 30 per cent of the total mass. Crude chemical analysis shows that of them about 60 per cent consists of protein, while the remainder consists of carbohydrates, fatty substances or lipids, other organic substances, and inorganic materials in varying proportions.

Ordinary chemical analysis kills the protoplasm, and there is always the possibility that in so doing it to some extent brings about chemical changes. Various ingenious methods have therefore been devised to find out how far the information obtained applies to the living material and not only to its dead counterpart. Some chemical reactions, for example with dyes, will go on with protoplasmic constituents without damaging them; material may be extracted, examined, and added back to restore its function; and physical means, such as measurements of refractive index, may, by comparison with artificially prepared mixtures, show probable identity of structure. By means such as these much has been found out about the particular proteins and other compounds present, and about their distribution in the protoplasm.

G 1.111. Proteins. When proteins are hydrolysed they give smaller molecules and eventually end as a number of nitrogen-containing acids. A few, such as proline, are iminoacids, of general formula HN=R—COOH, but the majority are aminoacids, of the general formula H_2N—R—COOH, where R is CH_2 or some other group with two spare valencies. Since H_2N— is basic and —COOH is acidic, such a compound is amphoteric, and should be able to combine with other similar compounds. This it will do (though not directly), the reaction being

$$H_2N \cdot R^1 \cdot CO\boxed{OH + H \cdot}N \cdot H \cdot R^2 \cdot COOH$$

$$\longrightarrow H_2N \cdot R^1 \cdot CONH \cdot R^2 \cdot COOH + H_2O.$$

The grouping —CONH— thus formed is called the peptide link, and a compound formed from two aminoacids and containing one such link is called a dipeptide. The molecule is still amphoteric, and so further condensations, leading to tripeptides and polypeptides, are possible.

The primary structure of a protein is thus a relatively simple chain. The actual aminoacids and their order have been worked out for several proteins. Two important, and opposite, results have come from this type of study. One is that a small alteration in the chain may greatly alter the ability of the protein to carry out its function. The disease sickle-cell anaemia, for example, which is often fatal, is due to the substitution of valine for glutamic acid at one position in the chain of 300 acids that make up the molecule of haemoglobin. The other is that most of the aminoacids in a chain are biologically inert, having little effect on the properties of the protein. The differing reactions of the various cytochromes, for example (section M 4.12), or of trypsinogen and chymotrypsinogen (section M 2.311), depend on only a few aminoacids.

The primary structure of a protein is that of classical chemistry, in which the bonds are covalent; the result is a chain which over all is straight, but which consists in detail of a regular zigzag depending on the angles of the bonds of carbon and nitrogen, thus

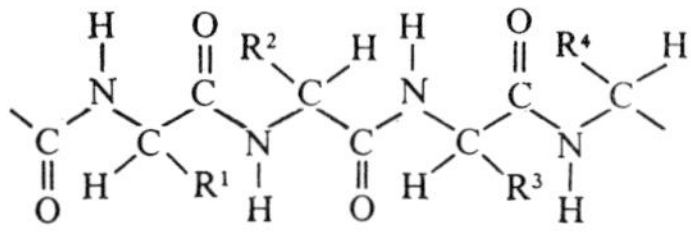

```
        H     O  R²      H  H      O  R⁴      H
        |     ||   \    /   |      ||   \    /
        N     C     C       N      C     C
      /   \  /  \  /  \    /  \   /  \  /  \
  \  C     C     N      C      C     N      \
     ||  H/  \R¹ |      ||   H/  \R³ |
     O           H      O            H
```

Other covalent linkages, such as the diacyl, —CO(—CO)N—, and the diamino-carbinol, —C(OH) (—NH—)—NH—, may perhaps be present. Many proteins have also a secondary structure, in which bonds are formed between the nitrogen of one peptide link and the oxygen of another; the hydrogen atom is shared between them to form a hydrogen bond. The effect of this is to pull the straight chain into a helix. Keratin (the protein of the epidermis), for example, normally has the structure shown in Fig. 1.1 in which the hydrogen bonds are formed between every third peptide link, giving a ring of thirteen atoms. When such a structure is pulled the hydrogen bonds break and the molecule is stretched, so that keratin is elastic. In collagen, of which white fibres are made, the hydrogen bonds are between aminoacid chains lying coiled round one another to form a cable, and the molecule is inelastic.

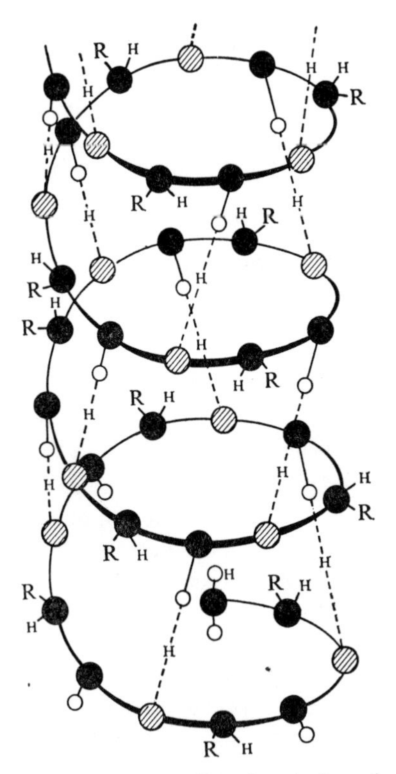

FIG. 1.1. Diagrammatic representation of part of a polypeptide α-helix to show hydrogen bonding. The black disks represent the carbon and the hatched disks the nitrogen atoms of the main polypeptide chain. R = an aminoacid less —$CH(NH_2)COOH$; H = hydrogen, those atoms that are involved in hydrogen bonds (dotted) being drawn larger than the others; O = oxygen of a carboxyl group. The curved bonds of the main chain are drawn thicker in the part of the helix nearer the reader. From any nitrogen atom of the main chain back to itself via the chain and one hydrogen bond there is a ring of thirteen atoms. From Thorpe, *Biochemistry for medical students*, Churchill (1964).

In many proteins there is a tertiary structure, in which globular and other shapes are formed by various types of link. One of the most important of these is the disulphide or cystine link, in which two atoms of sulphur are joined by a covalent bond, thus, —S—S—. Reduction of this will give two cysteine groups, —SH, and the reactive properties of the protein are often altered by such a change. Insulin, for example (section M 4.12) is active only in the oxidized form.

Some proteins polymerize, or aggregate together by weak electrostatic forces, to give a quarternary structure, which is readily destroyed by mild changes in the conditions. Mammalian haemoglobin, for example, breaks down reversibly into four similar subunits if the hydrogen ion concentration goes more than a short way from neutrality.

Many proteins are associated with a non-protein structure, called the prosthetic group, to give a complex molecule. The important properties of biological substances, such as haemoglobin and enzymes, often depend on the unique association of a protein with its prosthetic group.

Chains of up to fifty aminoacids, giving a molecular weight of about 6000, are usually called polypeptides; above that the word protein is used. The lengths of the bonds and the diameters of the electron-clouds of the various atoms are more or less constant, but the over-all dimensions of the molecule will obviously depend on the coiling and packing. Mammalian myoglobin (section M 3.41) has a molecular weight of 17 000 and is about $4 \times 3{\cdot}5 \times 2$ nm. Mammalian haemoglobin has a molecular weight of 68 000 and is $6{\cdot}5 \times 5{\cdot}5 \times 5$ nm. Fibrinogen, one of the proteins in solution in mammalian blood, has a molecular weight of 330 000 and is approximately $6 \times 6 \times 28$ nm.

With relatively small changes in temperature, acidity, or concentration, protein molecules may join together to form a solid network, so that there is a change from a colloidal solution of solid in liquid (sol) to one of liquid in solid (gel). Such gelation may be reversible, as in the setting of a gelatine jelly on cooling, or irreversible, as in the coagulation of white of egg on heating.

G 1.112. Fats. Definition of 'fat' is difficult. The usual statement of the chemistry books, that a fat is a glycerol ester of a higher fatty acid, is clearly tautologous, for the term fatty acid means simply an acid derived by hydrolysis of a fat. In an everyday sense, a fat is a greasy substance from an animal body, which melts at a temperature near or a little above that of the human body; the corresponding substances from plants, which are liquid at ordinary temperatures, are oils. These characteristics do not define the materials precisely and the word 'oil' is used also for substances of similar physical properties but of different chemical nature, the mineral oils. In biochemical literature, there is a tendency to use the word 'lipid' in a broad sense, and to restrict 'fat' to simple triglycerides, but since 'lipid' is derived from the Greek word for fat, and few if any authors are consistent in their usage, the simpler English term is used here for all fatty substances. A fat is always an ester, and the acid from which it is derived has a chain of several carbon atoms, usually in the teens. The base is usually glycerol, but other polyhydric alcohols occur. The term can be extended to cover compounds the main part of whose molecule is a fat, but which have other groups; of these phosphate is the commonest, and fats containing it are called phospholipids. As the length of the carbon chain and its degree of saturation rise, so does the melting point, so that there is a progression from what are ordinarily called oils through common fats to waxes. The last generally have bases other than glycerol in the molecule. The term 'lipid' is sometimes extended to include the acids and alcohols into which fats can be hydrolysed.

Fats, like proteins, enter into the fundamental structure of the living body, and are especially important in membranes (section G. 1.22).

G 1.113. Carbohydrates. Carbohydrates, which are quantitatively the third great class of organic substances in the living body, are also difficult to define. They have many hydroxyl groups and so are alcohols, but other groups are often present. They usually have a heterocyclic ring consisting of several carbon atoms and one oxygen, and in the typical members the formula is $C_x(H_2O)_y$, from which their name is derived. Carbohydrates in the strict sense are usually in solution or in the form of insoluble granules, and contribute little to protoplasmic structure, but the term is used in a loose sense to include similar substances with substituted groups such as $—NH_2$, and the mucopolysaccharides in which a protein is incorporated in the molecule. Many of these form important boundary layers.

G 1.12. Physical Structure

Since protoplasm is largely water, one would expect that this would determine its physical properties, but it is difficult to realize how liquid it usually is. The viscosity of a liquid is defined as the force between unit planes unit distance apart moving relative to each other with unit velocity, so that its dimensions are $ML^{-1}T^{-1}$. That of protoplasm from many sources has been determined in various ways, such as by calculations from the Brownian movement of contained particles and by the movement of introduced iron filings in a magnetic field. Most of the results make the viscosity between about 0·02 and 0·5 poises, while that of water is about 0·01 poises. (For comparison, the viscosity of glycerine is 8 poises.) Since the values for protoplasm depend on the size and density of small particles, neither of which can be measured accurately, they are not very precise, but at least they show that, in general, protoplasm is nearer to water than to glycerine. In some experiments the viscosity varied from place to place, that of the ectoplasm of *Amoeba*, for example, being higher than that of the endoplasm. Of the few high values that have been obtained some have been contradicted by later work, and others may be due to gelation of the colloidal proteins or to the presence of fibres.

The viscosity of a liquid is one of the properties that determine how fast it will flow through a fine tube. For most liquids viscosity is independent of the rate of flow, so that there is a linear relationship of rate of flow to pressure. For some liquids, which are called thixotropic, the viscosity decreases with increasing shear-stress in the moving liquid, so that once flow has started the pressure necessary to maintain it is reduced. (This phenomenon will be familiar to anyone who has ever used an ordinary house paint.) When a liquid is driven through a tube under pressure, the shear-stress is applied at the perimeter, where the liquid tends to stick to the walls; this means that the viscosity of a thixotropic liquid in these conditions will be less round the edges of the tube than in the centre. Observations of the velocity of particles in pseudopodia of the large amoeba *Chaos chaos* suggest that its protoplasm is of this type, with viscosities ranging from about 0·4 poises at the edge to 9 poises in the centre.

The hydrogen ion concentration is usually very near to neutrality, but since there is an imbalance of other positively charged ions (section G 1.331) protoplasm is nearly always electro-negative to the surrounding medium.

In addition to protoplasm in the strict sense, in which complex chemical reactions are going on so that it may be described as living, the body contains much material that is inert and can be separated unchanged from the active part. This dead or secreted part may be internal, as are the droplets of yolk (of varied nature, but often phospholipid) in eggs, or it may be external to the protoplasm, as are the fibres of collagen or elastin (both proteins) that make up much of connective tissue. Animal protoplasm is not always surrounded by walls of this type, but it may be. The keratin formed by the epidermis of all vertebrates except teleosts is a good example; though very different in chemical composition it is no different in position or function from the cellulose of plant cells.

We have noted that the viscosity of protoplasm may differ from place to place, and in general it appears to be stiffer or more solid on the outside than away from the surface. *Amoeba* has long been known to have an outer wall, the plasmalemma, which has somewhat the consistency of dough; it can be pulled out with a needle and shows some elasticity. Such properties could be given by a partially gelated colloid.

G 1.13. Organization as Cells

Contrary to the belief of many chemists who dabble in biology, cells are not necessary for life and are in no sense fundamental. For half a century or more, there have been only two definitions of 'cell': 'a mass of protoplasm containing a nucleus, the whole surrounded by some sort of wall', or words to that effect, and the same phrase to which is added 'specialized for some purpose'. Whichever definition one uses, Fungi, some Algae, many Protozoa, and some tissues are not cellular. Nevertheless, cellular structure is so prevalent that we must presume that it is valuable. It seems to have two selective advantages. The fact that protoplasm is essentially liquid means that both the size and shape of a mass of it are strictly limited. If small, it will be spherical, and if large (except in a medium of the same specific gravity as itself) it will spread out into a flat sheet, and in any case it will be easily fragmented. Cell walls, by confining the protoplasm, allow both large size and almost any shape to develop. Secondly, cell walls, by their spatial separation of parts, allow specialization of function. This is presumably based on suppression or reduction of some chemical processes and initiation or development of others. Nerve cells have developed special properties that depend on the common ionic differential across the surface, while gland cells can synthesize proteins or other substances that are not made, at least in measurable quantities, by other cells.

In view of the importance of the cell wall it is unfortunate that little attention has been paid to it in recent years. We have seen that it may be secreted protein, and that even in the cells generally described as naked, such as *Amoeba*, there may be some specialization of the surface. The surface, whatever it is, may be highly resistant to the passage of materials through it. The contractile vacuole of *Amoeba* appears to fill and empty very rapidly, but calculations show that water is entering at the rate of only 30 $mm^3 m^{-2} year^{-1}$, which represents a very high degree of water-proofing.

G 1.2. FINE STRUCTURE

G 1.21. Electron Micrographs and their Interpretation

With the invention of the electron microscope in the late 1940s there became available magnifications up to about 1 000 000, much higher than the limit of about 2000 set by the wavelength of light. Objects cannot be seen under the electron microscope, but micrographs comparable to photographic negatives and prints, and of the same nature, can be made. In the interpretation of these great care must be used. In the first place, the prints are technically bad, especially at high magnifications; in this they resemble, for example, photographs of bombed cities, taken from aircraft at high altitudes and then 'blown up' to show as much detail as possible. The interpretation of such detail was difficult, and, as we now know, not always correct. Secondly, before living material can be used in the electron microscope it must be cut into sections of thickness of the order of 10 nm, and it is obvious that the liquid protoplasm that we have described cannot be treated in this way. It must first be made solid, that is, coagulated and killed, or in the microscopist's language, fixed. The specimen must be placed in a vacuum, which means that it must be dried, and often for any detail to be seen it must be impregnated with heavy metals to provide something that is opaque to the electron beam.

All this means that what we see in the electron micrograph may or may not represent what is present in the living protoplasm, and interpretation is especially difficult when different methods of treatment give different appearances. The same difficulty occurs under the light microscope, but here many of the appearances can be checked by dark ground and phase-contrast observation of living material. In reconstructing the real thing from electron micrographs much stress is laid on consistency of appearance, and especially on constant spatial relationships.

G 1.22. Robertson Membranes

One of the most consistent observations of electron microscopy is that naked cells are surrounded by a membrane which is, after treatment, electron-opaque. Under higher magnification and some types of fixation it appears as two opaque layers, each about 2 nm thick,

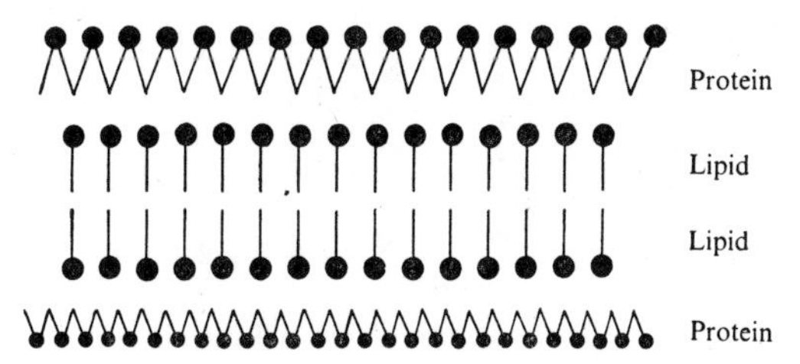

FIG. 1.2. Diagram of the structure of the cell membrane, based on electron micrographs and X-ray diffraction patterns.

separated by a transparent layer of about 3·5 nm. This membrane was called by Robertson, its discoverer, a 'unit membrane'. It has also been called a 'double membrane', but it is in fact triple (or perhaps more strictly quadruple, as we shall see shortly). All these names are unfortunate, confusing, and mutually contradictory.

When the cell surface is examined by X-rays it gives a diffraction pattern, which is similar to what would be given by two layers of lipid between two layers of protein, and the two proteins are different; possibly the outer layer contains carbohydrate. Lipid molecules are polar (that is, they can ionize at one end) and the diffraction patterns do not show whether their ionizable heads or their tails are together. It is possible to introduce water into the membrane, and when this is done each dark line in the electron micrograph is split into two. Since water attaches to the heads of the lipid molecules this means that they must be facing outward. The structure of a Robertson membrane is therefore as shown in Fig. 1.2.

It is interesting to compare this with the model of the cytoplasmic membrane proposed by Danielli and Davson on the basis of observations of the way in which various substances pass through the cell wall. This was first put forward in 1935; a later modification is shown in Fig. 1.3. The only differences between this and the Robertson membrane are that the latter has two different types of protein and that neither layer shows regular gaps. The possible existence of these is referred to in section G 1.331.

Electron micrographs sometimes suggest a further layer outside the Robertson membrane. It is probably largely mucopolysaccharide, and may be the light microscopist's 'basement membrane'. Where naked cells are closely packed, as in many epithelia, each has its own

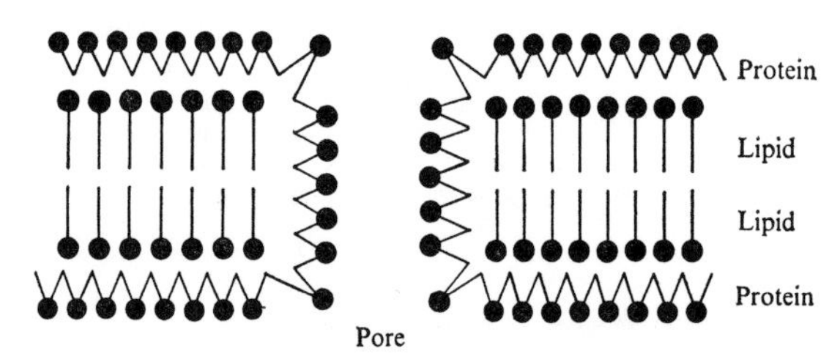

FIG. 1.3. Diagram of the structure of the cell membrane, based on its permeability.

Robertson membrane, and adjacent cells are separated by a layer that is transparent to electrons. Across this may stretch in places material that is slightly electron-dense. If it appears as short lengths in the electron micrograph (presumably representing cylinders) they are called desmosomes; if it appears to make rings round the cells, they are called terminal bars. At the terminal bars the outer layer of the Robertson membrane is not visible.

Where the cells are actively absorbing, as in parts of the intestine and kidney, their free surface is often drawn out into fine filaments called microvilli, about 1 μm long and 0·1 μm in diameter. In the intestine of the mouse there are about 600 per cell, or 5×10^7 per mm^2, and they are very regularly packed (Fig. 1.4). A surface with microvilli is called a brush border.

A special case of the Robertson membrane is given by the myelinated nerve fibre. The axon of an unmyelinated fibre is nearly surrounded by a Schwann cell, as in Fig. 1.5 (*a*), and each wall of this shows in an electron micrograph the dark–light–dark appearance of the typical membrane. In the myelinated fibre the Schwann cell has wrapped itself spirally round the axon; an early stage in this is shown in Fig. 1.5 (*b*). The cytoplasm of the Schwann cell disappears, and the sheath comes to consist of many layers of spirally wound Robertson membranes; hence its largely lipid nature. This type of structure, with layers of radially arranged lipid molecules separated by tangentially arranged protein molecules, was first shown by X-rays in 1935.

As we shall see in the next two sections, Robertson membranes bound many other structures, but in all except the cell surface the two protein layers appear to be similar.

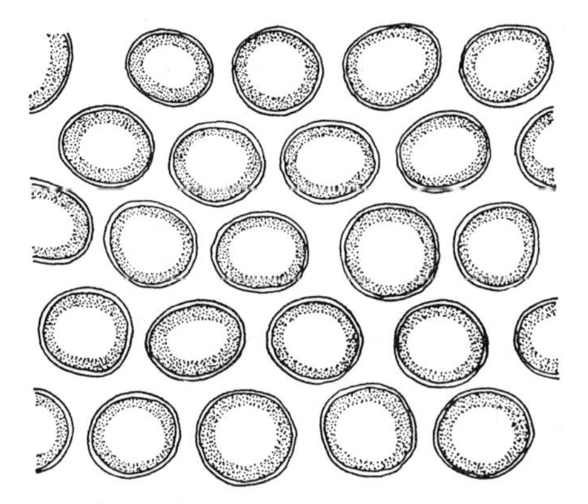

FIG. 1.4. Microvilli of rat intestine in transverse section, $\times 100\,000$. Notice the regular packing, and that each wall is a normal Robertson membrane. Redrawn from Haggis *et al.*, *Introduction to molecular biology*, Longmans (1964).

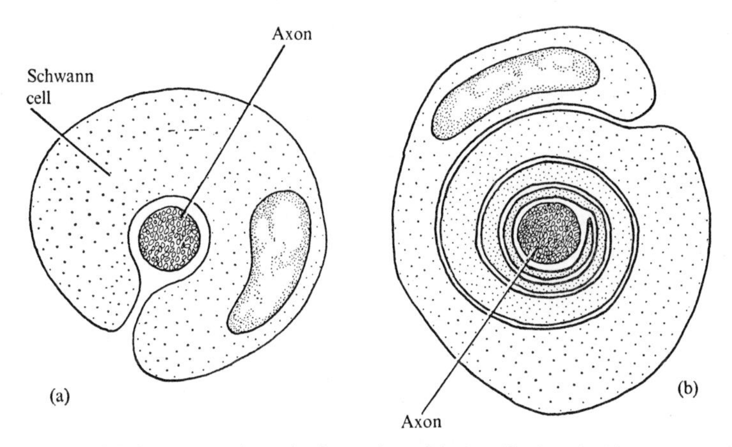

FIG. 1.5. Diagrams to show the formation of the myelin sheath. (*a*) a single axon in transverse section, nearly surrounded by its Schwann cell. (*b*) the Schwann cell has grown spirally round the axon to form the sheath. Later, the cytoplasm of the sheath disappears, so that it comes to consist only of the walls of the cell, which have the normal structure of a surface membrane.

G 1.23. The Nucleus

The 'nuclear membrane' of light microscopists consists of two membranes similar to those of the cell surface, separated by a gap of about 10 nm; the outer bears ribosomes (section G 1.24). Both appear to have many pores, about 50 nm in diameter, but in embryos and some adult tissues, and possibly more generally, these are closed by diaphragms. In young cells still capable of division the nuclear membrane is often continuous with the endoplasmic reticulum (section G 1.24) and so perhaps with the cell surface, but this connection is seldom seen in mature cells. At nuclear division, the membrane breaks up into vesicles, which appear to re-aggregate to form the new membranes in the daughter cells.

The main part of the nucleus, the nuclear sap or nucleoplasm, appears not to be greatly different, in physical or chemical properties, from the cytoplasm. Its most important contents, the chromosomes, consist of deoxyribonucleic acid (section M 4.11) attached to proteins. The molecule of deoxyribonucleic acid is a double helix about 2 nm in diameter, while the smallest chromosomes visible in the light microscope are about 100 nm thick, and the finest sub-threads shown by electron micrographs are about 20 nm. There must therefore be some sort of packing or reduplication of the molecules. The role of the acid in the synthesis of proteins is discussed in section M 4.11.

The disappearance of chromosomes between cell divisions remains unexplained. Occasionally phase-contrast microscopy of the interphase nucleus shows fine threads, which may be the chromosomes, attached to the nuclear membrane, but more usually the only structure visible is a nucleolus, which is a mass of granules of ribonucleic acid. Chemical analysis shows that the amount of deoxyribonucleic acid, and so presumably the size of the chromosomes, increases between divisions.

G 1.24. Inclusions

Various types of living inclusions, as distinct from secreted particles such as yolk granules and oil droplets, have long been recognized in cytoplasm, and some are visible even in the living cell. The electron microscope suggests that almost all of these are bounded by, or even consist of, membranes similar to those of the surface.

Mitochondria are bounded by two Robertson membranes, the inner of which is often folded inwards to make partitions or cristae. They contain many enzymes, and when they are separated from the rest of the cell by centrifuging, and then disrupted by high-frequency sound and extracted in aqueous solution, the insoluble fraction, presumably consisting of bits of the membrane, is found to contain cytochromes and flavoproteins (section M 4.12). The soluble fraction, presumably derived from the matrix, contains dehydrogenases and other enzymes of the tricarboxylic acid cycle (section M 4.12). This distribution agrees with the observation that in mitochondria from tissues with a high oxidative activity, such as muscle, the numbers of cristae are high. Many of the enzymes are present in fixed molecular proportions, suggesting that they are present in the surfaces in definite patterns.

Mitochondria divide, and the number in any one type of cell is approximately constant.

The endoplasmic reticulum is a folded membrane that appears to be continuous with that at the surface. It is studded with ribosomes, particles made up of about 60 per cent ribonucleic acid associated with several different protein molecules; they take part in the synthesis of proteins. Other membranes are less regular.

Many electron micrographs show the cell as being more or less filled with a tangle of membranes of one sort or another, and it is difficult to reconcile this structure with the liquid nature of protoplasm discussed above. Some of the membranes may be artefacts of fixation, but this cannot be true of the mitochondria (which are visible in living material) or the endoplasmic reticulum. The membranes may account for the thixotropic properties of protoplasm, and the small forces applied by particles falling through it may be enough to disrupt them and so reduce the viscosity. There is evidence from other sources that the lipid molecules of a membrane may be themselves moving about, so that the term membrane, in so far as it implies something fixed and solid, may be misleading. There may be, rather, a condensation of similar molecules of no greater strength than that of a layer of lipid on the surface of water. There could well be gaps which particles could easily enlarge and so force their way through. That something of this sort happens is suggested by observations of particles in *Amoeba* when the animal is centrifuged. At first they do not move at all, but as the speed is increased they begin to do so more or less suddenly, and then continue in jerks.

G 1.3. CHEMICAL PROCESSES

G 1.31. Energetics

The concept of energy, as developed in the nineteenth century, though usually taken for granted, is philosophically difficult to understand. That a falling body possesses some property, which we agree to call kinetic energy, by which it can break or dent anything that it hits, is simple enough; that the same body, before it starts falling, has as much potential energy as it afterwards possesses kinetic energy, is not so obvious. That heat is a form of energy, though suggested by ingenious experiments in the 1840s, is even less obvious (Joule's paper announcing his discoveries was rejected by all the scientific journals to which it was sent). We can, however, say that if we make the assumption that energy (measured as the product of force times distance, or ML^2T^{-2}), is never destroyed, but only converted into some other form, we can make many predictions that come true, not only qualitatively but quantitatively.

The concept of free energy is less familiar and even more difficult, not least because the term was used in one sense by Helmholtz (who invented it) and in another by Gibbs (who developed it), and most authors do not specify which they intend. The general principle is that of all the energy contained in a system only some is available for doing work; this is its free energy. We cannot measure either the total energy or the free energy absolutely, but we can sometimes measure changes in them. The free energy depends on the conditions of temperature, pressure, and so on, so that these should be stated. Changes that will reduce the free energy of a system tend to occur, although, owing to constraints that may be applied, they may not take place, just as a brick lying on a beam will not fall unless the beam is removed. Changes in the opposite direction will not occur unless energy is put into the system. If, as the result of a reaction, the free energy of the system is reduced, the reaction is called exergonic; if the free energy is increased, it is endergonic. Most of the free energy that is in a system can usually be made to appear as heat, so that reactions may also be described by the more familiar terms exothermic and endothermic, according to whether heat is given out or absorbed. The changes in free energy determine whether or not a reaction will go on, but for many reactions in the body the changes in the heat content are what matter to the animal. Moreover, while heat changes are easily measured experimentally, the changes in free energy have to be largely deduced, and are impossible to obtain for complex reactions. We shall therefore, in general, use 'exothermic' rather than 'exergonic' in this book. The quantitative difference is usually small; the heat liberated in the oxidation of carbohydrate is about 2 per cent less than the fall in free energy. That part of the free energy lost in a reaction which does not appear as heat may be assumed to be absorbed by a rearrangement of atoms or electrons.

The rates of most chemical reactions increase logarithmically with temperature, doubling for a rise of 10 or 12 °C. Biological reactions are no exception over moderate temperature ranges, but their rates fall off at higher temperatures, presumably because of the destruction of the enzymes on which they depend. The rates of activity of such processes as beat of cilia or the heart beat of invertebrates show similar temperature dependence.

Many chemical reactions in the body are coupled, that is to say two (or more) go on together; under these conditions very little of the free energy need appear as heat, for that liberated by one reaction may be taken up by the other; an exergonic reaction may be said to drive an endergonic one. In this way substances of high free energy may be formed, which can afterwards break down, liberating the free energy as heat or in doing work. As we shall see in section M 4.12, most of the energy that animals use comes from carbohydrates (and some from fats and proteins) in the food, but they seldom or never use this energy directly. It is used first, inside the body, to make one or more compounds that store it in a convenient form. Most of these are phosphates of aminoacids, called high-energy phosphates, because on hydrolysis they yield much free energy, 40 kJ or more per mole (the exact figure depends on the compound and the conditions, and is very difficult to determine), compared with 12 kJ for most phosphates. This free energy, when liberated, may be used for useful work, as in the contraction of a muscle, or for further endergonic reactions by which energy-storing compounds can be built up.

G 1.32. Enzymes and Catalysts

Students often have difficulty in understanding enzymes because they are told that enzymes are catalysts, and they have learnt that catalysts alter the rate of a chemical reaction without taking part in it. It is obvious that one of these statements is wrong. A lump of sugar sitting on the table is not, so far as any evidence shows, slowly turning into carbon dioxide and water, and even if we throw it into the fire, when it does burn to form these compounds, it undergoes quite different chemical changes from those which, under the influence of enzymes, it suffers in the body, as we shall see in section M 4.12.

A food such as sugar has more free energy than its equivalent in carbon dioxide and water, so that the reaction by which it changes into these is exergonic and in principle can go on. In fact, under ordinary circumstances it does not, so that the sugar is comparable to the brick on the beam mentioned above (p. 14). The enzyme is comparable to a piece of string that can be attached to the brick to pull it sideways and so allow it to fall.

There is now plenty of evidence that many enzymes act by combining with their substrate to form a compound that can then split up in a different way. By a series of steps, often involving several enzymes, the full reaction is carried out and the enzymes are usually restored. It is probable that this method of action is universal. For some reactions complexes of enzyme and substrate have been isolated. In other reactions the rate, with excess of substrate, is directly proportional to the amount of enzyme present, while when the amount of enzyme is constant and the amount of substrate is gradually increased, the rate increases to a maximum. Both these results suggest some form of quantitative association between enzyme and substrate, which is most likely to be chemical. This combination of enzyme and substrate is often called 'activation', the word being put in quotation marks, presumably to indicate that it is meaningless.

If we leave the common definition of 'catalyst' undisturbed, we must agree that enzymes are not catalysts. If, on the other hand, we use a definition such as that of Hinshelwood: 'whenever the addition of a new substance to the system offers the possibility of an alternative and more speedy reaction route, what occurs is catalysis', the new substance being a catalyst, then an enzyme is one. In practice the majority of the reactions that enzymes bring about are not reversible.

Enzymes are proteins, often of relatively low molecular weight, but they may also have additional or prosthetic groups of varying complexity. Many biological reactions are carried out not directly by enzymes but by coenzymes, which act in the same way but are simpler non-protein compounds (see, for example, Chapter 4). These coenzymes are themselves attached to or associated with proteins, so that the distinction, which arose when the nature of the substances was little understood, is not of great importance.

Since enzymes are proteins, and so amphoteric, they are greatly affected by hydrogen-ion concentration. The rates of their reactions depend on this, and also rise, as usual, with temperature. But since the protein is coagulated at a temperature that may be as low as 20 or 25 °C they are also destroyed as the temperature rises, so that there is an optimum temperature at which the rate of reaction is a maximum, and this optimum depends on the time over which the experiment is run.

Many enzymes, especially perhaps those that attack proteins, are secreted in an inactive form called a proenzyme, and have to be treated in some way before they can do their job. Three types of activation are known. In the first, part of the molecule may have to be removed by another enzyme, a process called unmasking; in the second, called de-inhibition, there is a simple internal molecular change such as the reduction of a cystine link to cysteine; and the third is the addition of a metallic ion that may possibly enter into the structure of the molecule.

If enzymes can be extracted at all they will carry out their reactions in solution in test-tubes, and some, such as the vertebrate digestive enzymes, work in the spaces of the body in a comparable way. Many intracellular enzymes are specially associated with membranes, such as the walls of the mitochondria, and some of their special properties perhaps depend on their ability to enter into these and displace some of the structural molecules of the membrane.

Since an enzyme must first combine with its substrate we need not be surprised that the reactions that it can bring about are limited. Some enzymes, such as arginase, are specific to one substance; others, such as most esterases, to a particular type of molecular linkage; and yet others to a particular optical isomer.

Enzymes are now, by international agreement, divided into six main classes, according to the type of over-all reaction that they bring about. Oxidoreductases carry out oxidations; transferases remove a radical, such as the methyl group, from one compound to another; hydrolases bring about hydrolysis; lyases add a group to a compound, or remove one from it; isomerases carry out isomerization; and ligases (or synthetases) join two molecules together, and at the same time break the phosphate bond of adenosine triphosphate. There are many subdivisions; thus oxidases are oxidoreductases that can use molecular oxygen; transaminases are transferases that take an amino group from one compound to another; esterases are hydrolases that split esters, and so on. Sometimes an enzyme can be put in more than one group; for example some hydrolases are also transferases.

8541112 C

Each individual enzyme has a name that may indicate the class or sub-class to which it belongs, or may merely show the substance on which it acts. All modern names of enzymes end in -ase.

G 1.33. Movement

G 1.331. Physical Movement. The physical movement of molecules would be at its simplest in an ideal gas, in which the molecules have no effect on each other except for the kinetic energy that they share when they collide. Under these conditions diffusion is relatively simple; its rate is proportional to the pressure gradient and to the square root of the absolute temperature, and inversely proportional to the square root of the density (or, what comes to the same thing, of the molecular weight). Except when a gas is diffusing into a vacuum there is always in practice interference by the molecules already present, but in most gases the simple laws, like the other gas laws, are obeyed fairly well.

In liquids things are rather more complicated, for the molecules are always close enough to each other to interfere. Indeed, since solution always involves a change of temperature, it probably also includes some chemical reaction (although in the limit, where forces involving elementary particles are concerned, the distinction between chemical and physical change is meaningless). Whatever one calls it, the effect is that diffusion is not simple, and may be slowed or hastened according to the nature of the forces. Nevertheless, solutes in dilute solution obey the gas laws to a fair approximation; that is, they behave as if they were gases. In particular, their rate of diffusion between any two points is proportional to the difference in concentration (that is to the difference in pressure) between them, and small molecules diffuse faster than big ones. The rate of diffusion of glucose in water is about one-tenth that of hydrogen under the same conditions, and that of haemoglobin about one-tenth that of glucose.

The effect of this physical diffusion is to increase the uniformity, or randomness, of a mixture, as would be predicted by the second law of thermodynamics.

If an impermeable membrane is placed in the way of diffusing molecules they cannot get through, and are necessarily reflected or bounced off from it. The reversal of their momentum exerts a pressure on the membrane, and calculations for an ideal gas relate this precisely to the gas's density and absolute temperature. Similarly, if a semi-permeable membrane, that is one which obstructs solute but not solvent, is placed in a dilute solution, the solute molecules, obeying the gas laws, exert a pressure on it. The result is osmotic pressure, which is caused not by the movement of solvent molecules (which cannot exert a pressure on a wall they pass through) but by the reflection of solute molecules.

Even with gases, the rate of diffusion through fine holes is not the same as in open space, and experiments show that the wall is having an effect. There are obviously plenty of possibilities of chemical or electrical interaction between the diffusing molecules and those of the wall. It is simplest to think of a semi-permeable membrane as a sieve with holes of molecular dimensions, and this is to some extent justified, since the degree to which substances are obstructed by it is roughly proportional to their size. The observed osmotic pressure is always less than the theoretical (that is, the sieve leaks), and from the ratio of the two and the known diameters of molecules it is possible to calculate the diameters of the pores. For red blood corpuscles it is about 0·42 nm; that of the hydrated potassium ion is 0·396, of the hydrated sodium ion 0·512. The fact that cations are obstructed more than anions can be explained if the pores are themselves lined with positive charges. The increases in permeability that occur for example in the propagation of the nervous impulse (section MV 6.31) may be explained by the removal of substances, such as calcium ions, that are blocking the pores.

We have already suggested that the molecules of the fatty protoplasmic membranes are probably highly mobile, and this would make it possible for other molecules to pass between them. The degree to which this could happen would be expected to depend on the various electrical forces between the solute and water on the one hand, and between solute and fat on the other. These can be expressed as the partition coefficient for a given solute between fat and water, which is the ratio in which the solute distributes itself between the two solvents when both are available. For many substances, the rate of diffusion across protoplasmic membranes from various sources is roughly proportional to the partition coefficients, which confirms that they are probably dissolving in the membrane before passing through it. Other substances, notably water itself, are anomalous, passing through more rapidly than their partition coefficient would suggest, so that they are said to undergo facilitated diffusion. The simplest explanation would be that there are gaps in the lipid layer, and there are arguments for holding that at these the protein layers are turned in as in Fig. 1.3. The polar groups that would thus be facing each other would prevent the enlargement that surface tension would be expected to bring about in a water-filled pore. Whether such protein-filled plugs in the fatty membrane are called pores or not is a matter of choice.

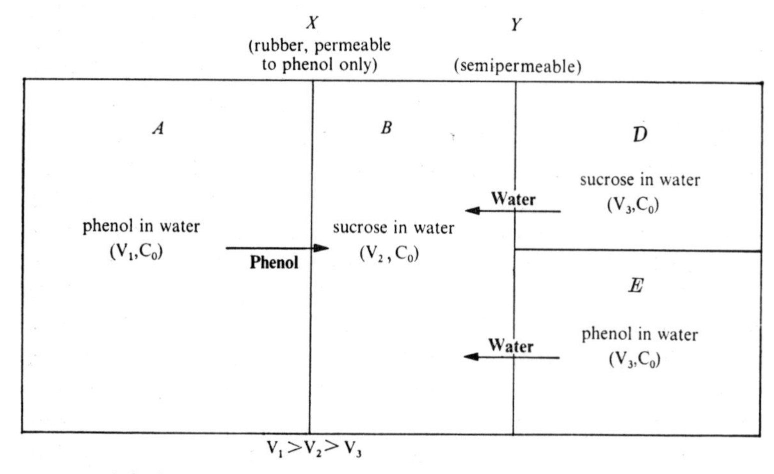

FIG. 1.6. A system in which solutions can be concentrated by osmosis. See text.

Another possibility is that a solute with facilitated diffusion passes through the membrane attached to a carrier of high lipid solubility. Glucose and glycerol probably enter red blood cells in this way, since their rapid entry is inhibited by cuprous ions and other substances, which presumably act by preventing the formation of the link with the carrier.

If only one solute and one membrane are present the case is simple, and diffusion always goes from regions of high concentration to those of low, but where more than one solute, or membranes of different permeabilities, are present, complications can arise.

Consider the case shown in Fig. 1.6. The membrane X is of rubber, which is permeable to phenol but to none of the other substances present. It separates a large volume (A) of a solution of phenol in water at concentration C_0 from a smaller volume (B) of a solution of sucrose in water at the same concentration. Phenol will pass from A to B until the concentration of phenol in B is C_0, and the total osmotic pressure in B is therefore doubled. Beyond Y, which is an ordinary semi-permeable membrane, are separate and smaller volumes of sucrose (*D*) and phenol (*E*) in water, initially at C_0. Water will pass from them into *B* until their concentrations are $2C_0$. Simply by physical means therefore the solutions of sucrose and phenol in *D* and *E* have been concentrated.

More complicated cases occur where ions are present. Suppose we have a solution of hydrogen chloride in which is a membrane permeable both to H^+ and to Cl^-. At equilibrium there is uniform distribution and, using square brackets to indicate concentration and subscripts to show the two sides, we can write

$$[H^+]_1[Cl^-]_1 = [H^+]_2[Cl^-]_2 = [H^+]_2^2 = [Cl^-]_2^2.$$

Now add to the first side a salt RCl, whose ion R^+ cannot pass through the membrane. The concentration of chloride is increased, for

$$[Cl^-]_1 = [H^+]_1+[R^+].$$

For equilibrium

$$[H^+]_1([H^+]_1+[R^+]) = [H^+]_2^2 = [Cl^-]_2^2;$$

hence

$$[H^+]_2 > [H^+]_1,$$

so that hydrogen ions must pass through the membrane even though their concentrations on the two sides are initially equal. More complicated cases can be worked out with reference to the activities of the ions, and give what is called a Donnan equilibrium, from the name of the physical chemist who investigated the problem. Both these anomalous types of diffusion are probably important in living tissues. The energy for the redistribution comes either from electrical work or from the kinetic energy of the molecules. In the second case there will be a fall in the temperature of the system.

G 1.332. Chemical Movement: Assisted Transport. In the cases that we have considered so far the final result is an equilibrium that could be predicted, if all the facts were known, by the application of simple physical laws. Even with facilitated diffusion, although enzyme reactions may take part in the formation of the carrier-complex, the concentration of solute in the end becomes equal on both sides of the membrane, just as it would without facilitation. In many parts of the animal body things are quite different; movement of solutes takes place against a concentration gradient and with no known electrical forces available, so that there is said to be assisted or active transport. In all the cases that have been investigated it has been found that interfering with some energy-providing process in the cells interferes with the movement. This agrees with the prediction that to decrease the randomness of the system, as is done by concentrating a solute, is to go locally against the second law of thermodynamics, and so will require the provision of energy.

Assisted transport of this sort is widespread, but is most obvious in the uptake of sugars and other products of digestion from the gut. In most animals that have been investigated one or two sugars, for example (usually including glucose), are taken up to a much greater extent than the others. The mechanism is obscure, but two types of reaction seem possible. In one, a compound such as a glucose phosphate is formed on one side of the membrane and as rapidly destroyed on the other, so that although the glucose, or whatever it is, is going against the concentration gradient, the temporary compound goes with it. Uptake of sugar from the gut has been shown not to depend on phosphorylation by any known enzyme, so the other hypothesis, according to which there is some sort of combination of solute and membrane, again with formation on one side and breakdown on the other, seems more likely. It follows from what we have said above, that when one substance has been taken against the concentration gradient by this chemical or primary assisted transport, there is a possibility of a passive secondary redistribution of other substances, some of it also against the gradient. Water in particular is probably often carried in this way, as in reabsorption by the kidney (section MV 5.1). Water can apparently be taken up from strong sugar solution by the gut and cuticle of insects by primary assisted transport.

Assisted transport may be followed and helped by a differential permeability of the two walls of the cell. Thus it seems likely that fructose, after passing into the intestinal absorbing cell, is converted to glucose, and that the permeability of the serosal surface to this is higher than that of the mucosal surface. Removal of the glucose by the blood will therefore be facilitated, and its return to the intestinal lumen reduced.

G 1.333. Pinocytosis. Larger particles are taken in by many cells by a process which, superficially at least, is comparable to the ingestion of food by an amoeba, or phagocytosis by a white blood cell. Large protein molecules, and experimentally such things as particles of colloidal gold (especially if negatively charged), are surrounded in various ways by the cytoplasmic membrane, so that they come to lie in small vacuoles or vesicles, which are drawn deeper into the cell. The word 'pinocytosis' was originally applied to one particular method of ingestion, but is now used as the general term for the process, whatever the details. It seems that many vesicles so formed later aggregate together, and finally the membrane may dissolve, leaving the particle free in the cytoplasm. (The process, of course, cannot be observed with the electron microscope; the account just given is a reasonable interpretation of the stills that appear in electron micrographs.)

The secretion of proteolytic enzymes by the cells of the pancreas, and the liberation of acetylcholine at synapses (section MV 6.32), seem to be more or less the reverse of pinocytosis, for membrane-bounded vesicles containing the active material are carried to the surface and there burst.

2. Nutrition

Food, which by common consent is any matter, other than oxygen and water, that an animal takes into its body so as in some sense to become part of it, is used in three ways: as a fuel to supply energy; as building material to make new tissue (that is, in growth), and as materials to repair the old; or, with little or no change, to make essential reactants in the body, which are hardly part of its structure but without which it cannot live. The distinctions between these are often more formal than practical, for the same atom may be at one time part of the structure of the body, at another in a molecule that is undergoing a rapid reaction. This is obvious where the tissues waste away in starvation, or in such rapid developments as the change of a tadpole into a frog, but it occurs also, to a lesser degree, throughout life.

Most animals, unlike plants, actively seek their food, or, if they do not, they exercise some selection on that which is presented to them; the activities that make up such search and selection are called feeding. Most food needs some preparation, mechanical or chemical or both, before it can be absorbed into the fluids of the body; this is digestion. Feeding, digestion, and absorption, are treated in this chapter, while the subsequent chemical fate of the constituents of the food is dealt with in Chapter 4.

G 2.1. THE FOODSTUFFS: THEIR NATURE AND SOURCE

Chemical analysis of the food of all animals so far investigated shows that it consists chiefly of the elements carbon, hydrogen, oxygen, and nitrogen, and that these are for the most part combined in various mixtures of the three classes of organic compound called proteins, carbohydrates, and fats. We may safely conclude that this is a universal rule. Fine analysis shows the consistent presence of other elements and sometimes other compounds, but only rigorous feeding experiments and detailed analysis of the tissues of animals can tell whether such substances are essential or whether they are accidental contaminants of the food. Such work has been done on relatively few animals, but we can be sure that some elements other than the four mentioned above are always needed. Sulphur, for example, is a constituent of many proteins, phosphorus (in the form of phosphate) is needed in reactions that seem to go on in all animals, and iron is a constituent of the substance cytochrome, which is probably present in all animals, as well as in other living creatures. The universality of particular compounds, especially organic compounds other than the three main groups, is not so clear, but some are certainly widely necessary. All these constituents of food may conveniently be called foodstuffs, and they can usefully be classified into

Proteins	Other organic compounds
Carbohydrates	Inorganic substances.
Fats	

G 2.11. Proteins

Proteins are an essential part of protoplasmic structure, and in the normal way the food from which they are derived consists also of proteins. These are broken down to aminoacids before absorption, and it would therefore be expected that the latter could take the place of proteins in the diet. This is so, but since not all aminoacids can be converted into one another and some of them cannot be made by animals, it is necessary that certain quantities of particular acids, or of the proteins that contain them, be present. The simpler acids such as alanine, $CH_3 \cdot CH(NH_2) \cdot COOH$, can, in general, be synthesized by transference of the amino-group of other aminoacids to the breakdown products of carbohydrates, but other complex ones, such as tryptophan,

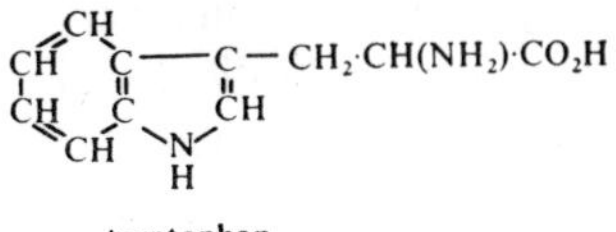

```
   CH
CH    C ——— C — CH2·CH(NH2)·CO2H
|     ||     ||
CH    C      CH
   CH    N
         H
```

tryptophan

with its ring-structure, cannot. They are, therefore, called essential. Some acids that can be synthesized cannot be made fast enough for optimal growth, and others can be made only if excess of one particular other aminoacid is present in the food, so that the list of acids regarded as essential is not easy to agree. It seems, however, that in rats the number can in no circumstances be brought below nine, nor in man below eight. Little critical work has been done on other animals except for some insects and Protozoa, where the list is probably the same as it is in mammals. The only exceptions seem likely to be animals such as cattle, which cheat by having in their gut bacteria than can synthesize a wide range of compounds, including aminoacids.

Proteins that are rich in the essential aminoacids are called first-class, and are mostly of animal origin. Casein, the protein of milk, is for some animals at least a complete source of aminoacids, but gelatin, although it comes from animals, is poor, since it lacks the essential acids methionine, tryptophan, and tyrosine. There is, therefore, no such thing as a minimum protein requirement unless the composition of the proteins is specified. For a man on a mixed diet, with what is in European civilization a normal proportion of meat, eggs, and milk, about 70 g of protein per day are required. On a vegetarian diet much more is needed, and on restricted diets, such as maize, which is deficient in tryptophan, not enough protein can be obtained however much food is eaten. More protein is needed when growth is rapid, as in children or during pregnancy and lactation. Where growth has completely ceased, as in the imagines of some Lepidoptera, no protein is needed or taken in the food at all. (A few imagines, such as the mayflies, have gone even further, and take no food whatsoever during their short life.) In carnivores protein is an important source of energy, and much more is taken.

Whatever may be their most appropriate taxonomic position, the photosynthetic green flagellates are physiologically plants, and are not dealt with in this book. Other animals, such as rats, can carry out a limited amount of synthesis of simple aminoacids from ammonia and carbohydrate residues, but if this occurs at all during normal life, which is highly unlikely, it must be on a very small scale. Many animals in several phyla have a symbiotic flora in the gut, and get some of their protein not directly from their food but after it has been metabolized by the symbionts. In ruminants (the animals that chew the cud), for example, there are Bacteria, some of which break down protein to ammonia, from which other species synthesize new proteins. Some of the ammonia is absorbed by the host and taken to the liver, where it is converted to urea. Some of this returns to the stomach in swallowed saliva, so that it can then be utilized for the production of protein. In this complicated way some 40 per cent of the aminoacids finally absorbed from the intestine come through Bacteria. Experimentally, urea can be substituted for protein in the original food.

The larvae of *Drosophila* have yeasts in the gut, and digest these to get their aminoacids. The only source of nitrogen the yeasts need in order to synthesize protein is ammonium tartrate, $(CHOH \cdot COONH_4)_2$, so that this, with sugar and mineral salts, makes an adequate food for the larvae. Neither in these insects, nor in the ruminants, is the food normally taken any different from that used by animals without an intestinal flora.

G 2.12. Carbohydrates

Carbohydrates are for most animals the chief source of energy, and so make up the bulk of the food. The ciliate *Tetrahymena* (= *Glaucoma*) *piriformis* can live on a diet of proteins (containing ten essential aminoacids) and vitamins only, but if carbohydrate is present it is used, and there is less breakdown of protein. This ability to do without carbohydrate seems to be exceptional, and in mammals its absence upsets the metabolism of protein and fat.

The chief carbohydrates of food, such as starch, are based on the six-carbon sugars called hexoses, of which glucose, fructose, mannose, and galactose are the commonest and probably the only ones that animals can utilize. Pentose-carbohydrates, based on a five-carbon sugar, are common in many plants, such as grasses, where they make up to 25 per cent of the dry weight, but little seems to be known of their value as food; some are probably indigestible.

Carbohydrates in the strict sense are not important structural constituents of animal cells, but combined with nitrogen in various ways they form some of the most important substances in the body. The mucopolysaccharides that are present in cell walls and are responsible for some immune reactions are based on the glucose molecule, and ribonucleic acid, which in various forms is involved in the synthesis of proteins, contains the pentose sugar ribose.

The amount of carbohydrate required in the food is, in man and many other animals, simply the expression of the energy that the body is using, since the possibility of supplying this from protein or fat is limited. This energy is used in three ways: in maintaining temperature, in doing internal work (secretion, the beating of the heart, and so on), and in doing external work. The first two are, in the resting state, and at a given ambient temperature, relatively constant, and constitute what is called basal metabolism. (In most animals which are cold-blooded no energy is used in maintaining temperature.)

External work obviously varies widely from time to time. Heat production to maintain a given body temperature will depend upon the body surface; the internal work will rise with increasing body weight. In practice it is found that for a given species the basal metabolic rate (BMR) is roughly proportional to the surface, and in man is about 47 W m^{-2} (40 kcal m^{-2} h^{-1}), less in women, more in children, and falls with age. Nomograms have been constructed connecting the height of man with his weight, from which his surface area may be read. For an average man standing 1·71 m (5 ft 7 in) and weighing 70 kg (11 stones) the surface area is 1·8 m^2, which gives a basal metabolic rate of 82 W (71 kcal h^{-1} or 1700 kcal day^{-1}). External work, which obviously depends on activity, will raise the total energy requirement to 120–200 W or exceptionally much more. An average figure of 155 W (3200 kcal day^{-1}) is generally taken, and so, to allow for loss in cooking, food enough to supply 14 200 kJ (3400 kcal) must be eaten in each 24 h. If, as is usual, a quarter of this is supplied by fat, and 9 per cent by 70 g of protein, about 550 g of carbohydrate must be eaten to supply the rest. This is a measure of dry weight, but most foods contain much water.

G 2.13. Fats

In mammals and many other groups, fat can be formed from carbohydrate. There is little doubt that this is general throughout the animal kingdom, so that fat is not an essential constituent of the diet in the way in which protein is. Nevertheless, many animals are unable to synthesize particular fats, which must therefore be present in the diet. Rats, for example, and other mammals, need either fats containing linoleic, linolenic, and arachidonic acids, or the free acids. All three are unsaturated, with double bonds in the chain. Linoleic acid, for example, has a double bond at $C_{12:13}$; the animal is thus apparently unable to insert a double bond at this point, although experiments have shown that rat liver can put one elsewhere in the chain. The larvae of the meal moth *Ephestia* and other insects also need linoleic and linolenic acids, but some insects that have been studied do not.

Some Protozoa are able to live on a diet containing no fat at all, but most animals normally eat some, and for mammals, and perhaps others, some proportion of the food must be fat, presumably because they cannot manufacture it from carbohydrate fast enough. It is also the normal vehicle by which the fat-soluble vitamins and some other essential substances are taken in.

G 2.14. Other Organic Compounds

When in the early years of this century it became clear that the classical constituents of a balanced diet—water, carbohydrates, fats, proteins, and mineral salts—were not enough to maintain healthy life in mammals and birds, the name 'vitamine' (from which the terminal 'e' was afterwards dropped) was given to the substances that were found to be necessary in minute quantities. As research goes on, the list of such substances increases, but the term vitamin is not usually applied to essential substances, such as linoleic acid, which can be assigned to one or other of the three main groups.

All vitamins are substances that take part in important chemical reactions in the body, or are simple precursors of such substances. In the second case the compound present in the food is sometimes called a provitamin, leaving the term vitamin for the reactant in the body, but this introduces the anomaly of making some vitamins constituents of the food but others not. Whether such a substance is an essential part of the diet for a particular species will depend on two things: whether the reaction that it catalyses is necessary in that species, and if so, whether the animal can make the compound from other materials. A complication arises in the many animals that have micro-organisms in the gut, for these can often make vitamins that their host cannot. In these cases the animal needs the vitamins, but does not have to be supplied with them in the food. Another complication is that an animal may be able to make some of a necessary substance, but not enough, so that we may speak of a 'partial vitamin'. Choline, for example, can be made by mammals from the aminoacid methionine, or from proteins such as casein that contain this, but man always needs some choline in his food. There may be a series of increasing ability to synthesize a given vitamin, so that in the limit one species can make all it needs and the substance ceases to be a vitamin at all. Ascorbic acid is probably needed by all mammals, but while men, monkeys, guinea-pigs, and marmots cannot synthesize it, or can do so only to a very limited extent, rats (and perhaps most mammals) can make all they need from glucose.

Before anything was known of their chemical nature vitamins were conveniently described by the letters of the alphabet, and to some extent this nomenclature persists. We can now make a more rational classification, based on the chemical nature of the compounds and their part in metabolism.

The substances generally known as the B-vitamins are mostly co-enzymes, or the prosthetic groups of enzymes, in the metabolism of carbohydrate, fat, and protein, and are therefore likely to be widely required. Most of the other vitamins are used in various ways in reactions that do not go on in all animals. Whether they rank as vitamins in a given species will depend therefore on whether the species needs to carry out the reaction. Vitamin A, for example, (though it has other functions) is a reactant in the chemistry of vision, and so is specially important in vertebrates, insects, higher crustaceans, and cephalopods, but may not be needed at all by most invertebrates. Some vitamins such as tocopherol (vitamin E) are still of somewhat doubtful function.

The distinction between water-soluble vitamins (B, C) and fat-soluble vitamins (A, D, E, K) is of no fundamental significance, but is important in assessing the likely adequacy of a given diet, since one that is deficient in fat will also be deficient in the vitamins that are soluble in it.

Historically, many of the vitamins were first recognized because of the symptoms caused by their absence, and there is still a tendency to think of them in terms of 'deficiency diseases'. In this book chief attention is given to their part in the physiology of the animal, but there are brief statements of the diseases associated with shortage of them, and of the foods that are most important as sources.

M 2.141. The B-Vitamins in Mammals. Eleven vitamins of this group are well established, and there may be others. They are wide-spread in animal tissues, but of more limited occurrence in plants, so that diseases caused by their deficiency occur most often in societies living on a vegetarian diet, especially where this is restricted to one type of food. All are needed by all mammals except those whose symbionts can synthesize them. The Bacteria in the rumen of artiodactyls can make all of them, and rodents get all they need by their habit of coprophagy, since the Bacteria in the faeces can synthesize all eleven. In other mammals there is a more limited synthesis by the flora of the gut. Further details of the reactions in which they take part will be found in sections M 4.11 (proteins), M 4.12 (carbohydrates), and M 4.13 (fats).

Thiamine or aneurin in its transport form has the formula

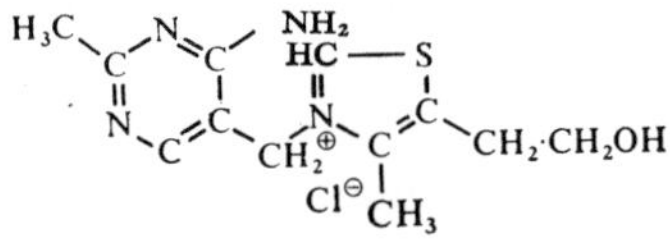

```
H3C    N       NH2
   \C/  \C/   HC—— S
    ||   |    ||    |
    N    C    N    C
     \C/  \CH2⊕ \C/  \CH2·CH2OH
            Cl⊖  |
                 CH3
```

thiamine

Its pyrophosphate is co-carboxylase, one of the coenzymes used in the oxidative decarboxylation of pyruvic acid and α-ketoglutaric acid (section M 4.12). The chief symptom of deficiency is the muscular wasting characteristic of the disease beri-beri. This is associated with a diet of polished rice, for when the embryo and pericarp of the seed have been removed there is little or no thiamine left.

Riboflavin, or lactoflavin, has the formula

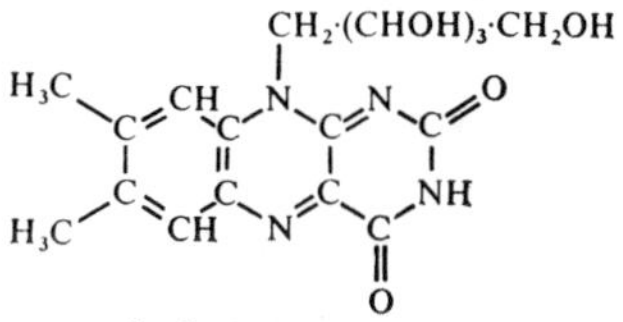

```
                     CH2·(CHOH)3·CH2OH
                     |
H3C      CH      N       N      O
   \C= /   \C/    \C= /   \C= /
    |        ||     |       |
    C        C      C       NH
H3C/ \CH /  \N /  \C /
                    ||
                    O
```

riboflavin

Its chief function is as a constituent of riboflavin adenine dinucleotide (FAD), which, when bound to protein to make a flavoprotein, is one of the most important hydrogen acceptors in the final energy-producing processes of protoplasm. It is active also in the oxidation of purines. Riboflavin is a constituent also of flavin mononucleotide, or riboflavin 5-phosphate (FMN), which makes another flavoprotein that is a hydrogen acceptor in the synthesis of fat from carbohydrate. Riboflavin is widely distributed in animal tissue, but mild degrees of deficiency, marked especially by inflammation of epithelia, are common.

Nicotinamide is easily derived from nicotinic acid; the formulae are

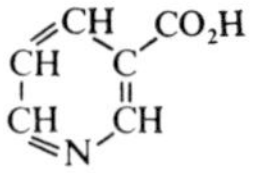

```
   CH   CO2H              CH   CONH2
CH   \C/               CH   \C/
|     ||               |     ||
CH    CH               CH    CH
  \N/                    \N/
```

nicotinic acid 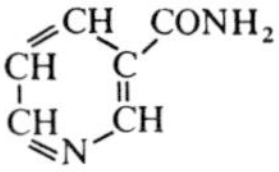   nicotinamide

It is the active constituent of nicotinamide-adenine dinucleotide (NAD, formerly called diphospho-pyridine nucleotide, DPN and coenzyme I), which is the chief hydrogen acceptor in the earlier stages of the oxidation of carbohydrates, and of phospho-nicotinamide-adenine dinucleotide (NADP, formerly triphosphopyridine nucleotide, TPN, or coenzyme II), which is the hydrogen acceptor in a few such reactions. Nicotinamide is widespread in animal tissues, and most mammals studied can synthesize it to some extent provided that their food has adequate quantities of tryptophan. Deficiency of it therefore occurs where there is not enough protein containing this aminoacid, and in fact is almost entirely a maize-eater's disease, since other cereals contain enough tryptophan. The disease pellagra, the symptoms of which are diarrhoea, dermatitis, and dementia, followed the introduction of maize, as a cheap food, into Europe and America.

Pantothenic acid, with the formula

$$HO\cdot CH_2\cdot \overset{CH_3}{\underset{CH_3}{C}}\cdot CHOH\cdot CO\cdot NH\cdot CH_2\cdot CH_2\cdot CO_2H$$

pantothenic acid

is a constituent of coenzyme A, which, in the form of acetyl-coenzyme A, takes part in many reactions, including the oxidation of carbohydrates and the synthesis of fats. It is found in most cells, both plant and animal, and deficiency in man is unknown.

Pyridoxal, or vitamin B_6, can be formed from pyridoxin or pyridoxamine; the formulae are

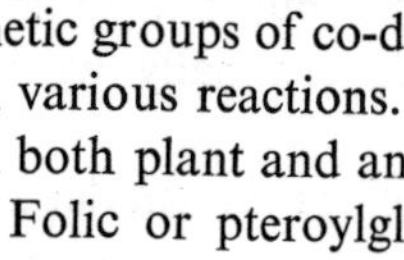

CHO; HO; CH_2OH; H_3C; N; CH — pyridoxal

CH_2NH_2; HO; CH_2OH; H_3C; N; CH — pyridoxamine

CH_2OH; HO; CH_2OH; H_3C; N; C — pyridoxin

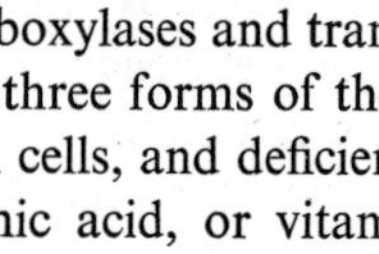

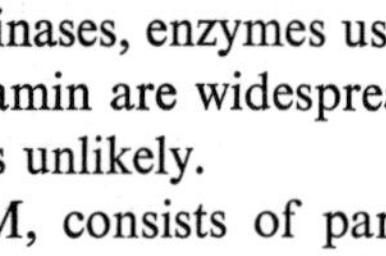

Both pyridoxal phosphate and pyridoxamine phosphate are the prosthetic groups of co-decarboxylases and transaminases, enzymes used in various reactions. All three forms of the vitamin are widespread in both plant and animal cells, and deficiency is unlikely.

Folic or pteroylglutamic acid, or vitamin M, consists of para-aminobenzoic acid linked at one point to glutamic acid and at another to a pterin,

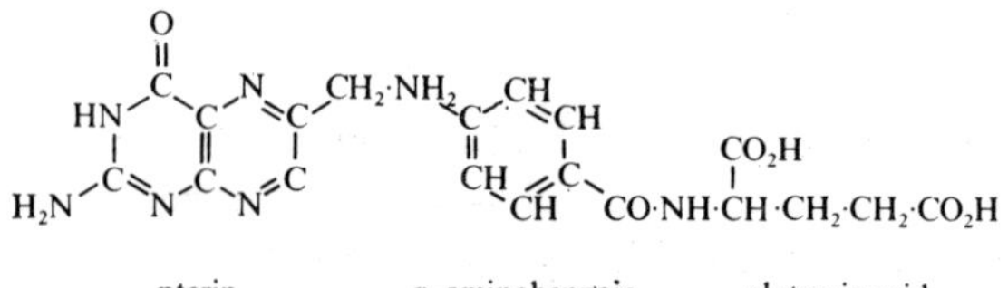

O; HN; H_2N; N; N; N; $CH_2\cdot NH_2$; CH; CH; CH; CH; CO_2H; $CO\cdot NH\cdot CH\cdot CH_2\cdot CH_2\cdot CO_2H$

pterin — *p*-aminobenzoic acid — glutamic acid

In the liver, with ascorbic acid as a catalyst, it is altered by changes in the side-chains to folinic acid, which is a coenzyme in the synthesis of many nitrogen-containing compounds, such as methionine, purines, choline, and deoxyribose nucleic acid. The chief sources of folic acid are liver and green vegetables, and deficiency causes sprue, a tropical disease characterized by an anaemia in which the red cells are enlarged.

Biotin (formerly called vitamin H) has the chemical structure

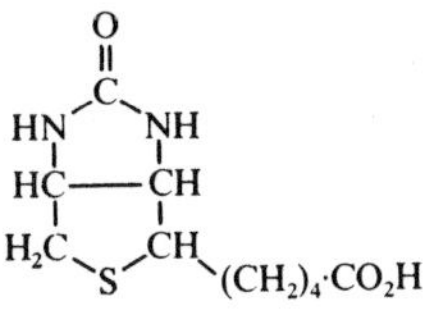

```
        O
        ‖
        C
      ╱   ╲
    HN     NH
    |       |
    HC──────CH
    |       |
   H2C      CH
      ╲   ╱   ╲
        S      (CH2)4·CO2H
```

biotin

It takes part as a coenzyme in a number of reactions, especially in carboxylations such as those by which carbohydrate is formed from propionate and fatty acids from carbohydrates. It is widespread in animal tissues and natural deficiency is unknown.

Inositol, or muscle sugar, is one of the stereoisomers of hexahydroxycyclohexane,

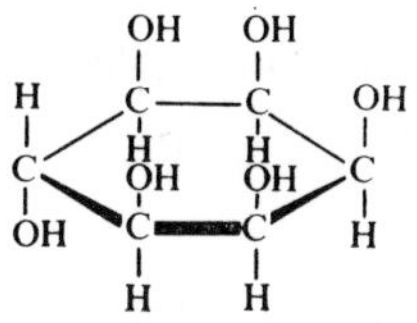

```
            OH     OH
            |      |
   H        C──────C       OH
   |      ╱ |      | ╲     |
   C        H      H       C
   |      ╲ OH     OH ╱    |
   OH       C━━━━━━C       H
            |      |
            H      H
```

inositol

Experimental deficiency in rodents causes symptoms such as loss of hair, but as it is widely distributed in fruits as well as animal tissue natural deficiency is unlikely. Its function is unknown, but it is a constituent of some phospholipids.

Choline is a relatively simple substance, being a substituted ammonia of the formula

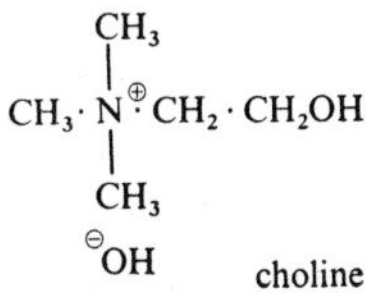

$$(CH_3)_3N^{\oplus}\cdot CH_2\cdot CH_2OH \quad {}^{\ominus}OH$$

choline

It is important as the substance from which acetylcholine, the chief transmitter substance of nerve, is formed (section MV 6.31), and in the constitution of phospholipids. Choline also supplies methyl groups for various reactions. It is present, usually in a combined form, in all cells, so that deficiency is unlikely. Most mammals studied can synthesize it from methionine and amino-ethanol, and its status as a vitamin in man is doubtful.

Cyanocobalamin, or vitamin B_{12}, has a complicated ring-structure of sixty-three carbon atoms, hydrogen, oxygen, nitrogen, and phosphorus, with one atom of cobalt attached to nitrogen atoms by co-ordinate bonds. Other closely related compounds occur, from which it is easily formed. It is involved widely in metabolism in ways that are not clear, but its most important functions are in the synthesis of the nucleic acids and in the formation of red blood cells. It is present generally in animal tissues but is absent from plants, so that deficiency occurs in strict vegetarians. More important is the disease pernicious anaemia, in which the vitamin, though present in the food, is not absorbed because of the absence of a substance called intrinsic factor, which is normally secreted by the gastric mucosa.

M 2.142. Other Vitamins in Mammals. Ascorbic acid, or vitamin C, is a simple derivative of the hexose sugars, and so far as is known all mammals, except primates, guinea-pigs, and marmots, can synthesize it, probably from both glucose and galactose. It exists in two forms, the formulae of which are

$$
\begin{array}{c}
\begin{array}{r}
O \\ \| \\ C \\ | \\ HO-C \\ | \\ HO-C \\ | \\ H-C \\ | \\ HO-C-H \\ | \\ CH_2OH
\end{array}
\quad
\begin{array}{c}
\xrightarrow{-2H} \\ \xleftarrow[+2H]{}
\end{array}
\quad
\begin{array}{r}
O \\ \| \\ C \\ | \\ O=C \\ | \\ O=C \\ | \\ H-C \\ | \\ HO-C-H \\ | \\ CH_2OH
\end{array}
\\
\text{reduced} \qquad\qquad\qquad \text{oxidized}
\end{array}
$$

(In each formula the first C and the H—C carbon are linked through a ring O.)

The ease with which one can be changed into the other suggests that it may help in dehydrogenation. It is needed for the formation of collagen and of the cement substance that binds cells together, so that it is probably required in the synthesis of mucopolysaccharides. Its chief sources are fruits and green vegetables, but its distribution is very erratic, even varieties of a species differing in the amounts they contain; for example the Mediterranean lemon (*Citrus medica limonum*) is a good source, the West Indian lime (*C. medica acida*) is a poor one. Since it is easily oxidized, cooking or exposure may destroy it. Deficiency is common, and causes scurvy, a disease characterized by haemorrhage of the limbs and gums.

The term 'vitamin A' may be used for a group of closely related fat-soluble substances; the active form in mammals is vitamin A_1 or retinol, which as its name implies has a hydroxyl group. It is easily formed in the intestinal mucosa by the hydrolysis of β-carotene, one of the constituents of the green pigment of plants; the formulae are

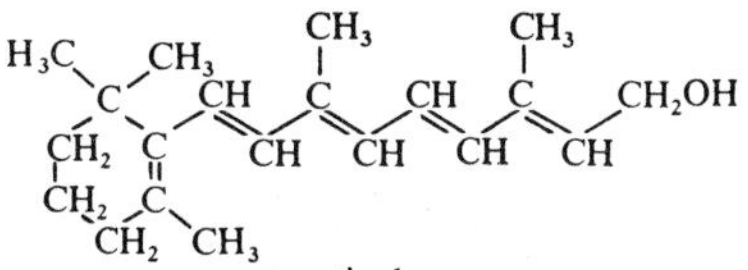

retinol

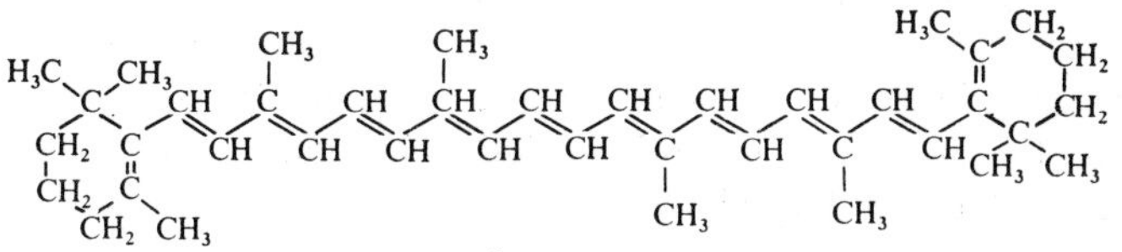

β-carotene

Although β-carotene is in effect two molecules of vitamin A joined end to end, only one is produced in the animal. There is some evidence that retinol may play some part in the oxidative processes that go on in or near biological membranes, including those of mitochondria, and if this is confirmed it seems likely that it and its relatives would have a more fundamental role in animal metabolism than is implied above. Its best known function is in vision, for retinal (retinene), the aldehyde corresponding to retinol, is the non-protein part of rhodopsin, the isomerization of which is apparently the change that starts the nervous impulse in the retina (section 6.46). Vitamin A is present in some form in the olfactory epithelia, and it seems likely that here also it is concerned with sensory reception. Vitamin A occurs in most animal fats, and carotene is present in all green vegetables, but in spite of its easy availability mild deficiency is common, the symptom being night blindness. More severe deficiency causes inflammation of the eyes, and defects especially in the formation of epithelia, which tend to become keratinized.

The tocopherols, known collectively as vitamin E, are present in the same sort of places in cells, such as chloroplasts and mitochondria, as vitamin A and, like it, they may be concerned in an oxidative process in membranes, perhaps the cytochrome *c* reductase system (section M 4.12). They probably also take part in the synthesis of ubiquinone, or coenzyme Q, a hydrogen acceptor used in the cytochrome reactions in mitochondria. The formula of the most active, α-tocopherol, is

$$\text{(ring: } H_3C,\ CH_3,\ HO,\ CH_3;\ O,\ CH_2,\ CH_2\text{)}\ \overset{CH_3}{C}\text{—}(CH_2)_3\cdot \overset{CH_3}{CH}\cdot(CH_2)_3\cdot \overset{CH_3}{CH}\cdot(CH_2)_3\cdot \overset{CH_3}{CH}\cdot CH_3$$

α-tocopherol

The others (β, γ, δ) differ in their side chains. Deficiency is not known in man, but experimentally in mammals it causes various defects, especially sterility.

Nine related forms of vitamin K occur in membranes and chloroplasts and in Bacteria. The active sort in mammals is called vitamin $K_{2(20)}$, the number in parentheses indicating the number of carbon atoms in the side chain; its formula is

$$\text{(naphthoquinone ring: } C{=}O,\ C{=}O,\ CH_3\text{)}\ (CH_2\cdot CH{:}\overset{CH_3}{C}\cdot CH_2)\cdot(CH_2\cdot CH{:}\overset{CH_3}{C}\cdot CH_2)\cdot(CH_2\cdot CH{:}\overset{CH_3}{C}\cdot CH)_3$$

vitamin $K_{2(20)}$

It is necessary for the formation of prothrombin, without which the blood will not clot, so that deficiency leads to prolonged bleeding from small cuts. More generally, the K-vitamins are used in the electron transport of phosphorylation.

Calciferol, or vitamin D_2, has a structure closely related to that of the steroids (section M 6.142), but the B-ring is broken between atoms 9 and 10. Cholecalciferol, or vitamin D_3, is a related substance. (The name vitamin D_1, given in error to an impure product, is not now used.) They are readily formed by the action of ultra-violet light on ergosterol and 7-dehydrocholesterol respectively, which takes place in the skin of mammals, including man, in sunlight. The vitamin is needed for the absorption of calcium from the gut, and there are other effects on the metabolism of calcium and phosphate. Deficiency therefore leads to malformation of bone—rickets in children and osteomalacia in adults. Rats can absorb enough calcium without the vitamin if the ratio of calcium to phosphate in the gut is not high, and can use D_2 and D_3 equally well. It seems likely that the vitamin acts not directly on the absorption but on the deoxyribonucleic acid in the nucleus to enable it to synthesise an enzyme needed for absorption. Some calciferol is present in most fats, but the best sources are the livers of certain fish, especially cod, halibut, and tunny. Ergosterol however is widely distributed, so that deficiency of the vitamin occurs only when a poor diet and little sunlight coincide. Rickets was notably a disease of poor living conditions and high latitudes, and was especially common in northern factory towns.

V. 2.14. Vitamins in other Vertebrates

It is probable that all vertebrates need the same coenzymes as mammals, but there are some limitations on the ability of particular groups to use the various forms in which a given vitamin occurs, and some variations in abilities for synthesis.

Fowls, like mammals, can synthesize some nicotinic acid from trytophan, and it seems likely that birds in general can synthesize ascorbic acid. Fowls are limited to cholecalciferol in their use of vitamin D, and need it whatever the ratio of calcium to phosphate.

Most of the vitamins have been shown to be present in the Amphibia, but little is known of their action. Ascorbic acid can be synthesized.

The distribution and function of the various forms of vitamin A in vision are discussed in section V 6.46.

J 2.14. Vitamins in Invertebrates

Very little is known of the nutritional requirements of invertebrates, except for a few insects and even fewer Protozoa. We may safely make the generalization that if a vitamin is an agent in a fundamental biological process it will be needed in the food unless it can be synthesized. But this tells us nothing, since we know no more of the distribution of chemical processes amongst animals than we do of their vitamin requirements.

A wide range of insects (Orthoptera, Coleoptera, Lepidoptera, and Diptera) have been shown to need six members of the B-group—thiamine, riboflavin, nicotinic acid, choline, pantothenic acid, and pyridoxal—in their food, but others do not require them. Some species need biotin and folic acid. The different requirements probably reflect not differences in fundamental chemistry, but varying abilities for synthesis of the vitamins by the flora of the gut. Many of those species that do not normally need the vitamins in their food have been shown to do so if their gut is first sterilized. Larvae of the blowfly *Lucilia* have Bacteria, and many blood-sucking insects (lice, bugs, tsetse flies) have yeasts or other fungi that can synthesize B-vitamins. In some, these microbes are present in special cells called mycetocytes, and there are special arrangements for infecting the eggs with them.

Some Protozoa need one or more members of the B-group, and so does the snail *Helix pomatia*, and the tapeworm of rats, *Hymenolepis diminuta*. On such rather scanty evidence it is customary to generalize and say that B vitamins, or at least the five important coenzymes (thiamine, riboflavin, nicotinamide, pantothenic acid, and pyridoxal) are needed by all animal tissues.

Ascorbic acid can be synthesized by all the insects studied, and by at least some Protozoa. The loss of the power to make it in some mammals may be a very special case. Insects are said not to need vitamin A in their food, but it certainly plays an important part in their vision, just as it does in the vision of vertebrates. Rather less certainly it plays a similar part in the eyes of crustaceans and cephalopods. Little is known of its possible synthesis, but in herbivorous species, as in mammals, there would be a ready source in carotene. Vitamin A_2 has not been found in invertebrates, but A_1 has been detected in a few nematodes (which perhaps merely absorb it from their hosts), possibly in echinoderms and annelids, and in many molluscs, crustaceans, and insects. The eyes of honey bees contain the aldehyde retinal. Molluscs, both cephalopods and snails, use their retinol more generally than in vision, as do the vertebrates.

Some insects need α-tocopherol, and most of those tested need cholesterol or a related substance, although they are unable to utilize calciferol. Some Protozoa also need cholesterol and *Helix pomatia* needs sterols. Of other animals and other vitamins we are almost completely ignorant.

Some invertebrates need substances in their food that are not required by mammals, and so have their own special vitamins. The predacious ciliate *Didinium* is unable to synthesize the peptidase necessary to digest the Paramecia on which it lives, but obtains it from its prey. *Trypanosoma* and other flagellates living in blood need haemoglobin or its derivatives in small quantities. The important part of the molecule is the haematin, which the flagellates apparently use for synthesis of respiratory enzymes. The mealworm needs carnitine, a fairly simple substituted amine. Some species of gnats will not lay eggs without a meal of blood.

G 2.15. Inorganic Substances

Besides the oxygen required for respiration certain other compounds and elements are necessary. Water, although it is formed in the body by oxidation of organic materials, is usually lost in large quantities by evaporation and in urine and faeces. The deficit must be supplied in food and drink, and in man four-fifths of the daily loss must be obtained in this way. Small birds can get all their necessary water by metabolism.

The majority of the elements which in small quantity are necessary for life and growth are probably supplied in the form of inorganic ions. Those of calcium help to form many skeletons, and are essential for the cell membrane. Chromium is possibly part of the permanent structure of the red blood cell. Cobalt is a constituent of vitamin B_{12}, but cattle and sheep, which need far more of it than do other animals, get it from their symbiotic organisms, which have themselves probably synthesized it from inorganic cobalt. Copper is a constituent of the haemocyanin present in the blood of most arthropods and molluscs (section G 3.4) and also of a number of vertebrate enzymes; in the fowl and several mammals it is necessary for the formation of the blood. In cattle and sheep it has an important action as an antagonist for molybdenum; on certain pastures there is an excess of this in the herbage, and the result is a disease that can be cured by the addition of small quantities of copper sulphate to the grass. Besides its well-known occurrence in haemoglobin, iron is present in several oxidative enzymes, including the cytochromes (section M 4.12). Magnesium is an activator of many enzymes. Manganese activates various oxidations, and the actions of intestinal aminopeptidase and of bone phosphatase, so that deficiency leads to malformation of the skeleton; birds need relatively large quantities of it. Molybdenum occurs in the prosthetic groups of some enzymes, including xanthine oxidase, which is important in the formation of uric acid (section M 4.11). Zinc is a constituent of carbonic anhydrase (section MV 3.42) and other enzymes.

By contrast, the non-metallic elements iodine, phosphorus, and sulphur are often obtained in organic form. Iodine occurs in the thyroid hormone; some of it is absorbed as iodide. Phosphate ions are important in many reactions and are universally present in cells; they are obtained from organic compounds such as nucleoproteins (section M 4.11), or directly from inorganic salts. Sulphate ions also are present in cells, and are derived from the sulphur of proteins by oxidation. Many ascidians contain niobium, tantalum, or vanadium, or more than one of these elements, attached to proteins; but their function is unknown. Selenium assists the action of tocopherol in forming coenzyme Q.

G 2.2. FEEDING AND FEEDING MECHANISMS

The method by which an animal obtains its food is intimately connected with the nature of that food, and it is possible to divide animals functionally into three or four main types:

Microphagous feeders	Detritus feeders
Macrophagous feeders	Fluid feeders.

Subdivision may be made on the basis of the type of structure used, which depends on the morphological possibilities and so on the taxonomic position of the animal.

G 2.21. Microphagous Feeders

Microphagous feeders are animals that live on small particles, usually plankton. It is a characteristic of these that they scarcely ever stop feeding, and they are often, though by no means exclusively, sedentary or sluggish. Perhaps the commonest method that they use is to draw a current of water to the mouth by means of cilia, food particles being abstracted and the water expelled. This is found in nearly all phyla from the Protozoa to the chordates (with the obvious exception of the nematodes and arthropods, which possess no cilia). The long cilia in the oral groove of *Paramecium* draw a cone of water towards the cytoplasmic mouth, and food particles thus brought into contact with this are ingested. Some sea anemones, such as *Metridium*, are covered with cilia which may beat towards the mouth, to which they convey plankton. For most of the time they beat in the opposite direction, so that waste products are removed, this being one of the few cases where reversal of cilia undoubtedly takes place.

In the lamellibranchs there is a complicated arrangement of cilia on the gills, palps, and mantle, by which a current of water is brought

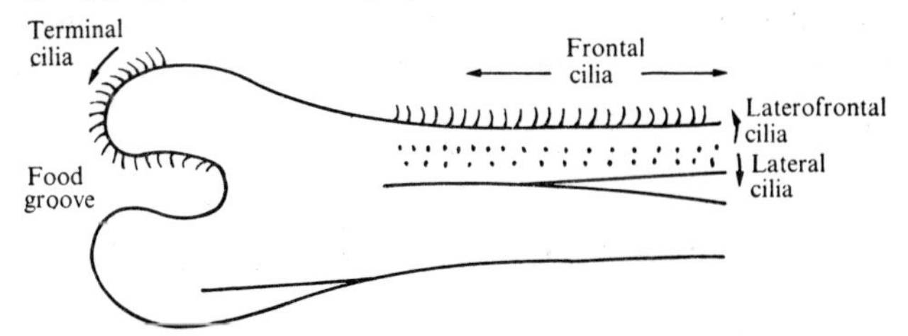

FIG. 2.1. Diagrammatic lateral view of gill filament of *Mytilus edulis*. Modified from Orton, *J. mar. biol. Ass. U.K.* **9** (1912).

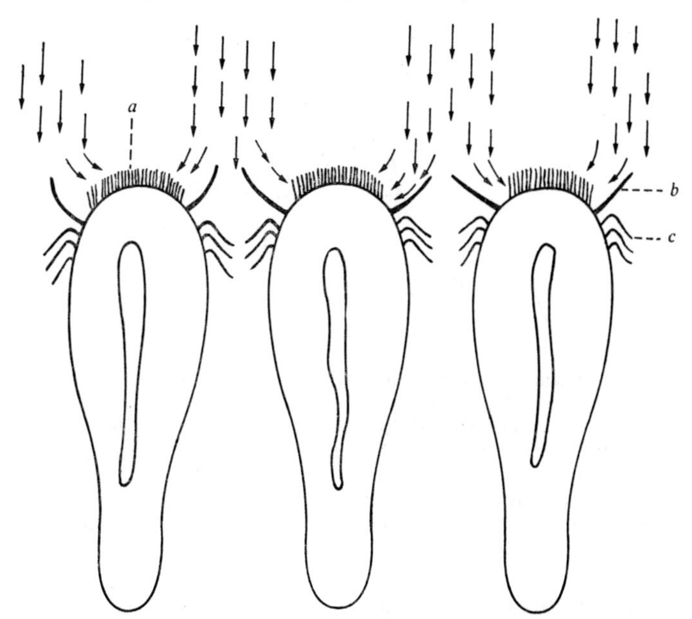

FIG. 2.2. Transverse section of three gill filaments of *Mytilus edulis*, showing the deflection of the vertical current on to the frontal cilia. *a*, Frontal cilia; *b*, laterofrontal cilia; *c*, lateral cilia. From Gray, *Proc. R. Soc. B* **93** (1922).

into the shell, particles of a certain size selected and carried to the mouth, and water and waste material ejected. In the sea mussel *Mytilus* an outline of what happens is as follows. The inhalant current is caused by cilia on the gills where they occupy the three positions shown in Figs. 2.1, 2.2. The lateral cilia cause an inward current directed on to the gills, and the particles carried in this are diverted by the laterofrontal cilia on to the edge of the gill filaments. Here the frontal cilia carry smaller particles dorsally to the base of the gill, while heavy or large particles are carried ventrally by other

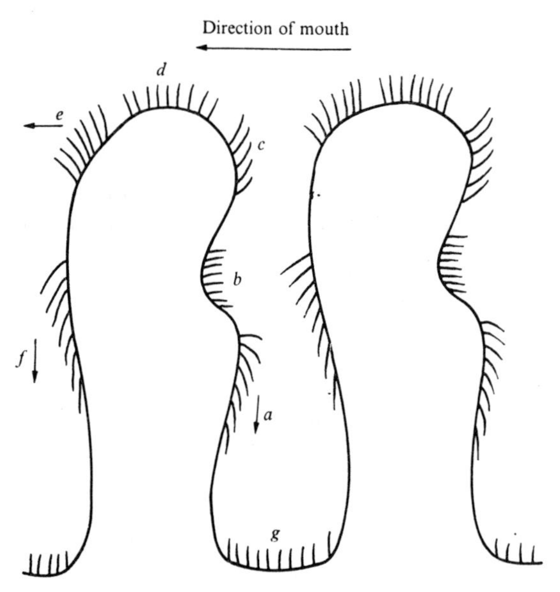

FIG. 2.3. Diagrammatic section through two ridges on the palp of an oyster, *Ostrea edulis*. The whole palp is covered with cilia, but for clarity these are shown in seven separate blocks according to the direction of their beat. Blocks *a*, *e*, and *f* beat in the direction shown by the arrows; block *b* through the plane of the paper toward the base of the palp; block *c* diagonally toward the mouth; and blocks *d* and *g* through the plane of the paper toward the upper margin of the palp. Reconstructed from Yonge.

frontal cilia to the groove in the free edge of the gill. These cilia beat only when stimulated by suitable particles, while the dorsally-beating cilia are active continuously. Both sets of particles are then carried forward to the palps, but many of the larger particles drop off and are lost. On the palps there is mechanical selection of particles by size. A simplified section of a palp of an oyster is shown in Fig. 2.3; light particles are thrown over the tops of the ridges and go to the mouth, while heavy particles fall into the grooves and are taken to the edge of the palp. Cilia on the mantle cause a current that takes these rejected particles, together with other waste products, to the *inhalant* siphon. In some species muscular contraction of the gills and palps helps in the transport of the particles.

The most advanced forms with ciliary feeding are the tadpoles of frogs and toads, which scrape off small particles with the buccal rasp. These and other particles are gulped into the mouth, and then taken to the oesophagus by cilia. In *Bufo bufo* and *Rana temporaria* there is probably no peristaltic swallowing. Well-developed ciliary feeding is also found in the ammocoete larva of *Petromyzon.*

There are other types of microphagous feeding. In some species of Crustacea setae on various appendages are used to create a current of water, and the larva of *Culex* feeds so efficiently in this manner that it is said even to be able to feed on colloidal particles, though it cannot get enough of them to grow if other food is not available.

Some active vertebrates, such as the tropical fish *Tilapia,* the sockeye salmon *Oncorhynchus nerka,* the basking shark *Ceteorhinus* and the whalebone whales, are microphagous, and get their food by swimming through the water and extracting plankton.

The secretion of mucus often helps in the transport of particles, and in some animals it has become all-important. *Urechis caupo* (Annelida, Gyphyrea), *Vermetus* (Gastropoda), and the larvae of *Chironomus plumosus* (Diptera) secrete threads or nets of mucus that float into the water, and are then withdrawn and ingested, together with the particles which they have trapped. A similar method is occasionally used by *Nereis. Ciona* and other tunicates use continuous mucous sheets secreted by the endostyle.

In many microphagous feeders, including molluscs, brachiopods, polyzoans, and tunicates, cilia in the gut twist the food and mucus into a rotating rod or bolus for which the name ergatula has been proposed.

The particles ingested by filter-feeders are usually a few microns in diameter. The mucus-feeders extract even smaller particles, but though a few can possibly retain haemocyanin molecules none can retain average proteins. Tadpoles and water-snails feed on the surface-film of water, which consists of a thin solid layer of protein probably derived from decaying plant material. A tadpole may take in 50 $mm^2\,s^{-1}$, which in stagnant water, where the film is thickest, means that it could in this way consume its own mass of protein in a day.

G 2.22. Macrophagous Feeders

Macrophagous feeders are those that feed on comparatively large masses of food. Examples, from the Protozoa to man, are fairly obvious. The Rhizopoda ingest particles of food in a number of ways. In some (e.g. *Amoeba proteus*), water is taken in with the food, so that a vacuole is formed at once, while in others (e.g. *Actinosphaerium*) the food is taken practically dry, and water is then secreted round it. The food particle may simply be drawn into the cytoplasm, the whole animal may flow round it, or two definite pseudopodia may be used to surround it. The last is the method normally used by *Amoeba proteus. Amoeba* can cut a paramecium into two pieces. All the Hydrozoa, although they move very little, are nevertheless macrophagous, animals that touch the tentacles being paralysed by the nematocysts. The tentacles carry the prey to the mouth, and in *Hydra* as soon as this is touched they withdraw so that the animal stretches itself over its food 'like a serpent or an automatic stocking'.

Since macrophagous feeders take their food only occasionally they need some specific stimulus. This seems generally to be chemical, or there may be a combination of chemical and mechanical stimuli. The formation of a protoplasmic food-cup on the surface of the heliozoan *Actinophrys sol*, normally induced by the contact of flagellates or ciliates, can be induced also by a jet of water containing a solution of egg albumen. In some species of *Hydra* swallowing is stimulated by glutathione released from the prey by the nematocysts, and the sensitivity is increased by starvation. In many mammals feeding depends on smell, in most birds probably on sight, but there is no satisfactory explanation of why a replete animal stops feeding. Certainly the central nervous system is involved, for prefrontal leucotomy in man and dogs, and destruction of another part of the brain in rats, lead to over-eating (and so to obesity).

In macrophagous forms the food has often to be broken up before the digestive enzymes can act on it, or even before it is taken into the mouth. Occasionally there is very vigorous external digestion. The rhizopod *Vampyrella* pours out a cellulase to dissolve the wall of the Alga *Spirogyra* on which it feeds and the ciliate *Glaucoma* and others liquefy gelatin in the medium, so that they must produce an external proteinase. The starfish *Asterias* extrudes its stomach over its prey, such as a mollusc, and a considerable amount of digestion takes place before the food is swallowed. A number of species inject a protease into living animals. Most of the cephalopods do this, and so do the larvae of the beetles *Dytiscus* and *Lampyris* (the glow-worm). The first of these insects lives on almost any living thing that it can catch—tadpoles, for instance—and the second on snails and slugs. Spiders likewise inject proteases into the flies that they have caught. The external digestion of the food is so complete that the material actually taken into the body is liquid, and any parts not dissolved are rejected. Blowfly larvae likewise liquefy their food before they swallow it, and since many of them are always present together they change the solid flesh, on the surface of which they begin life, into a semi-fluid pulp. Some of the enzymes concerned in this are present in the excreta.

More often the food is broken up by mechanical means. In decapod Crustacea such as *Astacus* food is first torn by the chelae, and then pulled to pieces by the endopodites of the second and third maxillipeds while being held in the mandibles. After it has been taken into the mouth it is further broken up in the gizzard, a special part of the foregut bearing internal teeth or pyloric ossicles. Insect jaws and palps are used to the same purpose as the mouth-parts of Crustacea, and species that feed on hard food (Coleoptera, Orthoptera, some Hymenoptera) have well-developed mandibles. The radula of such gastropods as the snail is used for cutting pieces of leaf of suitable size; for this purpose it works against a hard pad on the roof of the mouth.

For the most part the lower vertebrates swallow their food whole: this is so in the dogfish, where whole crabs may often be found in the stomach, and in the frog and snake. The egg-eating snake *Dasypeltis scabra* has an interesting special adaptation. The haemal spines of the cervical vertebrae project into the oesophagus, and as the egg is pressed against them while it is being swallowed, its shell is broken. Most birds also swallow their food whole, though parrots break it up very finely with the beak and tongue, and thrushes and some shore birds break the shells of molluscs before eating the soft parts. The gizzard, well-developed in game-birds and pigeons, is very muscular, but has no hard parts; instead, it contains small stones that have been accidentally or deliberately swallowed or fed to the chicks by their parents. Food is crushed by being worked against these. In mammals, the teeth, in addition to being used to seize the food, may also serve to crush it in the mouth. In accordance with this the type of food has a close connection with the form of the teeth, particularly of the cheek teeth. The molars of carnivores have cutting edges, but are not used for grinding: those of herbivores have ridges, and of omnivores rounded cusps, both being suited to chewing. Animals that live on small invertebrates, badly called insectivorous, have cheek teeth with sharp points. Piscivorous species do not chew their food, and have simple conical molars adapted for holding their prey.

G 2.23. Detritus Feeders

Detritus feeders, those that feed on the organic matter in soil and mud, are perhaps best placed with or near macrophagous forms. They include burrowing species like *Lumbricus* and other worms that suck in soil by the protruded pharynx, those that shovel sand into the mouth like the holothurians, and a few such as the sturgeons that suck mud with the lips. In order for the animal to get enough food a great deal of useless mineral matter always has to be passed through the body.

G. 2.24. Fluid Feeders

There remain several species that feed on food that is already liquid. Most of these suck fluid through some sort of mouth apparatus. The majority of them are insects. The chief free-living forms are Lepidoptera, bees, and blowflies, which have three different types of sucking apparatus developed from the mouth-parts. The ectoparasitic insects such as bugs, mammal-lice, and mosquitoes also mostly get their food in this way, but here the mouth-parts are capable not only of sucking but of first piercing through the epidermis of the host. Ectoparasites of other classes, such as ticks and leeches, use a similar method. Both of these last groups have been shown to produce an anticoagulin to prevent clotting of the blood while it is being sucked. A few endoparasites also use this method: the liver-fluke lives mainly on blood that it draws in through the mouth, which is surrounded by the anterior sucker, and some nematodes such as *Ankylostoma* browse on the mucous membrane of the gut wall and obtain much blood (Fig. 2.4).

The more extreme endoparasites, such as the Sporozoa, Cestoda, and *Sacculina*, absorb food over practically the whole body-surface, and the liver-fluke can absorb monosaccharides in this way.

G 2.25. Symbiosis

Many animals are assisted in their feeding by others, with which they are said to be commensal; the hermit crab feeds on particles dropped by the anemone that it carries, and the hawfinch on the stones of haws dropped from the trees by thrushes. In symbiosis there is a more intimate physiological connection between two species, and it is common for Metazoa to carry in their gut micro-organisms on which they are more or less dependent. Many of these are important in breaking down cellulose, which is otherwise difficult to attack, others manufacture vitamins. These aspects are dealt with in sections J 2.14 and M 2.312.

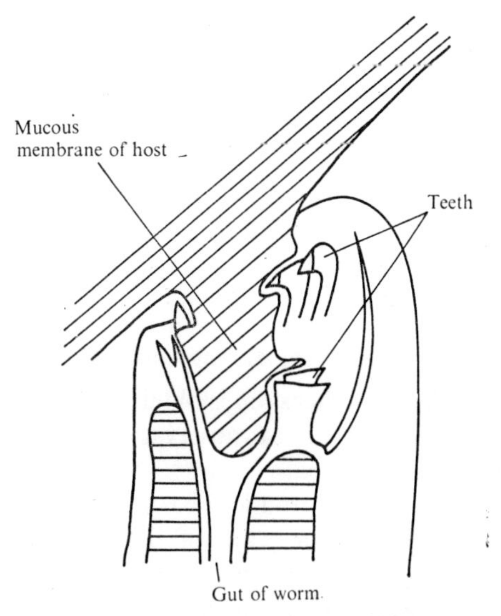

FIG. 2.4. Diagrammatic longitudinal section of the anterior end of *Ankylostoma* attached to the wall of the host's gut. In part after Brumpt, *Précis de parasitologie*, Masson et Cie, in part after Keilin.

G 2.3. DIGESTION

The complex organic substances that make up most of the food of animals are chemically relatively inert and often insoluble. They cannot be directly used for the metabolism of the body, but must first be broken down into simple diffusible products that can be absorbed by the cells. Such preparation of the food is called digestion. Even when the foodstuffs are soluble, they must still be broken down, because only so can they be rebuilt into the tissues of the body. A protein molecule can be compared to a model made of a constructional toy in which the strips of steel of different sizes represent different aminoacids. The same strips may build two different models, but one can be converted to the other only by being completely dismantled. Similarly, caseinogen and haemoglobin consist of the same aminoacids in roughly the same proportions, but a young animal can make its haemoglobin only after first hydrolysing the caseinogen in its food into aminoacids.

Intracellular digestion, in which particles of food are surrounded by cytoplasm and broken down within a cell, is normal in Protozoa, sea anemones, platyhelminths, lamellibranchs, lower gastropods, and arachnids, and is usually held to be primitive. There is some in annelids and in *Branchiostoma* and possibly a little still goes on in the small intestine of vertebrates, including mammals. In all the higher animals digestion is predominantly or entirely extracellular, the enzymes being liberated into the lumen of the gut. The advantages of this are that it reduces the internal area of the gut necessary for absorption and enables the enzyme-secreting glands to become specialized. In macrophagous forms it hastens digestion.

M 2.31. Digestion in the Mammal, especially Man

Most of our knowledge of digestion is obtained from experiments on dogs, with additions from clinical observations on man and experiments on a few other mammals.

The saliva in the mouth is produced partly by small buccal glands, but chiefly by three or four pairs of salivary glands—the submaxillary, the sublingual, the parotid or retrolingual, and the infraorbital, the last being absent from man. The saliva contains a protein called mucin, secreted chiefly by the submaxillary glands, and its chief function is to act as a lubricant. In man about a litre is produced in a day, and it is secreted at about pH 6·4–7·0 but it goes alkaline on standing through loss of carbon dioxide.

The tubular gastric glands of the stomach form gastric juice, which contains, besides mucus and enzymes, free hydrochloric acid. As secreted in man it has a strength of about 0·15N, but it is partly neutralized, so that 0·1N or 0·4 per cent, and pH 1·0, may be remembered as approximate values. Not all mammals have so acid a stomach as this; the high acidity may possibly be connected with the carnivorous habit, because the acid is a disinfectant, killing most, but not all, bacteria. Acid, mucus, and enzymes are produced by different cells.

The secreted contents of the intestine come from four sources. Brunner's glands, in the submucosa at the upper end only of the duodenum, produce mucus, the function of which is to protect the epithelium from the acid of the stomach. Succus entericus, the intestinal juice proper, is produced mainly by tubular glands in the Lieberkuhn's follicles (the crypts between the villi) of the duodenum. The pancreas produces pancreatic juice. The liver produces bile, which contains no enzymes, but which is nevertheless of great importance in digestion. After secretion it is in most mammals, including man, stored in the gall-bladder, where water is absorbed until the solids are ten times as concentrated as they were on arrival. The bile contains sodium chloride and bicarbonate, the pigments bilirubin and biliverdin (which are breakdown products of haemoglobin), lecithin and cholesterol, and the characteristic bile salts. These vary in different animals, but are always salts of acids closely related to cholesterol. In man the chief are sodium taurocholate and sodium glycocholate.

The secretions of Brunner's glands and the pancreas are alkaline, the bile is nearly neutral, and the succus entericus is slightly alkaline in man but slightly acid in the dog. Although the pH of the pancreatic juice is not above 8·9 it represents about 0·1N base, since the alkalinity is caused by bicarbonate, which acts as a powerful buffer. The acidity of the food passing into the intestine from the stomach is partially neutralized, but actual alkalinity is probably never reached. The pH of the duodenal contents is 4·5–5·1, that of the ileal contents 6·0–6·5.

M 2.311. Digestion of Proteins. There is no digestion of protein in the mouth. In the stomach the acid breaks up the collagen that forms the white fibres of connective tissue, and so liberates the individual cells, which are then more readily acted on by the enzymes. The elastin of elastic fibres is not attacked. The chief enzyme is a proteinase called pepsin, which hydrolyses most edible proteins to proteoses and peptones and a few aminoacids, all of which are soluble. Pepsin attacks those peptide bonds that are formed from the α-carboxyl of an L-dicarboxylic acid and the α- amino-group of an L-aromatic acid, and very few others. The chief acids that it liberates are consequently L-tyrosine and L-alanine. By this attack protoplasmic cell walls are broken down, and their contents liberated. The droplets of fat from the food give a milky appearance to the stomach content, which is then called chyme. Peptic digestion goes on at pH 4·5 or less; in man, calf, and pig there appear to be two pepsins each with two optima. The enzyme produced chiefly near the pylorus has optima in man at pH 1·5 and 3·2, the other, produced chiefly near the fundus, at 2·0 and 3·6. Pepsin is secreted as a pro-enzyme, pepsinogen, which is unmasked by the removal of a polypeptide by hydrochloric acid, or by pepsin that is already present.

In the stomach of some young mammals there is another enzyme, rennin, which has somewhat similar properties to pepsin. It is secreted as prorennin, which is activated by acid and then hydrolyses the same peptide bonds as does pepsin, but with a pH optimum of about 4. It is best known for its action on milk, which is also shared by pepsin. The soluble protein caseinogen (or casein) is converted to casein (or paracasein), which is precipitated as the calcium salt, so that the milk flocculates, although the acid of the stomach may bring the casein back into solution as the hydrochloride. The physiological function of all these changes is obscure. Rennin is probably absent from man, but is the chief enzyme in the abomasum of foetal calves (and perhaps lambs), where it is gradually replaced by pepsin after birth.

In the duodenum proteins are attacked by a battery of enzymes which hydrolyse them progressively to polypeptides, dipeptides, and aminoacids. The whole active principle from the pancreas was formerly called trypsin, but that name is now used in a more restricted sense for the pancreatic proteinase. The collection of proteases (or proteolytic enzymes—those that assist at any stage in the breakdown of proteins) in the succus entericus was formerly called erepsin, but that term has now gone out of use. The constituents into which 'trypsin' and 'erepsin' have so far been analysed are shown in Table 1. It is possible that some of the final stages of proteolytic digestion may be intracellular.

Nucleoproteins (section 4.11) are hydrolysed by the acid in the stomach, and the protein is digested in the ordinary way. The nucleic acids are split to simple nucleotides by nucleases from the pancreas, and then nucleotidases from the intestinal mucosa split off phosphate, and nucleosidases from the same place separate the pentose from the bases.

Much of the protein in the food of artiodactyls is converted to ammonia by the Bacteria of the rumen (section M 2.312), which can similarly digest non-protein nitrogen, including urea. Although some of this ammonia is absorbed, taken to the liver, and there converted to urea, it is not, so far as is known, used by the animal. The Bacteria also build up their own proteins from ammonia, and some of this can come from non-protein nitrogen. When the animal digests the proteins of the Bacteria in the ordinary way it is then getting its nitrogen from non-protein sources.

TABLE 1. *Proteolytic enzymes of the small intestine of mammals*

Place of secretion	Enzyme	Activation	pH optimum	Substrate	Point of application	Products
Pancreas	Trypsin	Trypsinogen unmasked by enterokinase, or by trypsin already present	8–11	Proteins Proteoses Peptones	Peptide links at ends of chain and in middle, formed from carboxyl group of lysine or arginine	Polypeptides, and aminoacids, especially lysine and arginine
,,	Chymo-trypsin	Chymo-trypsinogen unmasked by trypsin	7·5–8·5	Proteins Proteoses Peptones	Peptide links at ends of chain and in middle, formed from carboxyl group of an aromatic acid	Polypeptides, and aminoacids, especially tyrosine and phenylalanine
,,	Carboxy-peptidase (probably many)	Procarboxy-peptidase unmasked by trypsin, activated by Zn^{2+}	*c.* 7·3	Polypeptides	Terminal peptide link adjacent to free carboxyl group	Dipeptides and aminoacids
,,	Elastase	Proelastase unmasked by trypsin	8–9	Most proteins except keratin and collagen, but especially elastin		Soluble proteins
Small intestine	Amino-peptidase (probably many)	By metallic ions—Mg^{2+}, Mn^{2+}, Zn^{2+}	8·0	Polypeptides	Terminal peptide link adjacent to free amino-group	Dipeptides and aminoacids
,,	Dipeptidases (probably many and specific)	By metallic ions—Mg^{2+}, Mn^{2+}, Zn^{2+}	7·5–8·0	Dipeptides, e.g. glycyl-glycine	Amino- and carboxyl groups, and peptide link	Aminoacids

M 2.312. Digestion of Carbohydrates. Starch grains, the commonest form of polysaccharide in food, consist of a central core of amylose, making about a quarter of the mass, and a husk of amylopectin. Both these substances consist of chains of glucose units, but in amylopectin lengths of chain containing about twenty-four glucose units are joined to form a branching chain. These branches are not separated by any digestive enzymes. Glycogen has a similar structure to amylopectin, but the branches have only about twelve glucose units. The saliva of most mammals is free, or nearly free, from enzymes, but that of Primates, pigs, elephants, and some rodents contains an endo-amylase called ptyalin (optimum pH 6·2–6·8), which attacks the simple links of both amylose and amylopectin and splits off a series of maltose molecules leaving dextrins, as may be represented in a simplified form thus:

$$\text{amylose} \rightarrow \underset{+\text{maltose}}{\text{erythrodextrin}} \rightarrow \underset{+\text{maltose.}}{\text{achroodextrin}}$$

Maltotriose molecules (containing three glucose units) are apparently also formed in the proportion of one for every 2·3 molecules of maltose. Glycogen is hydrolysed similarly, also to maltose, but most of that present in liver and muscle is likely to be already broken down before the food is eaten.

The activity of digestive amylases is greatly reduced, and their pH optima altered, in the absence of chloride ions. The extent to which salivary digestion goes on depends on the time during which the food remains in the mouth. When the swallowed food is made strongly acid in the stomach the action of amylase is stopped, but the acid takes some time to diffuse into solid food, and it may be as long as 40 min before hydrolysis ceases. By this time very little starch is left. Ptyalin does not attack raw starch grains; they must first be broken down, either in the preparation of the food as by boiling, by chewing, by Bacteria, or by enzymes contained in the food itself.

The gastric juice contains no enzymes that act on carbohydrate, but the duodenal contents regurgitate into the stomach. The acid hydrolyses cane sugar, and also inulin, which, since it is the normal storage compound of the Compositae, will occasionally occur in the food of man in such things as the Jerusalem artichoke and salsify. Since vertebrates have no inulase, the action of the acid is the only way in which inulin can be digested, unless it is attacked by symbionts.

Starch is digested in the small intestine by another amylase secreted by the pancreas. It has a similar action to ptyalin, but it can attack raw starch grains and can split maltotriose to maltose and glucose; it is probably a mixture.

The small intestine contains also several enzymes called glucosidases or disaccharases, which hydrolyse various disaccharides to the appropriate hexoses. In man there appear to be four maltases (one of which comes from the pancreas and one of which can split sucrose as well as maltose), trehalase, and lactase. The last, which splits milk sugar, or lactose, to glucose and galactose, is prominent in young animals and is the only amylolytic enzyme produced by the calf for the first month of its life. Trehalose, which on hydrolysis gives two molecules of glucose, occurs in Fungi and in the blood-sugar of insects and some other invertebrates, but it must be an unimportant food in man. Maltose gives two molecules of glucose, and sucrose (which is the storage or transport sugar of many food plants) gives one of glucose and one of fructose, so that these two hexoses must be the commonest in the intestine. There is also a little mannose, split off from proteins, such as those of egg white, with which it is in combination.

The cell walls of plants are generally said to be made of cellulose or of lignin, but these names apply to materials that have different compositions according to the species from which they are obtained. Cellulose is basically a carbohydrate material; the sugar unit of α-cellulose is glucose, with some xylose and mannose and smaller amounts of galactose and pentoses; hemicellulose also contains various hexoses and pentoses. Lignin is a relatively simple compound based on linked benzene rings, and is not a carbohydrate. Neither substance can be digested directly by mammals, but cellulose, which is not only an important constituent of the food of herbivores but also a barrier between the animal and the cell-contents in its food, is often attacked by enzymes produced by symbionts. These are best known in cattle and sheep. The stomach rudiment in the embryo gives rise to four parts (Fig. 2.5). Food is swallowed without being chewed and is circulated through the rumen and reticulum, where the cellulose is largely broken down, and the half-digested stuff is then regurgitated into the mouth and chewed. On being swallowed a second time it goes straight to the psalterium or omasum, where there is some mechanical trituration, and then into the abomasum, which is the only part of the stomach bearing gastric glands. About half of a meal leaves the rumen within 24 h of entering it.

The rumen contains many millions of Bacteria of several species, some of which break down cellulose, other carbohydrates, and proteins to simple substances. The products depend to some degree on the food, but there is little difference between sheep and cattle.

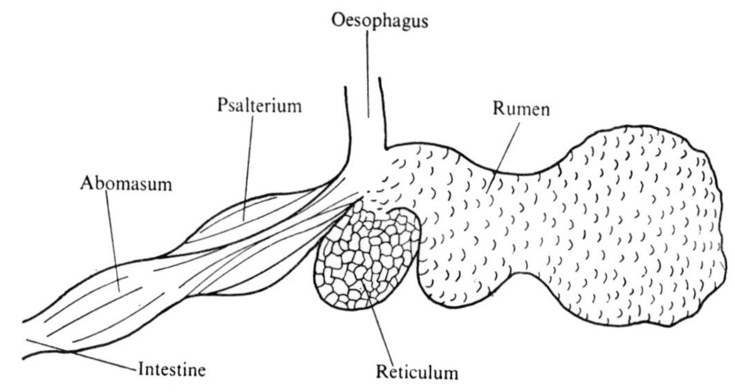

FIG. 2.5. Stomach of sheep, cut open and extended.

Usually 50–60 per cent of the product is acetic acid, about 25 per cent propionic, 10 per cent butyric, 5 per cent other acids, with small quantities of other substances such as ethanol. About 7 or 8 per cent of the energy of the food is lost as methane.

There are also in the rumen more than 100 species of ciliates of the Spirotrichida and Holotrichida. Their value to the animal is unknown, since if a calf is reared in isolation they are not present, but the animal survives. Some can attack cellulose, and as they store carbohydrates and proteins within their cells they may form a reserve on which the ruminant can live for a short time when grass is not available.

A comparable symbiotic digestion of cellulose occurs in many other mammals. In the kangaroo *Setonix brachyurus*, Bacteria act in an enlarged stomach and in the sloths (*Choloepus*) there is a compound stomach comparable to that of ruminants; in both, Bacteria convert cellulose to volatile acids. In the horse the Bacteria are mostly in the caecum and ventral colon, and in the rabbit in the caecum. The last, with other lagomorphs, many rodents, and the common shrew, has developed the odd habit of reingestion, or pseudorumination. When food is first eaten by the rabbit it passes rapidly to the caecum, and after 1 or 2 days is voided as soft faeces. These are eaten, and remain in the cardiac stomach while freshly eaten food passes straight through as before. After digestion in the stomach and intestine, the twice-swallowed food does not enter the caecum, but is voided as hard faecal pellets.

Similar ciliates to those in ruminants are found in the large intestine of the African rhinoceros, the tapir, the guinea pig, the chimpanzee, and the gorilla.

Chitin, which covers the bodies of arthropods, is, like cellulose, of varying and ill-defined composition. It is, however, always based on units of glusosamine, a 2-aminoglucose, so that it is essentially a carbohydrate. Chitinase is said to be produced by the gastric glands of some insectivores, bats and rats, and of the pig, and also the pancreas as well in some of these animals. All the mammals in which it has been found are insectivorous or omnivorous.

M 2.313. Digestion of Fats. There is a fat-splitting enzyme or lipase in the stomach, but it is unimportant as it has only slight activity at the acidity of the gastric contents, and that chiefly on finely divided fats such as those of egg yolk. In the small intestine there are two lipases, one from the pancreas and one from the mucosa. Both are strongly unspecific, attacking a wide range of glycerol esters to form diglycerides, monoglycerides, and free glycerol with one, two, or three molecules of free fatty acid respectively. Probably the terminal acids are removed first and separately from the triglycerides, giving two different diglycerides, and then the middle acid is removed, perhaps by a different enzyme. The ease with which the action is carried out, and the pH optimum, depend on the length of the fatty acid molecule. The pancreatic lipase, which is the more important of the two, attacks most readily triglycerides of long chain-length, and has a pH optimum ranging from 7 to 9 as the length increases. The digestion of fat is greatly helped by the presence of bile, the salts of which lower surface tension, so breaking up the droplets of fat to a finer emulsion. Since the intestinal contents are never alkaline, soaps cannot be formed. An investigation of the digestive enzymes of the horse failed to show any lipase. In the rumen of artiodactyls fats are hydrolysed by the Bacteria, and the glycerol is converted to propionic acid.

M 2.314. Absorption. The chief organ of absorption is the small intestine whose surface area is increased by projections called villi (Fig. 2.6). They contain both blood capillaries and small branches of the lymphatic system, called lacteals from the milky appearance of the chyle, or emulsion of fat, that they contain. They are in constant motion, both swaying and pumping, so that the fluid in contact with their surface is continually renewed and the lacteals are regularly emptied. The free surface of the absorptive cells has a brush-

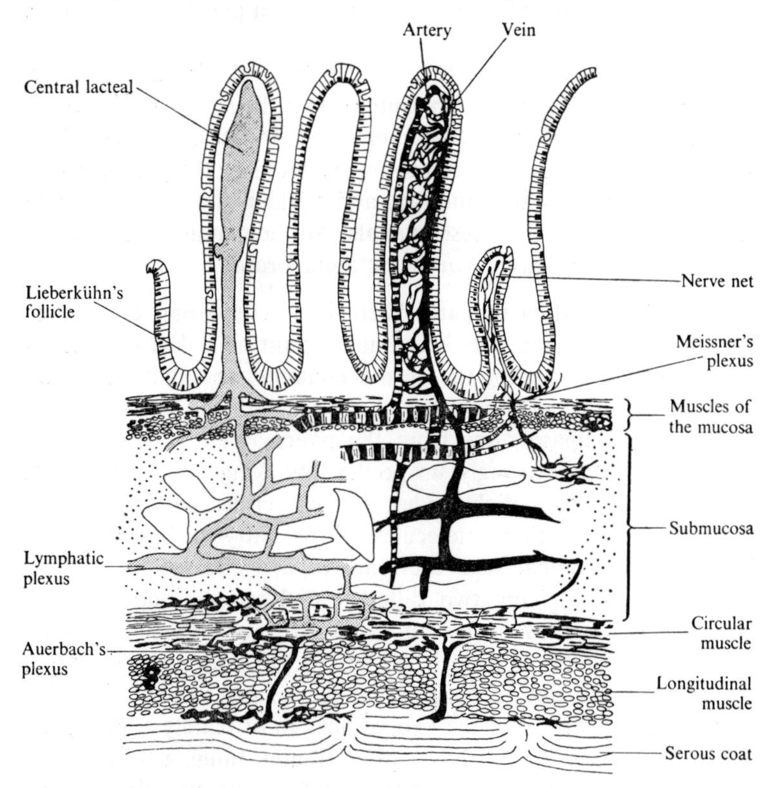

FIG. 2.6. Diagrammatic section through the wall of a mammalian small intestine. From Yapp, *Vertebrates; their structure and life*, Oxford University Press, New York (1965).

border, especially where absorption of solutes is most active. This is usually the upper part of the duodenum, but some goes on throughout the small intestine, and there is much specific variation in the rates in the different parts.

Absorption of the digested products of the food goes on both passively and by assisted transport, but there is much specific variation. That the second is used for at least some aminoacids is shown by the fact that the L-acids are taken up faster than the D-acids, but little is known of the mechanism. Since one aminoacid (e.g. leucine) may interfere with the uptake of another (e.g. valine), they presumably share a common carrier. The intestine of the rat can take up diglycerides, which undergo intracellular hydrolysis. It is probable that a few proteins can be taken up directly, presumably by pinocytosis. This is of no importance in ordinary nutrition, but new-born ruminants, horses, and pigs absorb antibodies (which are globulins) from their mother's milk, passing them into the cells of the intestine and then to the lacteals; the ability is lost after about 24 h. There is also some similar absorption of antibodies by rodents. In all the mammals where it is known there is delayed development of the gastric glands, so that the antibodies are not destroyed by pepsin.

Some monosaccharides, such as fructose, mannose, and pentoses, are absorbed without assistance, but others, notably D-glucose and D-galactose, are absorbed very rapidly and without relation to their concentration in the medium. This assisted transport needs sodium ions and phosphate, and also, in the adult, oxygen, but the fetus can make do with anaerobic glycolysis (section 4.12). The picture is complicated by the observation (in isolated hamster intestine) that glucose accumulates in the cell more rapidly when the sugar in the medium is sucrose, maltose, or glucose-1-phosphate than when it is glucose itself. Hence it appears that disaccharides can be taken up and that hydrolysis and absorption are somehow linked. Sodium ions are necessary to activate the enzyme that splits sucrose, and it is possible that sucrose and sodium ions go together into the filaments of the brush border, where the sugar is split, and some of the sodium is pumped out by an active process in which phosphate is, as usual, needed to supply energy. Glucose passes through the rest of the cell, by assisted transport, with sodium. The concentration of disaccharides is especially high in the upper region of the duodenum, where the brush border is most prominent.

Although pancreatic lipase is capable of completely hydrolysing triglycerides, much of the fat in the food (in some experiments on rats about four-fifths) is split only to monoglyceride. The resulting mixture of fatty acid, monoglyceride, and bile salts gives a clear solution, which is absorbed, chiefly in the lower part of the duodenum, even in the presence of metabolic poisons, so there is presumably no assisted transport. In the cells there is a monoglyceride lipase that splits some of the monoglyceride to a fatty acid and glycerol, but the chief process is a resynthesis of triglyceride from the absorbed acids and glucose. The glycerol from digestion does not take part in this. The process is inhibited by poisons that interfere with oxidative processes. The reconstructed fats pass into the lacteals.

The extent to which triglycerides can be absorbed is still disputed. Some workers claim that in the presence of bile and monoglycerides they can be broken down to particles small enough to be taken up by the cells, but others deny this. Some electron micrographs show globules in the channels between the cells, apparently pushing the opposed cytoplasmic membranes apart. In the absence of bile much of the fat in the food (in man about one-half) is not absorbed. Short-chain fatty acids are absorbed into the portal system, but the basement membrane of this is impermeable to larger molecules.

Absorption of some ions is assisted; that of calcium depends not only on vitamin D and the secretion of the parathyroid but on the presence of oxygen. Vitamin D itself is absorbed chiefly in the jejunum, with the help of bile salts.

Some water (as well as other simple molecules) is absorbed by the stomach, but the chief site where it is taken up is the small intestine. The material that enters the colon contains mucus, dead Bacteria, ions, and some fatty acids; only when there is much cellulose in the food is there any appreciable quantity of carbohydrate. After removal of ions and water the remains enter the rectum as faeces.

The ruminants have a very different pattern of absorption from most mammals. About four-fifths of the food is digested in the rumen, and the resulting volatile fatty acids are absorbed straight away into the capillaries of the portal system or into those of the omasum; a very small proportion goes on to the duodenum. The small quantities of glucose present, inorganic ions, and some of the ammonia and other gases are also absorbed by the rumen.

M 2.315. The Movement of the Food and Control of Digestion. Digestion is a complicated process and there is an intricate system of control which ensures that the food meets the right enzymes at the right time.

In man chewing is voluntary but it can be reflex, each closure of the jaws acting as a stimulus for their reopening. Swallowing also may start as a voluntary movement, but its completion is a chain of reflexes, involving the closure of the nasal and tracheal openings and the cessation of breathing, as well as the movement of the food; the trigeminal, glossopharyngeal, vagus, and hypoglossal nerves are all concerned. Once in the oesophagus, the food is moved chiefly by peristalsis, a type of muscular contraction in which the circular muscles of the gut contract successively in a wave that moves from the oral to the anal end. The stimulus for the contraction is normally the slight stretching caused by the presence of food at any point, and in the stomach the response is myogenic, and contraction is followed by relaxation. The effect of peristalsis is to move the food along the gut aborally, but this movement is interfered with at some points by the sphincters or tight circular muscles. The pylorus, at the junction of stomach and duodenum, is such a muscle, and as it is normally closed the effect of peristalsis on the fluid contents of the stomach is to cause peripheral onward movement and axial reflux, giving good mixing. When the pylorus relaxes, which after the stomach has been filled for some time it may do at every wave, some of the fluid is driven into the intestine. Control of the pylorus is probably through the central nervous system, and while fluids pass through the stomach in a few minutes, carbohydrate stays longer, and fat longer still, up to 3 or 4 hours.

Peristalsis continues in the small intestine, but is here controlled by the deep nerve plexus (Auerbach's). There is also some anti-peristalsis, and simultaneous contraction of the circular muscles at various points, the segmentation movements, both of which mix the gut contents. The sphincters between ileum and colon and between colon and rectum relax in a comparable way to the pylorus, but they are controlled by spinal and sympathetic nerves. The entry of food into the stomach may cause both to relax, so that gut contents enter the rectum and there is a call for defecation.

The time taken by the food to pass through the gut varies much, but the first marked constituents of a meal taken on an empty stomach (barium sulphate, which is opaque to X-rays, is generally used) reach the colon in $4\frac{1}{2}$ h and the rectum 14 h later. Indigestible matter may take 4 days or more to pass through.

The secretion of saliva begins from the taste-stimulus of food in the mouth, and is mediated by the fifth and ninth cranial nerves. Soon after birth conditioned reflexes are formed, so that there is salivation at the sight of food or on the receipt of any stimulus, such as a dinner gong, regularly associated with meals. Production of pepsinogen begins in a similar way and may become conditioned. The acid of the gastric juice is secreted under the action of gastrin, a hormone secreted and liberated into the blood by the stomach wall in the region of the pylorus when it is mechanically stimulated or comes into contact with peptones, and of another hormone similarly produced in the duodenum.

Both the intestinal juice and the mucus from Brunner's glands are produced by mechanical stimulation, Meissner's plexus being apparently involved in the former, but there are possibly also two further duodenal hormones, enterocrinin causing the production of juice and duocrinin that of mucus. The pancreas is stimulated to produce enzymes by the vagus, and also by a hormone called pancreozymin which is formed in the duodenal mucosa when proteoses or peptones make contact with it. Another hormone, called secretin, is formed similarly when acid comes in contact with the mucosa, and causes the pancreas to secrete bicarbonate. The neutralization of the stomach contents is therefore under automatic control, for the acid acts as a stimulus for the production of base, and when the intestinal contents are nearly neutral no more secretin, and so no more bicarbonate, is produced.

Bile is at least in part an excretion and its production by the liver is continuous. It is, however, increased by the presence in the intestine of protein, fat, bile salts, or dilute acid, or by the presence of secretin in the veins. Bile is stored in the gall-bladder where it is concentrated by absorption of water and salts and receives mucin and possibly cholesterol. The gall-bladder is innervated by the vagus, in which motor impulses that cause contraction are induced by the sight of food or the act of eating; the gall-bladder is also stimulated by a hormone called cholecystokinin, formed by the duodenal mucosa on contact with digested food, particularly when this is rich in fats, which need bile salts for their digestion. In those mammals, such as perissodactyls and elephants, that have no gall-bladder, the passage of bile into the duodenum is presumably continuous.

V 2.31. Digestion in Other Vertebrates

The general course of digestion seems to follow the same lines in other vertebrates as in mammals. Salivary digestion is rare, but an amylase is present in the saliva of frogs and toads, and of the fowl. The stomach secretes a pepsin-like enzyme acting in acid medium, the intestinal proteases are distributed as they are in mammals, and the intestine contains amylase, disaccharidases, and lipase. In bony fishes, however, both amylase and maltase are secreted predominantly by the pancreas. Small amounts of gastric lipase occur in some birds and fishes. Little work has yet been done on the identity of the enzymes in the different classes, but the individual proteases are very similar, if not identical, in fish, Amphibia, reptiles, and mammals. Salmon pepsin has been shown to be different in specificity from that of mammals and of the fowl. Chitinase is claimed to be present in goldfish, a lizard, and some passerine birds, but not in a tortoise or in pigeons. The birds called honey guides (*Indicator*), which feed on bees-wax, do not produce their own esterases but are dependent on their microflora. Bile salts are present in all vertebrates that have been examined and in several mammals, a bird, some reptiles, a toad, and a teleost, have been shown to be made from cholesterol, which has twenty-seven carbon atoms. Different species have different bile acids, but there is a general evolution towards C–24 acids, which are the predominant compounds in amniotes and teleosts but are present only in small quantities in lower forms. The significance of this is unknown.

The stomach contents are seldom so acid as they are in man and the dog, pH values from 2·5 to 4·5 being common in fish, Amphibia, and birds. In some rays and bony fish the stomach is alkaline, even though pepsin is present, but by contrast sharks have been reported as having an acidity twice that of man. The stomach is absent from Holocephali, Dipnoi, and many teleosts. Since the protochordates do not possess a stomach and do not produce pepsin, it is possible that the organ developed within the vertebrates, parallel with the change from microphagy to macrophagy. It may originally have been a storage organ, which is very necessary in carnivores, and high acidity and pepsin may have come later to kill the prey and give the possibility of preliminary digestion of protein. In Amphibia, including the frog, more pepsin is produced by the oesophagus than by the stomach.

In the fowl there is preliminary digestion in the crop by autolysis and by Bacteria, with production of lactic acid. The assisted uptake of the products of digestion in birds is broadly similar to that in mammals, again with specific variation. There is no evidence either for nervous or for hormonal control of gastric secretion in selachians. In *Rana*, the mudpuppy (*Necturus*), the tortoise *Testudo graeca*, and two lizards, stimulation of the vagus is not necessary for the production of pepsinogen, which is secreted following local mechanical stimulation of the stomach. In the toad *Bufo* there is some nervous control. There is some indication of the presence of gastrin in

Amphibia. Secretin is present in the intestine of the fowl, tortoise, frog, salmon, dogfish, and skate, but its natural activity in these animals is still obscure.

J 2.31. Digestion in Invertebrates

For a long time the only method available for investigating the presence of enzymes in invertebrates was to make an extract of the gut or of a gland, and observe its effect on various substrates *in vitro*. This has been done for most of the phyla, and it has been found that nearly all animals possess, in greater or less degree, proteases, carbohydrases, and lipases, only a few groups being lacking in one or other of the last two. Since the food of animals consists of proteins, carbohydrates, and fats, and of little else, these results merely give us proof of what might with little risk have been assumed *a priori*.

The relative concentration of the enzymes in the gut often bears a general relation to the type of the food. For example, the closely related flies *Glossina* and *Calliphora* both contain protease, carbohydrase, and lipase, but while the blood-sucking *Glossina* has much protease and little of the other two, conditions in the vegetable-feeding *Calliphora* are the reverse. The more recent and valuable work on enzymes has involved careful preparation of the extract, exact quantitative following by analytical methods of the change of the substrate to its scission products, and meticulous attention to the hydrogen-ion concentration and other physical properties of the medium.

In the 1930s Willstätter and Waldschmidt-Leitz introduced adsorption methods that enabled enzymes to be separated from one another. This for the first time made it possible to compare the enzymes of different animals. Such methods have been applied to only a few invertebrates. In the crab *Maia squinado* and in the marine snail *Murex anguliferus* the gut contains four proteases resembling those present in vertebrates, namely proteinase, carboxypeptidase, aminopeptidase, and dipeptidase. Moreover, the pH optima are of the same order as those for the vertebrate enzymes. Since the exact values of these optima depend on the degree of purification, exact equality is hardly likely to be obtained, but the proteases of these two animals are certainly very similar to those of vertebrates. The *Maia* proteinase does not need activation, but after it has been purified it becomes inactive, and can be reactivated by enterokinase. This suggests that the proteinase of *Maia* is similar to or identical with that of vertebrates, but that in the crab it is secreted along with its activator, which is separated in the purification by adsorption. Experiments on insects, on *Helix pomatia*, on the cuttlefish *Sepia officinalis*, and on a coral, have shown that in all these the protease consists of several components. In Orthoptera, beetles, lepidopteran caterpillars, and other insects the protease consists of a tryptic proteinase, a carboxypeptidase, an aminopeptidase, and a dipeptidase. For *Sepia* the pH optima are about the level of those for trypsin, and the proteinase requires activation by an extract of the caecum of the animal. This extract can be replaced by enterokinase, and can itself activate vertebrate trypsin: there is therefore little doubt of its close similarity to enterokinase. The general conclusion which may be drawn from all these results is that certainly in the molluscs, crustaceans, and insects, and probably in the coelenterates, digestion of protein is carried out by the same series of enzymes as are present in the intestine of vertebrates.

A few invertebrates, such as the annelid *Pheretima* and larvae of *Calliphora* and other meat-eating flies, have proteases that work in a slightly acid medium and may be compared to pepsin.

An interesting case is that of the larva of the clothes moth *Tineola bisselliella*, which lives largely on keratin and related compounds, which are not in the ordinary way attacked by enzymes. The pH of the gut is 10, and under these conditions the keratin is reduced by a system present in a special part of the gut to products on which the animal's proteinase can act. This proteinase cannot act on proteins *in vitro*, but it does act on keratin which has been attacked by a weak reducing agent. Dermestid larvae have a similar method, except that the pH is not above 8·2.

The carbohydrases of invertebrates have not been so fully investigated as the proteases, but something has been made out as to their nature and distribution. Amylase from a number of groups has been shown to be very similar to vertebrate amylase: it hydrolyses starch to maltose, but cannot attack pure starch grains, and usually has a pH optimum slightly on the acid side of neutrality. In the oyster activation by salts is necessary, just as with ptyalin. In most cases the amylase has not been separated from its associated maltase, but in the oyster they occur separately.

The three common disaccharides—maltose, sucrose, and lactose—can be hydrolysed by many invertebrates, and some can split many others, the snail in particular being able to deal with at least seventeen sugars and related substances. Since many disaccharidases (including some of the maltases) are not specific, but act on a group of sugars with similar linkages, the number of enzymes present cannot be determined without careful analysis. Some invertebrates can digest other polysaccharides besides starch. The snail and some others have an inulase that hydrolyses the inulin of the Compositae into fructose. Cellulase has been found in a few Protozoa, earthworms, a thysanuran, some (but not all) wood-boring beetles, and the wood-boring bivalve *Teredo* amongst others, and it is possibly present in the crayfish. The ability of the snail to digest cellulose (and chitin) is now said to be due to the presence of symbiotic Bacteria.

Chitinase (really two enzymes, a mucopolysaccharidase and a β-acetylglucosamidase) has been found in soil amoebae, in coelenterates, nematodes, earthworms, polychaetes, and insects, which are all animals that take arthropods as food. Some cellulases act weakly on chitin.

Alginic acid, which replaces cellulose in the cell walls of brown Algae, is built up from uronic acids, which are simple oxidation products of hexoses. It is digested by many molluscs and echinoderms, and by *Arenicola.* Freshwater and terrestrial molluscs, in which the action is weak, presumably inherited the necessary enzymes from their marine ancestors.

Lipases as a group are much less specific than proteases and carbohydrases, and all glycerol esters seem to be hydrolysed to some degree by any enzyme of this class. Invertebrate lipases differ in a number of points from those of vertebrates; they are differently affected by ions, they are inhibited in different ways, and they differ in the ease with which they attack different substrates. Many of them attack esters of lower fatty acids more readily than oils or fats, so that they are more properly called esterases than lipases. The distinction, however, is a fine one, and it seems that even in the same species the action is sometimes stronger on fats, sometimes on esters. The wax moth, *Galleria,* digests bees-wax, partly with the help of Bacteria, and partly, it seems, by its own enzymes.

In the vertebrates, the different constituents of the enzymes acting on proteins and carbohydrates are very largely produced by different glands, and to some extent act in different parts of the alimentary canal. It was therefore easy to recognize that more than one enzyme of each sort was present, even before the modern work on separating them by adsorption was begun. On the other hand, all the enzymes of an invertebrate are often produced by a single gland, and they nearly always act together in one part of the alimentary canal. This, while it means that the recognition of separate entities is difficult, also means that anatomically the gut of invertebrates is relatively simple. It is difficult to generalize about such specialization as there is, so that each of the major groups must be taken separately. In many echinoderms and lamellibranchs there is a special form of intracellular digestion; large food particles may be taken up and ingested by wandering amoebocytes, which in lamellibranchs may even go outside the animal and ingest food in the mantle cavity. They can attack all three classes of food and are probably of some importance.

Little is known of how the digested food is absorbed, but there is presumably at least some assisted transport. In insects a simple form of this is present, for hexoses are converted to the disaccharide trehalose as soon as they reach the blood. Surface-active agents, which might have a similar action to that of bile in emulsifying fat for digestion, have been found in annelids, echinoderms, crustaceans, and molluscs. In the crab *Cancer pagurus* they consist of fatty acids and aminoacids, and have no resemblance to the steroid bile salts of vertebrates.

Digestion in Protozoa. In the Protozoa digestion goes on in vacuoles. In all the genera that have been carefully investigated the vacuoles rapidly become acid (they reach a pH of 1·4 in *Paramecium*) and some digestion goes on under these conditions. The acidity may be due at least in part to the cytolysis of the food, as the prey is always killed before there is any great departure from neutrality. Later the hydrogen-ion concentration gradually returns to normal, more digestion takes place, and defecation occurs somewhere between pH 5·0 and pH 7·0. Protozoa seem to find proteins the easiest food to digest, many having difficulty with carbohydrate and fat. *Amoeba proteus* digests fat to glycerol and fatty acids, which pass into the cytoplasm and are there resynthesized, but *A. dubia* takes some days to digest olive oil and *Typanosoma evansi* seems to contain neither lipase nor carbohydrase. A number of soil-living rhizopods have both cellulase and chitinase, which are presumably useful if miscellaneous organic particles are used as food.

An account of protozoan digestion would be incomplete without a reference to the fact that the green flagellates are photosynthetic and sometimes (e.g. *Euglena*) saprophytic as well. Photosynthesis is, however, typically a phytological phenomenon, and it would be out of place to discuss it here.

Digestion in Coelenterata. In the coelenterates there are no special digestive glands, but in the Scyphomedusae the glandular cells are concentrated on the gastric filaments and in the Anthozoa on the mesenterial filaments. The enteron contains a proteinase which starts the hydrolysis of proteins. The further breakdown of these, and most of the digestion of fats and carbohydrates, are intracellular. Before the food particles are absorbed they may be moved about by peristalsis or ciliary currents or both. Indigestible parts of the food are extruded by the mouth, and may be removed by the ciliary currents. The green Algae that live symbiotically with many corals supply them with glycerol, and those of *Chlorohydra* supply maltose.

Digestion in Parasitic Worms. Most of the Turbellaria that have been examined resemble the coelenterates in that only the preliminary breakdown of proteins is extracellular, but in some digestion goes on in the meshes of a temporary syncytium formed by processes from amoeboid cells. In some of the Rhabdocoelida digestion is mainly or perhaps entirely extracellular. The pharyngeal glands of the Turbellaria secrete mucus, and in some species of Rhabdocoelida their secretion contains enzymes which are used for external digestion. The liver-fluke, *Fasciola hepatica*, feeds largely on blood, and perhaps also on pieces of solid tissue, so that it presumably has a proteinase, but it can also absorb glucose, fructose, galactose, and maltose, but not lactose or sucrose, through the general body surface; only the first of these sugars is normally available to it. Some tapeworms have been shown to be able to absorb only glucose and galactose; carbohydrases seem to be generally absent from these parasites. *Ascaris* can absorb glucose, fructose, sorbose, maltose, and sucrose, not mannose, galactose, or lactose; but these sugars are taken in through the gut, not by the general body surface. Some nematodes, including *Ascaris*, have amylases, maltase, protease, peptidase, and lipase, and so are not, like the tapeworms, completely dependent on predigestion by the host. Some phytophagous species have cellulase and chitinase, and an enzyme that splits mucopolysaccharides is present in larvae, such as those of *Ancylostoma*, that migrate through the tissues. Most of the enzymes are produced chiefly in the anterior part of the intestine.

Digestion in Annelida. All types of food are used and, in general, the three main classes of enzyme are produced by the wall of the intestine; but proteases are produced also by the pharynx of earthworms. Cellulase and chitinase are present in the intestinal wall of several species, and to a lesser extent in the pharynx. Little work has been done on digestion in polychaetes. In terebellids it is probably entirely extracellular, and absorption is into cells with a brush border. Lipase, proteinase, and amylase are all present, and the last two appear to be secreted by different types of cells. In *Arenicola* much of the digestion is intracellular. In serpulids and sabellids the transport of the food is by cilia throughout.

Digestion in Arthropoda. There is no intracellular digestion in the Crustacea, and all the movement of the food is by muscles, since cilia are entirely absent from the group. In the decapods the food is well broken up in the gastric mill, and is at the same time attacked by a protease which is sent forward from the digestive diverticula. At the entrance to the latter there is a complicated filter that allows only fine particles to pass. All the further digestion and most of the absorption take place here, the main midgut region being extremely short. Material is taken into the diverticula by the contraction of longitudinal muscles, and expelled by the contraction of circular muscles. Some decapods and copepods have a peritrophic membrane similar to that of insects.

In some insects the labial glands contain an enzyme, for instance amylase in the cockroach, sucrase in honey bees and moths, and a protease in the larva of *Corethra* (Diptera), and they nearly always form a liquid that softens the food. Glands seldom open into the crop, but digestion may take place here by enzymes sent forward from the midgut, as for example in the cockroach, where fat is digested and absorbed in this region. In this animal and many others the crop is the chief site of digestion, but this is carried out mainly by yeasts and Bacteria, which are subsequently themselves digested by their host. Midgut digestion is by enzymes in the normal way, and it is also from this region that the proteases used in external digestion by beetles are obtained. Blowfly maggots liquefy the meat in which they live by passing proteases in their faeces. When an insect is feeding on solid food, the bolus on leaving the foregut is enclosed in a thin sac of chitin called the peritrophic membrane (Fig. 2.7). This is permeable to both enzymes and digested food, and absorption takes place through it. It is absent from many fluid-feeders, such as bugs, adult Lepidoptera, fleas, lice, and tabanid flies, and from many carnivorous beetles, in which the midgut cells break down completely.

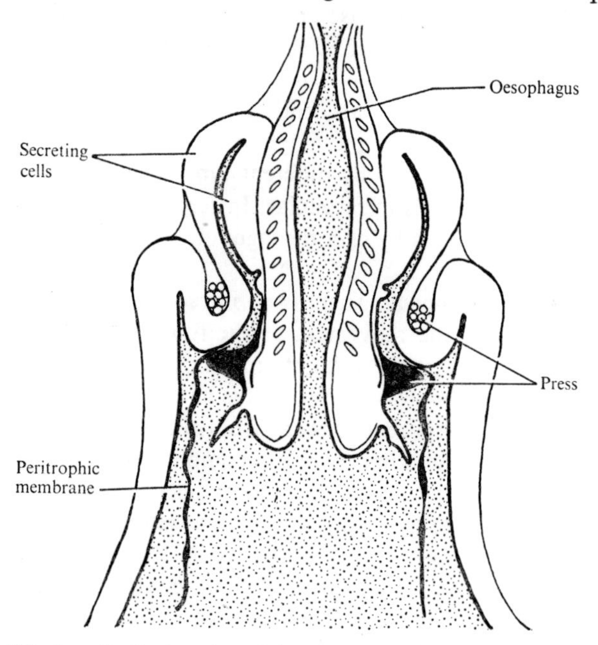

FIG. 2.7. Longitudinal section of the gut of an earwig, *Forficula*, to show the formation of the peritrophic membrane. It is squeezed out as a tube by a press formed by two rings of muscle. Redrawn from Wigglesworth. *Q. Jl microsc. Sci.* 73 (1930).

The chief function of the hindgut is to absorb water, loss of which is very dangerous to insects, but in some beetle larvae digestion of symbionts and absorption of the products takes place there.

Where there is a marked metamorphosis there is often a complete change-over in the enzymes produced corresponding to the change in food. Many adult butterflies and moths have no digestive enzymes except sucrase, and some, such as *Lymantria*, which has vestigial mouth-parts and does not feed, have none at all. Even when sucrase is present and sugar is taken, it seems not to be essential, for many moths can lay just as many eggs if they are fed on water as if they are given syrup. Many male adult Diptera and some females do not need protein, but in other species the females must have a meal of protein before they can lay eggs.

Digestion in Mollusca. The Lamellibranchia are peculiar amongst relatively advanced animals in that most of their digestion is intracellular. The only enzyme certainly set free in the gut of most of them

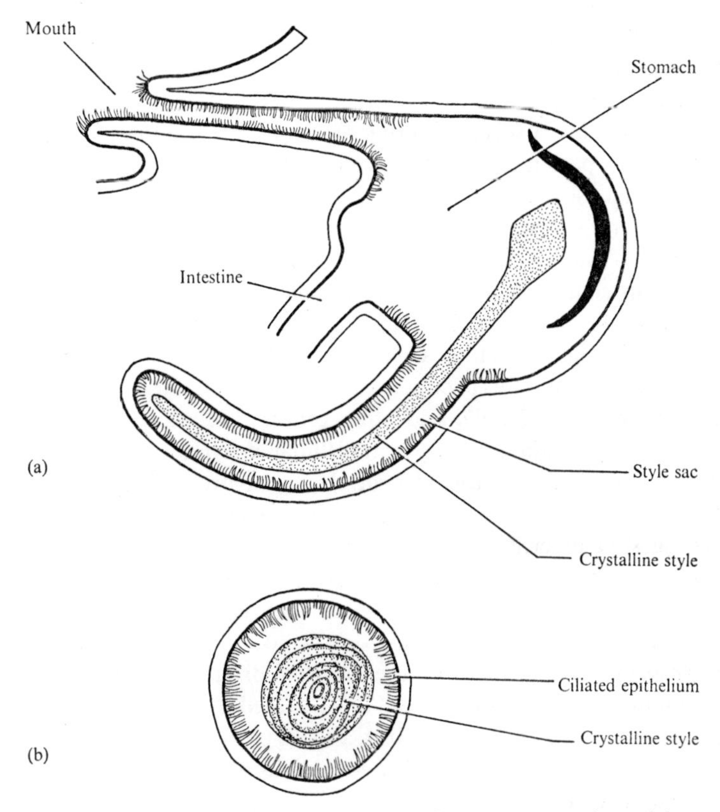

FIG. 2.8. Diagrams of (*a*) longitudinal, and (*b*) transverse, sections of the gut of *Donax*, to show the crystalline style. Redrawn from Borradaile and Potts, *The Invertebrata*, Cambridge University Press (1932). After Barrois.

is the amylase formed in the crystalline style, and in some species there is a lipase. The style is a rod of material consisting largely of globulin (a protein), but containing food particles and the enzyme as well (Fig. 2.8). It is secreted in a style sac in the intestine, and is rotated by means of cilia, its speed in the young oyster being about 70 rev./min. It is gradually worn down in front so that the amylase is liberated and at the same time brought into very close contact with the food. The function of the protein is to maintain the pH of the gut at about 5·5, which is optimal for the amylase. The style is perhaps derived from the ergatula (section G 2.1). Some authors have maintained that the cells of the digestive diverticula break down and so liberate a protease and a lipase into the gut, and this now seems to be generally accepted for *Lasaea rubra*. In the primitive *Nuccula sulcata* there appears to be no intracellular digestion. In many genera the stomach has a special filtering mechanism that allows only small particles to pass into the digestive diverticula, in the cells of which further digestion takes place. Particles are also taken up and digested by wandering amoebocytes. Muscle is almost entirely absent from the gut, so that all the movement of the food is by cilia, which maintain a circulation through the diverticula.

The more primitive Gastropoda closely resemble the lamellibranchs in their methods of digestion. Some, such as *Crepidula*, possess a style, and this produces the only extracellular enzyme, which is an amylase. In *Patella* and other genera that are without a style the amylase, again the only extracellular enzyme, is produced by lateral diverticula of the foregut. In carnivorous forms, such as *Murex*, there are extracellular proteases, which may be formed by the digestive diverticula or by buccal glands or both. In the Pulmonata, such as *Helix*, most of the digestion is extracellular, but it seems that the digestion of protein goes on only in the cells, though the evidence is conflicting. In all gastropods most of the digestion, as well as absorption, takes place in the digestive diverticula. The radula is an important feeding organ, and is lubricated by the secretion of the buccal glands. In *Dolium* and *Murex* these produce also the sulphuric acid for the solution of the calcareous shells of the bivalves on which these animals feed.

In the cephalopods digestion appears to be entirely extracellular. The digestive diverticula of *Sepia* and *Octopus*, but not of *Loligo*, are absorptive as well as secretory.

3. Respiration and Transport

By derivation, and in ordinary usage, the word 'respiration' means simply sucking air into the body and blowing it out again: it is in fact synonymous with breathing. Zoologists early recognized that the object of breathing was that the animal might absorb oxygen and give off carbon dioxide, and the meaning of respiration was extended to cover the gas exchange even where, as in the earthworm, it takes place without any special bodily movement. But further it became obvious that this taking up of oxygen from the atmosphere and giving up of carbon dioxide were simply the beginning and end of a process common to all animals (and indeed to all living matter) by which energy was made available for use when oxygen was taken to the tissues and entered into chemical reactions. In botany, since plants do not breathe, respiration has long meant merely the exothermic chemical changes by which energy is supplied, even when oxygen takes no part in them. It could logically be taken to include comparable processes in animals, and animal respiration would then be an exothermic chemical reaction (or series of reactions) by which energy is supplied for the use of the organism. Where oxygen is required for these reactions the whole mechanism whereby the gas is supplied would be included in the term, and where carbon dioxide is a by-product of the reaction its removal would be included. The word has never been used in connection with the supply of reactants other than oxygen, nor with the removal of products other than carbon dioxide. A distinction would be made between external respiration, which covers the supply of oxygen and removal of carbon dioxide, and tissue respiration, which is the chemical processes in the cells.

Animal physiologists have never been entirely happy about this extension of the meaning of the word, and many of them have compromised by not using respiration unless oxygen comes in at some stage. This is illogical and inconvenient, since oxygen only enters into the very end of the chain of reactions that supply energy and many of them are the same whether oxygen is present or not. Our increased knowledge of the metabolism of the body has now made any extension of the term respiration to include chemical processes both awkward and confusing, since, as will be shown in Chapter 4, the reactions that are used in the process of building up tissues and reserves are inseparable as a class from those that are used in providing energy. It seems best to revert to the older zoological sense of the word, and in this book respiration is taken to mean 'the provision of oxygen and removal of carbon dioxide'. For the wider processes the phrase 'provision of energy' must be used until an acceptable shorter term is invented.

G 3.1. THE RESPIRATORY QUOTIENT

The ultimate fate of the oxygen taken in when an animal respires is to combine with the carbon or hydrogen in the foodstuffs and form carbon dioxide or water. Water is taken up and given off in various forms, and is thus difficult to measure, but the gas exchange is easily measured with some form of manometer. The ratio

$$\frac{\text{volume of carbon dioxide given out in time } t}{\text{volume of oxygen absorbed in time } t}$$

(both being measured under the same conditions) is the respiratory quotient, usually abbreviated to R.Q. The volume of a gas, under given conditions of temperature and pressure, is proportional to the number of molecules that it contains, so that the respiratory quotient is also the ratio of the number of molecules of carbon dioxide and oxygen taking part in the over-all reaction by which the carbohydrates, fats, or proteins are oxidized. It is obvious that this is different for the three classes. For most carbohydrates the equation is

$$C_x(H_2O)_y + xO_2 = xCO_2 + yH_2O,$$

so that the R.Q. is unity. Tristearin, a common fat, gives

$$\begin{array}{l} \;\;CH_2\text{—}OOC_{18}H_{35} \\ \;\;\,| \\ 2CH\text{—}OOC_{18}H_{35} + 163O_2 = 114CO_2 + 110H_2O, \\ \;\;\,| \\ \;\;CH_2\text{—}OOC_{18}H_{35} \end{array}$$

which makes the R.Q. 114/163, or 0·70. Other fats give similar values and, though the figure for protein is more variable, it is always about 0·8 or a little more.

Knowledge of the respiratory quotient can be very helpful in suggesting what is going on in an animal or a tissue, but great care must be used in its interpretation. Tissues with quotients of 1·0 and 0·7 obviously have different chemistry, but two with values of 0·8 are not necessarily similar, for while one may be using protein the other may be using a mixture of carbohydrate and fat. There may also be incomplete combustion, and errors occur where carbon dioxide is retained within the body, as in hibernating mammals. An R.Q. of more than 1·0 probably always means that fat is being formed from carbohydrate, as in the fattening of beasts for market and in birds before migration.

G 3.2. RESPIRATORY ORGANS

The essential feature of respiration is the passage of oxygen from the surroundings of the animal into its cytoplasm. In the latter the oxygen is always in solution; in the environment it may or may not be. If it is not, the first step is to dissolve it, and this means that the respiratory surface must be kept damp. In a small animal, provided that it lives in water or in a damp environment where its surface will not dry out, the general body surface may be large enough to absorb all the oxygen that is needed, without any special apparatus. This is the state of affairs in earthworms (the largest of which may be more than 6 ft long), and since they are covered with a layer of water in which the oxygen dissolves there is little distinction between aerial and aquatic respiration. Some species (e.g. *Eiseniella tetraeda*), which live near the margins of lakes, seem equally at home whether they are above or below the level of the water, and even the more terrestrial genera, such as *Lumbricus*, can be kept in the laboratory under well-aerated water.

Most of the higher animals have some special region or respiratory organ that is used for the uptake of oxygen. A part of the body may be considered as such if it has a greater rate of gas exchange per unit area than has the general body surface, or if it seems to confer no other advantage on the animal than an increase of surface for gas exchange. Often we must guess, and some structures that have been claimed as respiratory organs have been shown to have other functions. The 'anal gills' of gnat and midge larvae are mainly organs for water exchange, while the 'gills' of lamellibranchs are feeding organs, oxygen being taken up chiefly by the mantle.

Respiratory organs are necessary for large or active animals because while the energy (and so the oxygen) needed by an animal increases as its weight, and so as its volume or the cube of its length, the rate at which it can take up oxygen increases as its surface, or the square of its length. Beyond a certain size, therefore, enough oxygen cannot be obtained without a respiratory organ. The magnitude of this critical size is less the more active the animal. Respiratory organs have probably also conferred an advantage in evolution because they allow for specialization of the body surface. An armoured cuticle, for example, or a surface impermeable to water, can rarely be developed unless there is some special region for taking up oxygen.

Where special circumstances have allowed it, occasional species have abandoned the respiratory organs used by their ancestors and reverted to the use of the skin. The semi-terrestrial hermit-crab *Coenobita*, which feeds on coconuts, has no gills; there are some Apterygota without tracheae, and a few aquatic insect larvae (*Simulium*, *Chironomus*) where they are not functional. In all the amphibians there is much gas exchange through the skin; some salamanders have neither gills nor lungs, and the axolotl, though it has both, can live perfectly well when they are cut off.

Respiratory organs may be classified by their physiology or by their morphology. If they work in water they are gills, if in air, lungs. They may project outwards, or they may be invaginations. Many animals have taken to using not the general body surface, but the front and back ends of the alimentary canal which communicate easily with the surrounding air or water.

M 3.21. The Lungs and Breathing of Mammals

The lungs of mammals are diverticula of the pharynx which have become very finely divided, so that they end in several millions of alveoli, about 0·1 mm in diameter. These have a thin and very vascular wall, lined with a thin layer of a lipoprotein of very low surface tension which presumably helps to keep the walls damp. Breathing is complex, but in man consists of three types of movement; the ribs, normally directed forward and slightly downward, are raised; the diaphragm, which at rest is convex upward, is contracted and so flattened; and the vertebral column is slightly extended. The parts played by these components depend on the depth of breathing. All have the effect of increasing the volume of the thoracic cavity, and so reducing the pressure within it. The lungs, at atmospheric pressure, must therefore expand; their volume is in turn increased and their pressure lowered, and so air moves in from the atmosphere and inspiration occurs. In expiration the various muscles relax, and the elasticity of the lungs, aided by the contraction of some intercostal muscles, reduces the volume of the lungs and forces air out.

Various adaptations prevent the passage of food and air into the wrong tube. The entrance to the oesophagus is normally closed by a sphincter, but opens for swallowing, at which time the larynx closes. In some mammals, such as the horse and cattle, the larynx is extended into the pharynx, so that there is a lateral food channel down which food can pass while breathing continues.

The lungs are never completely emptied, and the maximum tension of oxygen in them is only about two-thirds that in the air. The respiratory movements are partly controlled reflexly; impulses arising in the lungs when they are stretched pass by the vagus nerves to the medulla, whence efferent impulses go to the various muscles. But breathing is much more complicated than this, and depends largely on the composition of the blood. An increased hydrogen-ion concentration of the blood stimulates receptors in the aortic wall and in the carotid bodies, and high acidity in the cerebrospinal fluid stimulates receptors in the medulla itself, so that the depth of breathing and its frequency are increased. Acidity is normally produced by an accumulation of carbon dioxide in the blood. Receptors in the aorta and carotid body respond similarly to low oxygen concentration in the blood, but they are active only under exceptional circumstances, such as at high altitudes. In healthy men not taking exercise rapid breathing does not become noticeable until heights of about 4250 m (14 000 ft) are reached, and at this altitude the tension of oxygen in the air is just about three-fifths of the normal.

V 3.21. Breathing of other Vertebrates: Lungs

Lungs homologous with those of mammals are present in some fishes, the adults of most amphibians, and in all reptiles and birds. They are simpler than those of mammals, with little or no branching except in birds, where there is a complicated arrangement, not fully understood, in which some of the air-passages or bronchi lead to non-vascular air sacs. The effect of this is that there is little dead space, and the oxygen in contact with the respiratory surface is at almost the same tension as in the atmosphere. Ribs are sometimes thought of as being primarily concerned with breathing, but since the earliest amphibians and reptiles had them in the abdomen and tail as well as in the thorax, this is improbable; rather, the vertebrates were slow to make use of them for breathing and only the birds and mammals do so fully. Mammals alone have a muscular diaphragm. Birds use their intercostal muscles in a similar way to mammals; expiration is more active than inspiration. Control is similar to that in mammals. Some reptiles use their intercostal muscles in breathing, but more use the method of the force pump, in which air is pushed into the lungs when the floor of the mouth is raised with the mouth and nostrils shut. Since the volume of air in the mouth is less than that in the lungs this must be inefficient. Frogs and urodeles also use this method, but they take up much of their oxygen elsewhere than in the lungs—through the buccal cavity and the lungs in both, and by external gills in urodeles. In the salamander *Ambystoma maculatum* the proportion of the oxygen taken up by the lungs increases with rising temperature, which corresponds to the common observation that an active frog fills its lungs much more often than one at rest. The method by which lung-fish breathe is not entirely clear, but it is certain that they take in air through the mouth, not the nostrils.

Air-breathing vertebrates that dive and catch their food under water have developed devices to meet the difficulties that this leads to. In seals and ducks there is a reflex that produces cessation of breathing. In ducks the stimulus can be contact of hot or cold water (but not that between 30 and 40 °C) with the nares, or even holding the head in a downward-pointing position. The carotid and similar receptors must be relatively insensitive to carbon dioxide.

V 3.22. Breathing of other Vertebrates: Gills

The gills of fish are expansions of the wall of the pharynx situated on what are usually called slits; in the elasmobranchs these are expanded pouches carved out of the body wall, while in the teleosts, where they are more slit-like, the gills project into the outer world but are covered by an operculum. The effect is the same; the gills are situated in a chamber that has two openings, one to the pharynx and one to the exterior. The mechanism of breathing is essentially the same in the two groups, and is a combination of force pump and suction pump. In the selachians the two hemibranchs of a gill slit meet and effectively divide the pouch into two chambers. When the hypobranchial muscles contract and lower the floor of the pharynx, the external openings of the gill slits being shut, water is drawn in through the mouth and spiracles. These then close, and adductor muscles contract and draw the epi- and ceratobranchials closer together, increasing the pressure in the pharynx and forcing water through the lattice-work of the gills. This is the force pump. In the suction pump, the outer gill chambers are first enlarged by the contraction of the adductor muscles and of muscles in the gill septum. While this is happening the external openings are closed, so that water must be drawn through the gills from the inner chambers. The outer chambers are emptied to the exterior by the contraction of superficial muscles, back-flow into the inner chamber being prevented by the high resistance of the small channels through the gills. The two mechanisms may work independently or in conjunction with each other, and one or both may be passing water over the gills for nine-tenths of the time. By the suction pump the fish can breathe with its mouth continuously open.

In teleosts there is a similar buccal force pump and a suction pump worked by the operculum. In many species there are two velar folds of tissue in the mouth, which form a valve just behind the front teeth; in inspiration they open and the opercula close, in expiration the folds meet and the opercula are lifted away from the sides of the body; by these means the fish can breathe with its mouth continuously open. Similar valves are present in the lung-fishes *Polypterus*, *Amia*, and *Lepidosiren*.

Some fast-swimming fish, such as mackerel and sharks, probably breathe little or not at all at speed, but simply swim with the mouth open so that water flows over the gills.

There is undoubtedly some control of breathing, chiefly through the medulla. In the dogfish there are proprioceptors in the pharyngeal wall, innervated by the vagus and comparable to those in the lungs of mammals. When they are stimulated by inflation of the pharynx breathing is inhibited, and other receptors situated on projections from the gill-bars behave in the same way when they are deflected. These reflexes would maintain a rhythm. Elasmobranchs in general do not respond to changes in the concentration of oxygen or carbon dioxide in the blood, but teleosts do, increasing the depth or frequency of the respiratory movements, or both, with rising carbon dioxide tensions or falling oxygen.

The external gills of a few fish, such as *Polypterus*, and of amphibian larvae and the perennibranchiate urodeles, are vascular expansions of the skin which, since they project into the surrounding water, need no special mechanism for their ventilation. They are often capable of some movement.

J 3.2. BREATHING AND RESPIRATORY ORGANS OF INVERTEBRATES

Many invertebrates have structures that may be called gills. In the polychaetes, the marine gastropods, and crustaceans such as crayfish they are external vascular expansions of the skin. The aquatic larvae of Trichoptera, Ephemeroptera, Odonata, a few Lepidoptera, and *Simulium* (Diptera) also have structures called external gills, but they are supplied with fine tracheae (section J 3.7) instead of blood vessels. They do not always absorb oxygen, and in the nymphs of the mayflies some connection can be seen with habitat. In *Ephemera vulgata*, which burrows in mud, they absorb oxygen and also help by their beating to ventilate the surface of the body, even at high oxygen tensions. In *Chloeon* they do not absorb oxygen, but they beat actively when the oxygen tension is lowered; in *Baetis*, which lives in fast-flowing, and so well-aerated, streams, they seem to be functionless.

The alimentary canal is an alternative site of gas exchange. The echiuroid worm *Urechis caupo* pumps water in and out of its rectum, and some aquatic oligochaetes (e.g. *Nais*, common in British waters) have an ascending ciliary current beginning at the anus, often assisted by antiperistalsis; the rectum is also used for oxygen uptake by many primitive crustaceans such as water fleas, and to some extent by the crayfish. Dragon-flies of the sub-order Anisoptera have rectal tracheal gills, similar except in position to those of other aquatic larvae. The holothurians breathe by respiratory trees, which are branched diverticula of the rectum; one species, *Holothuria tubulosa*, which lives in stagnant water, rises to the surface and draws in air to oxygenate them.

There is often some special arrangement for renewing the layer of water in contact with the respiratory surface. *Spirographis spallanzanii* and *Sabella pavonina* are polychaetes, with external gills, that live in tubes made of mucin. Through these they maintain a current of water by rhythmical contractions of the body; a swelling appears at the hind end of the worm, completely filling the tube, and then moves forward. As a result, the worms can live in water containing little oxygen, while if they are removed from their tubes they die unless the water is well aerated. Their natural tubes can be successfully replaced by glass ones. The external gills, which project from the mouth of the tube, have been shown in other species to account for only a third to a half of the total uptake of oxygen.

In most of the higher crustaceans some of the limbs maintain definite currents over the gills. In the decapods, such as the crayfish and shore crab, the exopodite of the second maxilla is modified as the scaphognathite, which flaps and draws water forwards through the gill chamber about once a second. Other mouth parts assist in the process, some of them ensuring that the current actually passes over the gills, and the epipodites of the three maxillipedes brush the surface of the gills to prevent the accumulation of foreign matter. The scaphognathite reverses for a few strokes every now and again. The effects of the concentration of oxygen and carbon dioxide on the beat of the respiratory limbs of several Crustacea have been investigated. They fall into three groups: in the first, represented by the shore-crab (*Carcinus maenas*), the beat is independent of the concentration of the two gases. In the second, of which the amphipods *Gammarus pulex* and *G. locusta* are examples, the beat (in this case of the pleopods) is increased by falling oxygen tension and by increasing carbon dioxide tension. In the third, typified by the crayfish (*Astacus fluviatilis*), the beat is increased by oxygen deficiency but is unaffected by carbon dioxide. The distribution of these three groups certainly does not follow the classification of the Crustacea, and neither does it seem to agree with the habitats or habits of the animals. It appears that the power to regulate the respiratory current in accordance with needs has been independently evolved on several occasions.

In the lamellibranchs the ciliary food currents (see section G 2.21) also serve to bring in oxygen.

There are few successful and active terrestrial invertebrates other than the insects, and lungs are correspondingly rare. The mantle, developed to enclose a hollow chamber, is used as such in the pulmonate gastropods. They renew the air in their lung at fairly frequent intervals, since all the gas is expelled when the animal contracts into its shell. The freshwater forms, such as *Planorbis* and *Limnaea*, frequently come to the surface, push the pulmonary aperture into the air, open it, raise the floor of the mantle cavity, and lower it again. This brings in fresh air. Scorpions and spiders have two or more small openings or stigmata, usually on the ventral surface, which lead into chambers into which project several vascular and leaf-like projections of the body wall called lung-books.

The tracheae of insects and some other arthropods are in part functionally comparable to the vascular system of vertebrates, and this side of their physiology is dealt with in section J 3.7. The replacement of air in them is surprisingly rapid; if a cockroach is kept in pure oxygen until its tracheae are full of this gas, and is then returned to air, the tracheae contain 80 per cent nitrogen after 1 min. In small insects replacement is by diffusion only, but in large ones it is assisted by some sort of pumping, either of the large tracheae themselves (dragon-flies, moths, beetles) or of the abdomen (*Hymenoptera*, flies). In some there are air-sacs comparable to those of birds. In locusts and grasshoppers there is a current of air, in at the anterior spiracles and out posteriorly, and there is a similar current in the opposite direction in the corixid bugs. At rest the spiracles of most insects are closed, or open only at intervals, but they are open continuously when the insect is active. Those of the rat-flea *Xenopsilla cheopis* and others can be induced to open either by low oxygen concentration or by high carbon dioxide, or by acidity, but those of an African dragon-fly open to low oxygen only and do not respond to carbon dioxide. Changing conditions cause a progressive increase in ventilation; more and more spiracles open, and then pumping begins.

Those insects that, like mosquito larvae and many of the beetles, live beneath the surface of water but come to the air to breathe, are of some interest. The surface of the stigmata is covered with non-wettable substances. These have a high angle of contact with water, which is thus prevented from entering, but oil will wet them, so that a film of it spread on the surface will enter as soon as the spiracles touch it. Once the tracheae contain oil no air can go in and so the animal must drown.

Many insects (such as *Dytiscus* and *Notonecta*) carry bubbles of air below the surface with them. This is used directly as a source of oxygen, and it also enables them to extract oxygen from the water in the following way. Oxygen is being used by the animal, so that its partial pressure in the bubble falls, and in time goes below that in the water. But as this happens the partial pressure of the nitrogen in the bubble must necessarily rise and will go above that in the water, for the total pressure in the bubble is determined simply by its depth below the water, and is constant. Equilibrium can be restored either by oxygen entering the bubble or by nitrogen leaving it, but the first goes on three times as fast as the second, so that little of the latter takes place. *Notonecta* was found to be able to live for 7 h below water saturated with air, but only 35 min below water saturated with oxygen. In the latter case the nitrogen in the bubble would rapidly dissolve in the water, and the conditions described above would not apply, so that although as much oxygen was present as in the first case, it was not available.

As the bubble of an animal such as *Dytiscus* slowly loses its nitrogen, it shrinks, and so the surface over which oxygen can enter it is reduced, and the rate of diffusion of oxygen inwards falls also. Eventually the bubble disappears, and the animal has to come to the surface to renew it. Some insects do better than this. Their body is covered with a complete layer of air, called the plastron, held by unwettable hairs. The bubble can never disappear, any more than air can disappear from the rigid tracheae of insects with no spiracles, and oxygen is always available if it is present in the water. The bug *Aphelocheirus* forms its gas layer below the skin just before ecdysis, so that when the skin is shed the bubble is complete and in position. If the water is well aerated, as it is in swiftly flowing streams, the bug need never come to the surface to breathe. The hairs are about 5 μm long, and there are two and a half million of them per square millimetre. *Corixa* has a thicker plastron, which is not quite so permanent, so that the animal can survive in oxygen-saturated water for only 40 h. Plastron respiration is found also in many beetles, in some ichneumons, and possibly in water spiders.

G 3.3. COMPARISON OF AQUATIC AND AERIAL RESPIRATION

Students sometimes ask whether aquatic or aerial respiration is the more efficient. The question is probably meaningless, since the conditions that have to be fulfilled are very different for the two. It has been suggested that since the teleost *Erythrinus unitaeniatus*, which can breathe completely through either gills or lungs, has a slightly greater area of the latter, aerial respiration is less efficient, but obviously no conclusions can be drawn unless full information is available about the activity of the fish. Nevertheless, there is some reason for saying that air-breathing is less efficient. Ultimately all respiration is aquatic, since the first thing the oxygen has to do when it reaches the surface of the lung is to dissolve in the surface film of water. The difference between lungs and gills from a physiological point of view is that the former are covered with a relatively thick surface film of water, while the latter may be swept by a current so that the stationary layer is only a few molecules thick. Although there is more oxygen in air than in water, it is not so easy for the animal to obtain it. The four species of snail *Littorina littorea, L. obtusata, L. rudis*, and *L. neritoides* form an ecological series of animals that live in progressively drier situations on the shore. The first lives in rock pools and is exposed only at low tides, but the others may be exposed for several hours. Associated with the habitats is the degree of vascularization of the mantle cavity, which increases in the same order as the dryness of the surroundings.

The rate of diffusion of oxygen in air is 450 000 times that in water, so that unless there were some means for continual renewal of the layer of water in contact with the gill, access of oxygen would be much slower than in air.

Land animals may drown because their respiratory surfaces become covered with mucus which prevents free access of oxygen, but necropsies show that in man the chief cause of death is absorption of water, with consequent damage to the red cells.

There seems no reason why aquatic animals should die in air provided that they are kept moist, and, as we have seen, earthworms can live equally well in damp air or in well-aerated water. Crabs have no difficulty in walking from one tidal pool to another. Where there are finely divided gills they may collapse and so expose a much smaller surface to the atmosphere when the animal comes out of the water, and this is no doubt one reason why fish die in air. Another is a rather special one connected with their haemoglobin (section V 3.41). In the presence of acidity produced by 2 per cent of carbon dioxide this may have a loading tension as high as 150 mm of mercury. In water the carbon dioxide is easily removed, and the acidity is in any case buffered by bicarbonates, but in air the carbon dioxide concentration will rise—there is 5 per cent in the lungs of mammals—and the haemoglobin will not be able to take up enough oxygen. It is noticeable that many of those fish that do live on land breathe, like the eel, through the skin, so that carbon dioxide is easily carried away. An eel on land, however, still gets one-third of its oxygen through its gills, and the branchial chamber is continually replenished with air.

G 3.4. TRANSPORT OF OXYGEN AND CARBON DIOXIDE BY A CIRCULATORY SYSTEM

Oxygen must be able to reach all the cells of the body, and in a small animal it can do this by diffusion in solution in the cytoplasm; but as it goes it is being used up and its concentration is falling, so that there must come a point at which there is not enough to sustain metabolism. The animal has two evolutionary choices: to grow no bigger, or to develop some method of transporting oxygen rapidly and without loss to the interior of the body. The size at which this becomes necessary will depend on the activity of the animal and the availability of oxygen; isolated frog muscle in aerated water can just survive without blood vessels if it is not more than 4 mm thick, and most animals of this size or larger have a circulatory system that conveys oxygen round the body. The exceptions are either inactive, or, like the Scyphomedusae, have a large bulk of non-metabolizing tissue.

For a circulatory system to work it must carry the oxygen more or less rapidly from the respiratory surface to the other parts of the body, which implies a branching system with some sort of pumping mechanism, and probably valves to control the direction of flow; it must be able to pick up oxygen at regions of high concentration and give it up at low. The water vascular system of echinoderms, which is open to the exterior and contains sea water, is respiratory, and oxygen is taken up by the tube feet, but this seems to be a secondary function. In other large animals (annelids, arthropods, molluscs, chordates) the vessels conveying the oxygen are not open to the exterior and there is a pump or heart (or more than one) which provides a circulation of the fluid or blood in them. Most of the blood system of arthropods and mussels consists of ill-defined spaces, the haemocoele, in which the organs are situated; the only distinct vessels are the heart and the main channels leading from it. In the annelids and chordates there is a closed system, in which large vessels or arteries leading from the heart communicate by fine capillaries with large vessels or veins leading to it. In all four phyla the blood contains a substance that increases its power of carrying oxygen; it is always coloured in at least one of its states (though this seems to have no importance) and so is called a respiratory pigment. It combines with oxygen by a reversible reaction, so that it takes up oxygen at the respiratory surface, where the concentration is high, and gives it up in the tissues where the concentration is reduced by metabolism.

Four groups of such pigments are known (Table 2). Haemoglobin consists of a base called protohaem or haematin, which is a ferrous ion porphyrin, united to a protein, globin. This differs in different species, and while in vertebrates it is usually small in invertebrates it may be large, giving a high molecular weight to the resulting haemoglobin. Chlorocruorin resembles haemoglobin but has a different base. Haemerythrin, in which the iron is attached to a protein, has a molecular weight of 66 000 or 120 000. In haemocyanin also, the metal, copper, is in the protein, which is very large, giving a molecular weight of about 2 000 000.

TABLE 2. *Respiratory Pigments*

Pigment	Contained metal	Colour oxygenated	Colour deoxygenated	Occurrence
Haemoglobin	Iron	Red	Purplish	Vertebrates and scattered invertebrates
Chlorocruorin	Iron	Green	Green	Some polychaetes
Haemerythrin	Iron	Red	Pale yellow	Some annelids and brachiopods
Haemocyanin	Copper	Blue	Colourless	Most molluscs (not lamellibranchs), decapod and stomatapod crustaceans, some arachnids

TABLE 3. *Carrying power of bloods (various authors)*

Fluid	Pigment	Per cent oxygen by volume when saturated
Sea water		0·7
Human blood	Haemoglobin	19·0
Blood of whales and seals	Haemoglobin	20–30
Blood of pigeon	Haemoglobin	20
Blood of alligator	Haemoglobin	*c.* 10
Blood of frog	Haemoglobin	1·6–5·1
Blood of cod	Haemoglobin	*c.* 7
Blood of mackerel	Haemoglobin	15·8
Blood of elasmobranchs	Haemoglobin	*c.* 4–8
Blood of lamellibranchs	Haemoglobin	1–5
Blood of *Arenicola*	Haemoglobin	5·7–9·7
Blood of *Sprirographis*	Chlorocruorin	9·1
Blood of *Cancer*	Haemocyanin	1·4–2·3
Blood of *Octopus*	Haemocyanin	3·1–4·7
Blood of *Limulus*	Haemocyanin	0·74–2·7

The quantitative ability of blood to take up oxygen, its carrying power, will depend on the concentration of pigment present and on the amount of oxygen with which unit mass of the pigment will combine; both, but especially the second, may change with changing circumstances. Some typical values of carrying power are shown in Table 3. The high carrying-power of the blood of vertebrates is due in part to the fact that the haemoglobin is contained in envelopes, the red blood corpuscles, or erythrocytes, which allow a high concentration of pigment with a large surface.

M 3.41. The Haemoglobin of Mammals

The best-known respiratory pigment is the haemoglobin of mammals. If we represent its reaction with oxygen by the reversible equation

$$\mathrm{Hb}+n\mathrm{O}_2 \rightleftharpoons \mathrm{HbO}_{2n},$$

where Hb stands for a molecule of haemoglobin and HbO_{2n} for a molecule of oxyhaemoglobin, we can apply the Law of Mass Action. Using the conventional square brackets to represent concentration, we have

$$k_1[\mathrm{Hb}].[\mathrm{O}_2]^n = k_2[\mathrm{HbO}_{2n}],$$

or

$$\frac{[\mathrm{Hb}]}{[\mathrm{HbO}_{2n}]} = \frac{1}{[\mathrm{O}_2]^n}.\frac{k_2}{k_1} = \frac{1}{k[\mathrm{O}_2]^n};$$

adding one to each side,

$$\frac{[\mathrm{Hb}]}{[\mathrm{HbO}_{2n}]}+\frac{[\mathrm{HbO}_{2n}]}{[\mathrm{HbO}_{2n}]} = \frac{1}{k[\mathrm{O}_2]^n}+\frac{k[\mathrm{O}_2]^n}{k[\mathrm{O}_2]^n};$$

reversing,

$$\frac{[\mathrm{HbO}_{2n}]}{[\mathrm{Hb}]+[\mathrm{HbO}_{2n}]} = \frac{k[\mathrm{O}_2]^n}{1+k[\mathrm{O}_2]^n}.$$

Now $[\mathrm{Hb}]+[\mathrm{HbO}_{2n}]$ is a measure of the total haemoglobin present in whatever state it may be, so that this equation gives us a relation between the proportion of the total that is in the form of oxyhaemoglobin, and the concentration of oxygen with which it is in equilibrium. In practice it is convenient to state the oxyhaemoglobin as a percentage of the total, and to measure the oxygen concentration by its partial pressure in millimetres of mercury. The equation must then be modified to read

$$100\,\frac{[\mathrm{HbO}_{2n}]}{[\mathrm{Hb}]+[\mathrm{HbO}_{2n}]} = 100\,\frac{k(cp)^n}{1+k(cp)^n}, \qquad (1)$$

where p is the partial pressure of oxygen and c is a factor relating this to the concentration in gram-molecules per litre.

The power n is the number of molecules of oxygen that combine with one molecule of haemoglobin, and this was determined in the following way. Analysis showed that 32 g of oxygen reacted with that mass of haemoglobin that contained 56 g of iron, so that for each

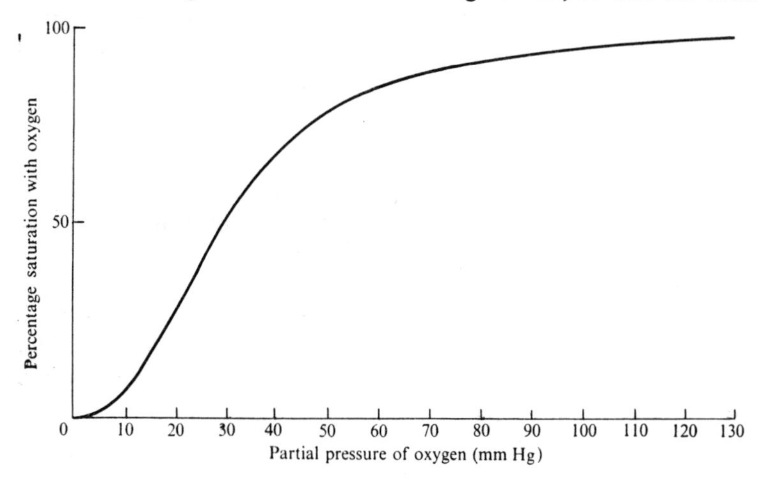

FIG. 3.1. Dissociation curve of human oxyhaemoglobin under 40 mm pressure of carbon dioxide at 38 °C. After Haldane and Priestley, *Respiration*, Clarendon Press, Oxford (1921).

atom of iron one molecule of oxygen was taken up. Measurements of the osmotic pressure of solutions suggested that the molecular weight of haemoglobin is about 68 000, and this was confirmed by measurement of diffusion rates and rates of sedimentation in the ultracentrifuge. In 68 000 g of haemoglobin there are 224 g of iron, so that one molecule contains four atoms of iron and $n = 4$.

When experimental values for the percentage saturation of haemoglobin (the left-hand side of eqn (1)) are plotted against p (the tension of oxygen with which they are in equilibrium) a curve of the form of Fig. 3.1 is obtained. This fits the equation only if $n = 2{\cdot}5$ or thereabouts. The hypothesis put forward more than 50 years ago to explain this was that the oxygen molecules are added one at a time, so that the equation represents something of an average of four different reactions. X-ray crystallography has now shown that this is so. A single oxygen molecule becomes associated as a ligand with an iron atom, which is not oxidized, but whose bonding changes from ionic to covalent. This leads to a change in the domain of π-orbital electrons, and so affects the ease with which the next atom of iron can react with oxygen. The first three steps go on relatively slowly, and then the last molecule of oxygen is added rapidly. It seems likely that more than four reactions go on, since the haemoglobin probably dissociates into its subunits, and a single atom of oxygen can be added not only to each of these (the monomers) but to all the possible dimers of molecular weight about 34 000.

Dissociation curves like that in Fig. 3.1 give one much information about the working of a blood pigment. It is obvious that complete saturation is approached asymptotically, and will seldom be attained; since in practice the oxygen tension at the respiratory surface is usually enough to give 95 per cent saturation, the pressure of oxygen in equilibrium with this percentage of oxyhaemoglobin is called the loading tension (t_L). Similarly, the haemoglobin in the centre of the body is seldom more than half dissociated, so that the pressure in equilibrium with 50 per cent oxyhaemoglobin is called the unloading tension (t_U) (Table 4). Between these pressures the pigment works; if one follows the curve of Fig. 3.1 from right to left, there is a rapid giving up of oxygen as the pressure falls over this range. Displacement of the curve to the right means a decrease in the affinity of the pigment for oxygen.

TABLE 4. *Loading and unloading tensions of bloods (various authors)*

Animal	Pigment	Temperature (°C)	t_U (mmHg)	t_L (mmHg)
Mammals	Haemoglobin	38	*c.* 27	*c.* 90
Pigeon	Haemoglobin	38	35	90
Fowl	Haemoglobin	40	*c.* 30	*c.* 120
Alligator	Haemoglobin	29	28	84
Frogs and toads	Haemoglobin	25	*c.* 30	*c.* 70
Sea-fish and trout	Haemoglobin	17	12–15	30–40
Carp, pike	Haemoglobin	15	2–3	10
Planorbis	Haemoglobin	12	1–2	15
Arenicola marina	Haemoglobin	17–20	2	5
Daphnia magna	Haemoglobin	17	3	..
Chironomus riparius	Haemoglobin	17	0·6	..
Spirographis	Chlorocruorin	10 (pH 8)	9	*c.* 26
Spirographis	Chlorocruorin	26 (pH 7·35)	29	*c.* 50
Sipunculus	Haemerythrin	19	8	35
Crustacea	Haemocyanin	15	14	33
Helix	Haemocyanin	11	4	17

The exact shape of the curve depends on the nature of the haemoglobin and on certain physical factors. The formation of oxyhaemoglobin is exothermic, so that a rise in temperature will dissociate it, and the curve will move to the right. Oxyhaemoglobin is a stronger

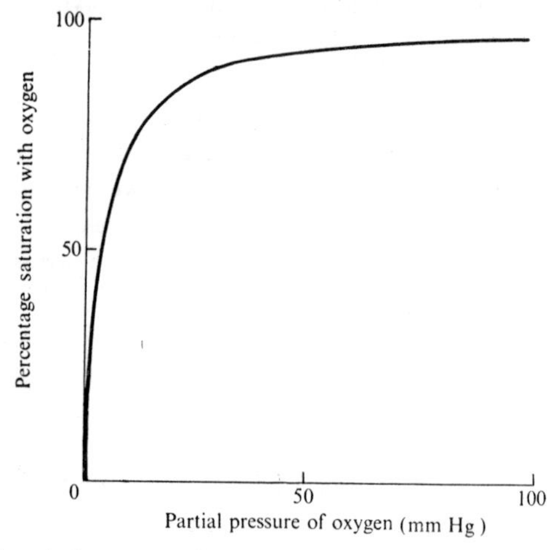

FIG. 3.2. Dissociation curve of muscle oxyhaemoglobin. Cf. Fig. 3.1. After R. Hill, *Proc. R. Soc.* **B120** (1936).

acid than haemoglobin, so that an increase in acidity will have the same effect as a rise in temperature. This, known as the Bohr effect, is important in life, for in the tissues the acidity (caused by carbon dioxide and lactic acid) is higher the greater the activity, and so oxyhaemoglobin dissociates most where it is most needed.

There is a gradient of oxygen tension from the atmosphere to the respiratory surface, from this to the blood plasma, from the plasma to the pigment in the red blood cells, from the pigment back again to the plasma in the capillaries, and from the plasma to the active cytoplasm. Oxygen necessarily follows this gradient. In muscles there is interpolated another stage, myoglobin, which resembles haemoglobin but has a hyperbolic dissociation curve (Fig. 3.2), suggesting that its molecule takes up only one molecule of oxygen, which is confirmed by its having a molecular weight of 16 800 and only one atom of iron in the molecule. The curve shows that it gives up its oxygen only at very low oxygen-pressures. These are probably reached only for short periods when a muscle, in its contraction, constricts its blood vessels and reduces its own oxygen supply. The myoglobin then dissociates and gives up its stored oxygen, and is recharged when the muscle relaxes.

Carbon monoxide and nitric oxide can act as ligands in the haemoglobin molecule, and have a very much higher affinity for it; they therefore displace oxygen from oxyhaemoglobin, and are rapid poisons.

V 3.41. The Blood of other Vertebrates

Most vertebrates possess haemoglobin, and are dependent on it for their existence. Three antarctic species of teleost have neither erythrocytes nor haemoglobin, and their blood has the carrying capacity of sea water; there is no haemoglobin in the leptocephalus larvae of eels, nor in the larvae of the sand-eel *Ammodytes*, and occasional specimens of *Xenopus laevis* without it have been described. Otherwise it is present in all the vertebrates that have been examined, although goldfish, carp, pike, eels, and other sluggish fish can live in water that has been saturated with carbon monoxide, so that they presumably use haemoglobin only when they are swimming fast. *Xenopus* also can live with its blood saturated with carbon monoxide.

Dissociation curves similar to those for mammals have been found for all the birds examined, for lizards, and for frogs, and the blood of all these animals presumably functions in the same way. The molecular weight of the haemoglobin of birds is the same as that of mammals. The curves for the fowl are slightly to the right of those of mammals, so that the blood has less affinity for oxygen and dissociates more readily. This would be expected, since birds have a higher oxygen tension in the lungs and in general a higher metabolic rate than mammals. The haemoglobin of fish, on the other hand, has a higher affinity for oxygen, and so dissociates only when the tension in the tissues has fallen below what would be tolerated by mammals. The more sluggish fish have the lower loading tensions (Table 4). Another general effect is that within a group the more active species, or those that need more oxygen, have more haemoglobin in a given volume of blood, and so a greater carrying power. Thus ducks have more than game birds, and lizards more than chelonians.

In a tortoise, the urodeles, aquatic anurans, many tadpoles, and in lampreys, the dissociation curve is a rectangular hyperbola, suggesting that only one molecule of oxygen is added to each molecule of haemoglobin. In accordance with this, the molecular weight of the haemoglobin of *Petromyzon* is only 17 000. If such a curve is examined (it is similar to that for myoglobin, Fig. 3.2), one can see that as the oxygen tension falls there is at first very little dissociation; then with a further fall there is a sudden emission of oxygen, until over a very small range of pressure the pigment is almost completely dissociated. A pigment of this sort cannot therefore function in the ordinary way. It will act rather as a store of oxygen, giving it up only at very low pressures, which possibly occur only occasionally. This is indeed the way in which myoglobin is assumed to act. The haemoglobins of some lower vertebrates, on the other hand, have high molecular weights, for example, 60 000, 140 000, and 290 000 in various amphibians and reptiles. In some fishes the dissociation curve is hyperbolic at temperatures of 0–5 °C, but sigmoid above 20 °C. There is a Bohr effect so strong that at high concentrations of carbon dioxide the blood can never become saturated with oxygen.

It is difficult to see any clear phylogeny in the vertebrate haemoglobins. As is so often the case, the mammals and birds seem to share the top of the evolutionary tree, although they have arrived there by different routes. The molecule with four atoms of iron, which takes up four molecules of oxygen and gives a sigmoid curve, is presumably more efficient for most purposes than the molecule with the single atom of iron and a hyperbolic curve, but whether this, as we see it in lampreys and urodeles, is primitive or whether, as with so many of their peculiarities, they have lost something their ancestors possessed, we do not know. The same is true of the sensitivity of haemoglobin to acid, which is not always present and seems to be found chiefly in active animals; it is present in the frog for example, but not in tadpoles. The haemoglobins of many species are electrophoretically heterogeneous, so that if the different forms have other different properties there is opportunity for natural selection to work.

J 3.41 The Blood of Invertebrates

Most of the more highly organized invertebrates, except the insects (which have other means of conveying oxygen) and the lamellibranchs (which are sluggish), have some pigment which can combine reversibly with oxygen and gives a dissociation curve broadly comparable either to the sigmoid curve of haemoglobin or to a rectangular hyperbola. Comparable pigments also occur in some invertebrates without a blood system.

Haemoglobin is found in some polychaetes (e.g. *Arenicola*, *Nephthys*), some oligochaetes (*Lumbricus* and other earthworms), some crustaceans (*Daphnia*), some insects (*Chironomus*), some lamellibranchs (*Anadara* = *Arca*), and some gastropods (*Planorbis*), as well as in *Paramecium*, *Ascaris*, and other non-vascular animals, and even in some plants. The molecular weight of haemoglobin from invertebrates is usually high, for example 1 630 000 in *Planorbis* and 3 000 000 in *Arenicola* and *Lumbricus*, but that of *Anodonta* is 72 000 and that of *Chironomus* is 17 000 or 34 000. Haemerythrin is present in a few polychaetes and some of the worms of doubtful affinity—sipunculids, priapulids, and *Lingula*. Chlorocruorin is confined to the polychaetes, but its prosthetic group has been recognized in some echinoderms. Haemocyanin, characteristic of gastropods, cephalopods, and the higher crustaceans (Stomatopoda and Decapoda), is found also in *Limulus* and some scorpions. The invertebrate pigments are usually in solution in the plasma, but haemerythrin is sometimes in cells of coelomic fluid, and haemoglobin is in corpuscles in a few polychaetes.

The very patchy distribution of these pigments suggests opportunism: the basic proteins and prosthetic groups must be widely distributed, and chance combination of them, which confers the power of reversible reaction with oxygen, has presumably been selected when it confers any advantage. It would be expected to do this when an animal is living in conditions where obtaining enough oxygen is difficult. There is some confirmation of this in the details of distribution. For example, freshwater snails of the genus *Planorbis* have haemoglobin and live longer under an atmosphere of 4 per cent oxygen than under air, while *Limnaea*, without haemoglobin, uses more oxygen and is adversely affected by 4 per cent oxygen. Two genera of notonectid bugs (*Anisops* and *Buenoa*) that have haemoglobin can dive for longer than other bugs without it. There are, however, difficulties. In nature, *Limnaea* is present in a wider range of habitats than *Planorbis* (e.g. in Windermere); while most polychaetes of the family Serpulidae, for example *Spirorbis borealis*, have chlorocruorin, *S. corrugatus* has haemoglobin, two species of *Serpula* have both chlorocruorin and haemoglobin, while *Spirorbis militaris* has no pigment, and no relationship can be seen with habit or habitat.

Although it seems certain that these pigments have some connection with oxygen supply, the way in which they function is seldom clear. For many of them the unloading tension is low, and it seems unlikely that in conditions of plenty of oxygen they can ever dissociate enough to play much part in transport; they may be useful in circumstances where the external oxygen pressure falls. It was suggested long ago that the lugworm *Arenicola*, which lives between tide-marks and has an unloading tension of 1·8 mm at 17 °C, uses its haemoglobin only when the tide is out and aeration of the sand is difficult; the pigment is perhaps acting as a store of oxygen. In contrast, chlorocruorin has a low affinity for oxygen, with unloading tensions of 9–30 mm, and can only work when the oxygen supply is good; we have already seen that *Sabella* (which has chlorocruorin) needs such a supply.

The matter is complicated by the fact that physical conditions greatly affect the properties of the pigments. The haemoglobins of *Lumbricus* and *Allolobophora* give fairly typical sigmoid dissociation curves at 20 °C, with unloading tensions of 8 and 6 mm respectively, but at 7 °C the curves are almost hyperbolic, and unloading tensions are 2 and 0·7 mm. With these properties the pigment might be of much greater use at one temperature than another, and its survival value could be in the protection it gives the animal under extreme conditions. In *Tubifex* and *Lumbricus*, which have haemoglobin, and in *Sabella*, with chlorocruorin, administration of carbon monoxide reduces but does not stop the consumption of oxygen at atmospheric pressure, so that, like sluggish fish and amphibians, they are probably not normally dependent on their pigments. In the more active *Nereis diversicolor*, also with haemoglobin, carbon monoxide completely stops the uptake of oxygen when the water is little less than half saturated, so that this species is probably dependent on the pigment under normal conditions. Haemerythrin could carry oxygen only at low concentrations, and haemocyanin is very variable in its properties; its oxygen affinity is lowest in the cephalopods, where it probably acts as an ordinary oxygen carrier, but in most other species the affinity is high and the carrying capacity is low, so that its function is somewhat doubtful.

The haemoglobin of *Ascaris* does not lose its oxygen even when the animal is kept anaerobically for several days, so that perhaps the function of the pigment is to keep the concentration of oxygen low, since moderate concentrations are fatal (section G 3.8).

Chlorocruorin shows the Bohr effect, but haemerythrin does not. The haemocyanins are irregular; some, such as those of the slug *Agriolimax* and lobsters, show it, but in others it is negligible or depends on the acidity. In *Limulus* it is reversed.

MV 3.42. Transport of Carbon Dioxide in Mammals and other Vertebrates

The blood system of vertebrates carries carbon dioxide as well as oxygen; the methods in mammals are these.

Carbon dioxide dissolves in the plasma in the ordinary way and there combines with water to form carbonic acid:

$$CO_2+H_2O = H_2CO_3. \qquad (i)$$

This reaction is slow, so that much carbon dioxide in simple solution diffuses into the red cells. Here carbonic acid is formed very rapidly under the influence of the enzyme carbonic anhydrase. Dissociation into hydrogen and bicarbonate ions occurs immediately and these react with haemoglobin, most of which is present as the potassium salt, also dissociated:

$$H_2CO_3+KHb = KHCO_3+HHb. \qquad (ii)$$

The chief ions that are present are therefore

in the plasma: Na^+ and Cl^-,

in the corpuscles: K^+, HCO_3^-, H^+, and Hb^-.

The walls of the corpuscles are permeable to anions, but not to cations nor to the large haemoglobin ion. Under these circumstances the chloride ions diffuse in and the bicarbonate ions out, to give a Donnan equilibrium (section 1.331). The result is that carbon dioxide is carried as sodium bicarbonate in the plasma, and that much more of the gas can be taken up than by mere physical solution or formation of carbonic acid. When the pressure of carbon dioxide is reduced by its diffusion out into the alveoli of the lungs, where the concentration is less than in the plasma, the reactions are reversed; the acid haemoglobin, whose concentration has not altered, drives back reaction (ii), so that bicarbonate enters the cells, and chloride diffuses out. Oxyhaemoglobin is a stronger acid than the reduced form, and so when it is present reaction (ii) does not go so far and less carbon dioxide can be carried; hence oxygen, by forming oxyhaemoglobin, displaces carbon dioxide from the blood.

There are subsidiary phenomena, of which the following are the chief. In the corpuscles, some of the buffering of the carbonic acid is carried out by phosphate,

$$H_2CO_3+K_2HPO_4 = KHCO_3+KH_2PO_4. \qquad (iii)$$

Some carbon dioxide combines directly with haemoglobin to form a carbamino-compound,

$$CO_2 + HbNH_2 = HbNHCOOH, \qquad \text{(iv)}$$

and a similar, but less important, reaction takes place with other proteins in the blood.

Of the carbon dioxide added to the blood in the tissues (the arterial-venous difference) about 5 per cent is carried in simple solution, 30 per cent as the carbamino-compound, and the rest as bicarbonate.

Other vertebrates use similar methods, but there are differences in detail. In teleosts and amphibians more is carried as bicarbonate, less in the red corpuscles. In a dogfish and a crocodile, on the other hand, the concentration is higher in the corpuscles than outside, but this seems not be to due to any combination between carbon dioxide and haemoglobin.

J 3.42. Transport of Carbon Dioxide in Invertebrates

Carbon dioxide is readily soluble in water, and diffuses easily through tissues and even through chitin; it therefore seems likely that invertebrates, with their small size and relative inactivity, would need no special mechanism for its transport. That this is so is confirmed when the amount of carbon dioxide that their blood will take up is plotted against its partial pressure, for many of the curves do not differ significantly from those given by sea water. Some active invertebrates such as cephalopods have blood that can carry much more carbon dioxide than can sea water, especially at low partial pressures. This is probably because the blood is highly buffered with bicarbonates and proteins, including the blood pigments. There is no evidence that haemocyanin combines directly with carbon dioxide as haemoglobin does, and carbonic anhydrase, though present in the gills of cephalopods, is absent from their blood.

G 3.5. OTHER FUNCTIONS AND PROPERTIES OF A CIRCULATORY SYSTEM

There can be little doubt that the circulatory system in its various forms was selected in evolution because of its ability to carry oxygen, but it is obvious that a system of vessels containing water can also be used for carrying substances in solution and suspension, and for conveying heat. Circulatory systems are used for all these purposes.

The soluble products of digestion—hexoses and aminoacids—are probably carried in the blood of all animals that possess it, and the same or similar substances are carried from storage organs to where they are needed. Similarly excretory products, not only carbon dioxide, are carried from where they are produced to the organ that eliminates them. All the chief phyla with blood are known to use it for carrying hormones, although these are developed in the vertebrates, and especially the land vertebrates, to a degree unknown elsewhere.

Fats are carried in suspension, and so in a sense are those breakdown products of the tissues which are ingested by amoeboid cells in the blood of metamorphosing amphibians and insects.

All metabolism produces some waste heat, and in a large animal the distribution of this and its conveyance to the surface so that there are no intense local rises in temperature must be important. The circulatory system performs this function just as does the water in the cooling system of a motor-car. The special function of the blood vascular system of contributing to the formation of a thermostat, which has developed in mammals and birds, is discussed in section 8.2.

The ease with which the circulatory system can carry substances in solution means that it can act as an intermediary between organs such as the liver and kidney and endocrine glands to help in maintaining a constant internal environment, but by itself it has no power to alter its composition. This aspect of life is dealt with in section 8.1.

Sporadically, natural selection has taken advantage of the ability of an enclosed fluid to transmit a pressure. Examples are the protrusion of the foot of bivalve molluscs, the stiffening of the labium in blood-sucking insects, and the erection of the penis of mammals. The colour of blood pigments is from the biological point of view accidental, although physically it stems from the system of π-electrons that is necessary for the carriage of oxygen. Animals have here and there taken advantage of it, the best-known example being the blushing and flushing of man. More often the colour conferred by the respiratory pigment, as in the haemoglobin that makes many small freshwater oligochaetes and insect larvae bright red, probably has no more function than the colour of the heart or other internal organs of vertebrates, 'arrased with purple' though they be.

In molluscs and arthropods the organs are directly surrounded by the blood and are said to be in a haemocoele. In vertebrates and others they are separated from the blood by the walls of the blood vessels, but are in contact with another fluid through which oxygen and other substances can diffuse. This is tissue fluid, largely derived from blood by passage through the walls of the capillaries. It is sometimes called lymph, but this name would perhaps be better confined to the same fluid after it has passed through the thin walls of a system of blindly ending vessels, the lymphatics. These are not known in invertebrates, cyclostomes, or elasmobranchs, but in other vertebrates they open into the venous system and sometimes have pulsating regions or lymph hearts. In some teleosts and amphibians the lymphatics are not blind, but arise from spaces in the tissues, so that the distinction between tissue fluid and lymph is not clear-cut.

Whether the blood system is open or closed there is a great fall in pressure in the region of the active tissues, that is in the capillaries or in the haemocoele. It might be expected that this difficulty would be met by the presence of local or auxillary pumps, but these are rare. They occur in cephalopods, where there are contractile branchial hearts interpolated between the capillaries of the system and the gills, and in *Myxine*, where there are several, the best known being in the portal vein and driving blood through the liver. In land vertebrates, and to some extent in Dipnoi, blood returns to the heart after passing through the lungs, so that the heart is in effect two pumps working in parallel.

G 3.6. CONTROL OF THE CIRCULATORY SYSTEM

One of the first necessities of the beating of the heart of an animal, at least one with a high metabolism, is that it should continue without intermission. The excised heart, whether of the frog or the mammal, will under appropriate conditions go on beating rhythmically for hours, although it is receiving no obvious stimulus. (The correct temperature must be maintained, and in the mammal oxygenated Locke's fluid must be perfused through the coronary blood-vessels that supply the heart.) It seems, then, that there must be some source of stimulation within the heart itself. The situation of this has been investigated by making cuts or ligatures that functionally separate the various parts of the heart. If the sinus of the frog be thus separated from the (auricles plus ventricle), it goes on beating rhythmically, but the auricles and ventricle stop for a time and then start again at less than the normal rate. If a ligature be applied between auricles and ventricle, the first continue beating but the second stops, and either never beats again or does so only very slowly. There is thus a gradient of excitability and of speed of rhythm, which shows that

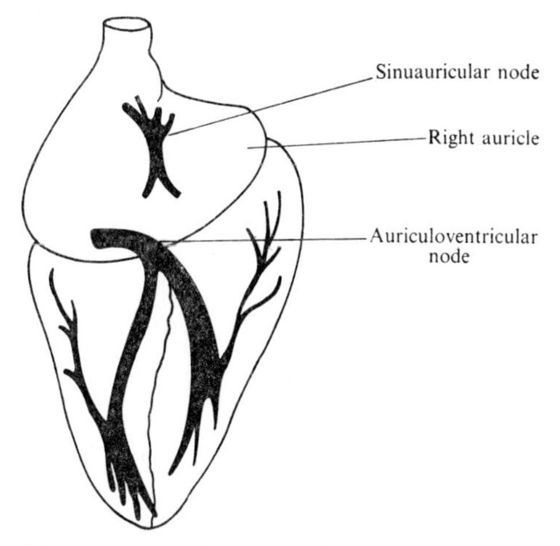

FIG. 3.3. Diagram to show the pace-maker and conducting tissue of the mammalian heart. From Yapp, *Vertebrates; their structure and life*, Oxford University Press, New York (1965).

in the normal heart the sinus must act as a pace-maker; the more slowly moving auricles and ventricle are quite unable to beat in their own rhythm, because they are in physiological continuity with the sinus and so are compelled to beat at the pace of this. The rhythm is entirely independent of nervous origin, for it goes on when the nerves are poisoned by drugs, and it starts in the embryonic chick before any nervous tissue is formed. Isolated cells of heart muscle grown in tissue culture beat rhythmically, so that rhythm of this sort, which has its origin in the muscle itself and is called myogenic, seems to be a fundamental feature of cardiac tissue.

The mammalian heart likewise has an automatic rhythm, which is more rapid at the venous end than in the ventricles. In addition to the complete separation of the pulmonary and systemic circulations there are certain anatomical modifications as well (Fig. 3.3): in particular the muscles of the auricles and ventricles are not in general histological continuity. The pace-maker is the sinu-auricular node, which is a piece of tissue lying between the right superior vena cava and the right auricle, and representing part of the sinus venosus. From this the contraction spreads over the auricles. At the bottom of the

8541112 H

auricular septum is another piece of tissue, the auriculo-ventricular node, which also represents part of the sinus venosus. From this starts a band of muscle-fibres known as the bundle of His, or the auriculo-ventricular bundle. It runs in the inter-ventricular septum, and divides into right and left branches for the two papillary muscles. Each branch subdivides so that its fibres are distributed to nearly the whole of the ventricular muscle. It is this bundle that provides the continuity between auricle and ventricle along which the wave of excitation can pass. The rate of conduction is ten times as fast in the bundle as in ordinary muscle, which means that the whole of the ventricle contracts practically simultaneously.

There is a similar arrangement of special fibres in birds, but in cold-blooded vertebrates the conduction is carried out by the ordinary fibres. Pace-makers have been found at various places in the ductus Cuvieri, the sinus venosus and the auricle in both elasmobranchs and teleosts, and at the base of the conus in elasmobranchs.

The peculiar heart of a tunicate, which is unique in driving the blood first in one direction and then in the other, has a pace-maker at each end, the two taking it in turns to control the beat, which is possibly myogenic, for about 1 min each. How they come to alternate is not known; possibly each in turn becomes fatigued by the accumulation of a metabolite.

Molluscs also have a myogenic beat with a pace-maker consisting of muscle, but most arthropods, including insects, have a nerve ganglion running the length of the heart or have nerve-fibres that control its beat, which is thus neurogenic. That this is a secondary system of control is suggested by the fact that nerves are absent from the hearts of embryos, which nevertheless beat, and from those of some of the lower crustaceans, such as *Daphnia*.

Whatever the pumping mechanism, it must be adjusted to the needs of the animal. The rapid increase in the supply of oxygen that is required in an organ as the activity of an animal increases can be met only by an increase in the amount of the blood arriving at any moment. This can be got by increasing the frequency of beat of the heart, or by increasing the volume of blood given out at each stroke. Both occur, and there are similar alterations for needs arising in other ways. In general, the system in well-developed hearts is that there are both accelerator and inhibitory (decelerator) nerves, reacting to different stimuli.

In mammals there are stretch receptors in the wall of the aorta, at the base of the innominate artery, and in the carotid sinus. These will be stimulated at systole, and they send impulses through the ninth and tenth cranial nerves to the medulla, which sends to the heart impulses travelling in motor or parasympathetic fibres of the vagus, which have the effect of slowing the heartbeat. Other receptors in the right auricle and the great veins, which will be stimulated at diastole, send to the medulla impulses that have the effect both of inhibiting the parasympathetic deceleration centre and of initiating accelerator impulses through the sympathetic system. In addition, chemoreceptors in the aorta and carotid body, and in the medulla itself, respond to an increase in acidity of the blood, so that the heart is accelerated. There is probably no reflex response to varying oxygen tension in the blood, but a reduced supply of oxygen causes a brief increase in heart-rate, probably by its direct effect on the sinu-auricular node.

It is within the experience of everyone that the heartbeat becomes deeper and more rapid in exercise and in certain emotional states, especially anger, fear, and love. While the effect of exercise is produced chiefly by the stimulation of the chemoreceptors, that of the emotions is due to the liberation of adrenaline from the adrenal medulla (sections M, V 6.41). The control is, however, not simple, for the heart rate may rise in a man who knows he is to take exercise before there can be any accumulation of acid in the blood. In exercise the rate of output of blood by the heart may treble: most of this increase is due to the increased rate of beat, and only about one-quarter of it, at most, to an increased emptying of the heart at systole.

One may guess that something similar must happen in other vertebrates, but little is known of such control. Birds have carotid bodies, much like those of mammals, which may be presumed to have a similar function. In fishes a rise in the pressure of blood in the gills slows the heartbeat, a reflex comparable to that initiated by pressure in the carotid sinus (which is derived from branchial vessels) in mammals. In elasmobranchs the heart beat maintains a simple relationship—1:2, 1:3, or 1:4—with the respiratory rate for long periods, and this seems to be based in part on a reflex from unknown receptors.

Many invertebrates are much less dependent on their circulatory system for their supply of oxygen, so that a fine control is not to be expected. In some, temperature is the chief influence, as can be very easily shown in *Daphnia*. The higher crustaceans have a nerve supply to the heart which includes both accelerator and decelerator fibres, and so there is some external control imposed on the pace-maker. In *Lumbricus* the rate of beating of the dorsal vessel and of the pseudo-hearts is increased by acetylcholine and by adrenaline (sections MV, J 6.32). Both of these substances occur in oligochaetes, but whether they act on the circulation in nature is unknown.

J 3.7. TRANSPORT BY TRACHEAE

The alternative to taking oxygen round the body in solution or in chemical combination is to carry it as a gas, and this method is used by eight groups of animals: insects, myriapods, isopods (Crustacea), four groups of arachnids (Phalangidae, Pseudoscorpionidae, Solifugae, Araneidae), and *Peripatus*. In all these there are tubes called tracheae running in from the surface; they branch and subdivide, and take air to the innermost parts of the body. Oxygen finally dissolves, either in cytoplasm, or, more often, in fluid that occupies the tips of the finest branches. The phylogenetic relationships are unknown, but it seems certain that there have been at least three separate evolutions, and all eight may have arisen independently.

The best-known tracheae are those of insects. There is in general a pair opening on each segment by apertures called spiracles or stigmata. As invaginations from the surface, tracheae are chitin-lined; they anastomose, and break up into finer branches until they end in tracheoles about 0·75 μm in diameter (Fig. 3.4). Each tracheole is formed from a single cell; it may be 400 μm long and may be branched. It gradually decreases in diameter, sometimes to as little as 0·2 μm, and then ends blindly. The tracheoles are lined by chitin, but this is not shed at ecdysis. They may form a felt-work around the muscles, or may penetrate deeply into a large cell or muscle fibre.

However oxygen may enter the tracheal system, calculation has shown that, once in, it will diffuse rapidly even to the inner ends of the tracheoles, and no other mechanism need be postulated to account for its arrival at the interior. The ends of some tracheoles are filled with fluid, and this can be caused to move up and down by varying the osmotic pressure. The same result is got by the activity of a muscle or deprivation of oxygen, which presumably produces metabolites that increase the osmotic pressure around the tracheoles.

This extracts water, so that air is drawn into the tracheoles when it is most needed (Fig. 3.5).

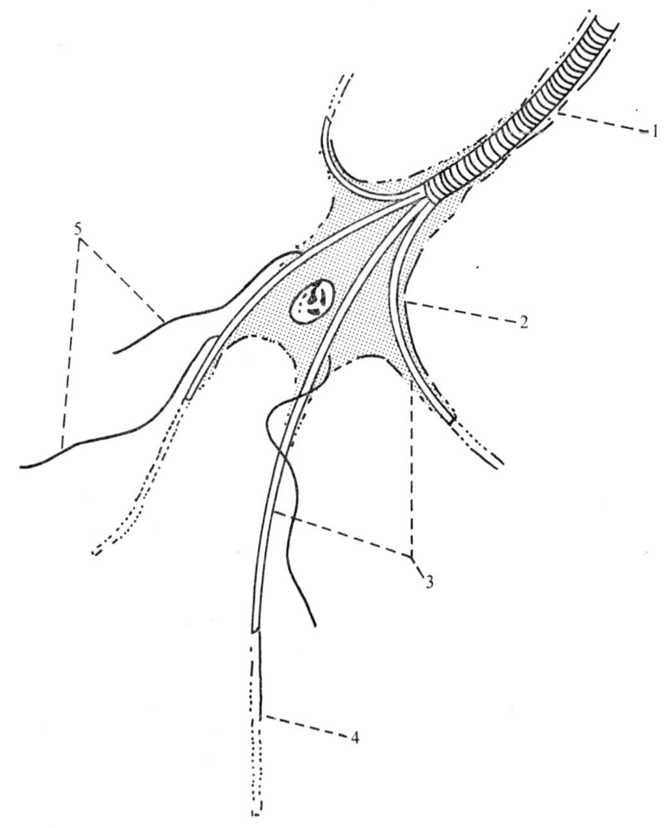

FIG. 3.4. Detail of tracheal ending. 1, trachea; 2, tracheal cell; 3, main tracheoles containing air; 4, main tracheoles containing liquid; 5, fine tracheoles containing air. From Wigglesworth, *Proc. R. Soc.* **B106** (1930).

G 3.8. LIFE IN LOW CONCENTRATIONS OF OXYGEN

From time to time species or families of animals have apparently changed their habits so as to live where there is much less oxygen than normal, and a few of the major groups live largely in such habitats. The change may involve alterations both to the respiratory system and to the circulatory system.

We have already mentioned in section J 3.2 the difficulties of insects and pulmonate gastropods that have returned to the water, and in section V 3.21 the reflexes by which diving amniotes cease breathing. Seals, diving ducks, and penguins have evolved a series

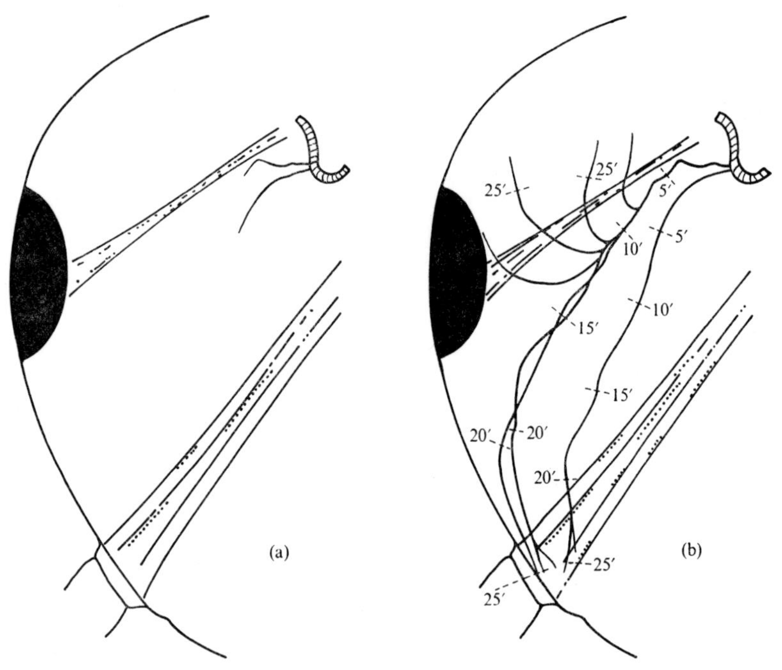

FIG. 3.5. Effect of asphyxiation on air in the tracheoles of a larva of a gnat. (*a*) larva at rest; (*b*) during asphyxiation. The figures in (*b*) show time in minutes, after onset of asphyxiation, at which air reached the points indicated, so making the tracheoles visible. From Wigglesworth, *Proc. R. Soc.* **B106** (1930).

of mechanisms for use under water that are much alike in the three groups. First, there is a high content of myoglobin in the muscles, so that they carry oxygen with them in the oxygenated form of this. There is enough in the seal *Phoca vitulina* to last for 5–10 min, and only when it is all used up does anaerobic work start. Secondly, the rate of circulation falls, either by a fall in the rate of heartbeat or by the use of an arteriovenous shunt or by both, so that little carbon dioxide and lactic acid enter the blood from the muscles, and other organs are not affected. This enables a large oxygen-debt to be built up without distress. Thirdly, there is a fall of body-temperature. This saves oxygen by allowing all that is available to be used for swimming, instead of much of it being used merely for heat production.

The physiology of whales is less well-known. They also have arteriovenous shunts and much myoglobin, but another problem is to avoid dissolving too much nitrogen at the great depths—1000 m or more—to which they descend. They probably achieve this by the shunt (which reduces the volume of blood to which the nitrogen has access) and by the reduction in volume and surface of the lungs caused by the high pressure.

The rate of heartbeat of alligators and frogs falls when they dive, and in the former, since lactic acid does not appear in the blood until after the dive, there is presumably either vasoconstriction or a shunt. Some turtles are said to be able to extract oxygen from the water and absorb it through the highly vascular pharynx. A teleost, *Leuresthes tenuis,* spawns on land, and while out of the water it has a lowered heart rate and little lactate appears in the blood.

Animals living at high altitudes have much less oxygen available than those at or near sea level, since at 2300 m (7500 ft) the tension of oxygen is only three-quarters of what it is at sea level. Fig. 3.1 shows that this is more than enough to saturate the blood, but it may not be enough to supply all that is needed for exercise. When a man goes to an altitude of 3000 m (10 000 ft or more) very little exercise produces an increased breathing rate, which supplies him with more oxygen but also removes more carbon dioxide; the kidney excretes more base, so that the acidity of the blood does not greatly fall. There is an immediate contraction of the spleen, which puts more red corpuscles into the circulation, and there is a rapid formation of more red corpuscles and more haemoglobin, so that after a week or so the amount of haemoglobin in the blood is at a new high level. With a longer stay, there is an increase in the size of the thorax.

Similar differences from the normal are seen in populations that live permanently in high mountains. Little is known about the effect of high altitudes on animals other than man. Some mammals that live in mountains, such as the llama, have higher red cell counts than most, and haemoglobin with a high affinity for oxygen; montane populations of an American thrush have been shown to have larger hearts and lungs than those from lower down. *Daphnia* can be induced to manufacture more haemoglobin by depriving it of oxygen.

It is sometimes said that parasites live in surroundings devoid of oxygen, but a little consideration will show that this is not generally true. The cells of the host are supplied with oxygen, and a parasite such as *Trichinella* living among them must be able to share it. The malarial parasite and others in the blood are very well supplied with oxygen indeed. The only situations where parasites live in which the oxygen concentration is low are the coelom and the gut, and even in these places it will seldom be zero, for there must be diffusion out from the wall, which is supplied, often richly, with blood. Measurements of the pressure of oxygen in the small intestine of the rat have shown it to be from 19 to 30 mm of mercury near the mucosa; in the sheep, as would be expected with the wider lumen, it is lower, from 4 to 12 mm. In the small intestine of the duck it ranges from 25 mm near the villi to 0·5 mm near the centre. The lowest values are likely to be found in the large intestine, where there are usually many Bacteria and much production of other gases. In the depths of the sea and in the mud of ponds and swamps, the concentration of oxygen sometimes falls too low to be measured, but it is not always so. On the whole it seems that life without oxygen, or anaerobiosis, is rare in animals.

Most small invertebrates are able to survive for some time without oxygen, but all that have been investigated use oxygen when it is available. *Paramecium*, for instance, normally aerobic, can survive in hydrogen or nitrogen for from 1 to 8 h, and the ciliate *Balantiophorus minutus*, which lives in dung, for from 5 days to 3 weeks. Most of the mud-living species are inactive forms requiring little oxygen, and probably able to survive for a time by one of the common anaerobic mechanisms that will be discussed in section M 4.12. Special attention has been given to the parasitic nematodes. The small forms *Nippostrongylus muris* from the rat, *Haemonchus contortus* from the abomasum of the sheep, and *Nematodirus* from its intestine, all increase their rate of oxygen consumption as its concentration rises. The maximum oxygen consumption corresponds closely to the maximum concentration observed in the intestine. The larger *Ascaris* uses less oxygen per unit mass than do the smaller forms; in complete absence of oxygen it produces carbon dioxide at only one-tenth or one-quarter the rate at which it does if oxygen is present, and its survival is not affected, but it is less active. Some species of nematode living in mud can be active for 35 days without oxygen, but others become quiescent.

Larvae of *Taenia taeniaeformis*, a tapeworm whose hosts are cats and rats, survive longer when oxygen is present than when it is not, and their consumption increases with the concentration. *Schistosoma* can survive and lay eggs when the oxygen uptake is reduced to 20 per cent of the normal. *Trichinella spiralis* needs oxygen for mobility, but not for survival. The larva of the warble fly, *Gastrophilus equi*, can live for about 3 weeks submerged in oil, but in life it appears to be an opportunist, seizing bubbles of air that pass down the host's oesophagus and storing the oxygen in haemoglobin. *Chironomus* larvae also can live anaerobically for a time.

The mechanisms by which such anaerobiosis is achieved are discussed in section J 4.1. While all animals probably use oxygen if they can get it, a few, such as *Ascaris* and some ciliates, are killed by moderate concentrations of it.

4. Metabolism and Cellular Physiology

All the soluble materials that the blood obtains from the small intestine suffer one of two fates: they may be built up into the tissues of the body, or they may take part in reactions the final result of which is usually to supply energy. In either case they undergo complex chemical changes, and in recent years, largely through the technique of introducing radioactive isotopes into the molecule as 'tracers', much has been learnt both of the details of the reactions and of their speed. There is an extremely rapid turnover, many atoms changing their position in the body and their state of chemical combination several times a day. Thus in 24 h the mass of iron removed from human blood plasma is 32 mg, which is eight times that present at any one time, and half of this is used for the manufacture of haemoglobin in the bone marrow. The haemoglobin in the blood contains about 3–4 g of iron, so that all of it must be renewed once in 100 days or so. The proteins of the rat liver have a half-life of only 6–7 days, and those of the skin of even less. The great lability of the atoms means that, from a chemical point of view, it is impossible to draw sharp dividing lines at particular points in a series of reactions that begin with those of assisted transport as the molecules enter the cells, and end only when the waste products are finally expelled from the body.

G 4.0. ENERGY AND SYNTHESIS

As we shall see later on in this chapter, most of the metabolism in the body takes place in quite small steps, so that we should not expect any one of them to involve a big energy change. But in fact some steps do mean a very much bigger change than others; these nearly always involve high-energy phosphates. Compounds of this sort are conventionally written R∼P, or R—Ⓟ, or R∼ph, where P or Ⓟ or ph represents —$PO(OH)_2$, R is the rest of the molecule, and the tilde (∼) shows that this bond has a large chemical potential energy so that when it is broken much work can be done. The commonest high-energy phosphate is adenosine triphosphate (or ATP),

which may be written

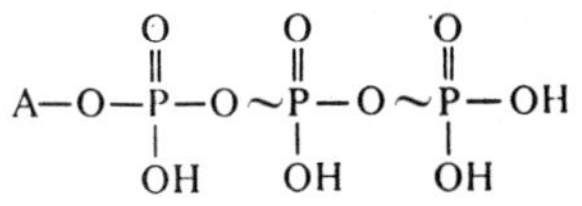

$$\mathrm{A{-}O{-}\overset{\overset{\displaystyle O}{\|}}{\underset{\underset{\displaystyle OH}{|}}{P}}{-}O{\sim}\overset{\overset{\displaystyle O}{\|}}{\underset{\underset{\displaystyle OH}{|}}{P}}{-}O{\sim}\overset{\overset{\displaystyle O}{\|}}{\underset{\underset{\displaystyle OH}{|}}{P}}{-}OH}$$

It has two high-energy bonds, but in reactions where phosphate is transferred to another chemical compound usually only the last one is broken, adenosine diphosphate (or ADP) being left. Occasionally two phosphate groups are lost, usually as pyrophosphate, leaving adenosine monophosphate (adenylic acid, AMP) which has no high-energy bond left.

Much metabolism consists of the formation of high-energy phosphates with the help of energy supplied by other reactions, and the subsequent use of the energy so stored in synthesis and other reactions. The exergonic reactions of metabolism correspond broadly to those that constitute katabolism, or breaking down, the endergonic reactions to anabolism, or building up. The two series are closely interlinked.

M 4.1. METABOLISM IN MAMMALS

M 4.11. Metabolism of Proteins and Nitrogen

The products of digestion of protein are aminoacids, and experiments with isotopically labelled compounds have shown that there is a continuous breakdown of proteins to aminoacids in the tissues as well. The metabolism of proteins is therefore largely the metabolism of aminoacids: their synthesis to proteins and their degradation into other compounds. These processes go on in all or nearly all tissues, but quantitatively the liver is the most important site; in man it accounts for more than half the breakdown and re-synthesis of the total of about 100 g tissue protein per day, and is responsible for the major part of the destruction of aminoacids.

Most of the aminoacids derived from the food are destroyed by chemical reactions known as transamination and deamination, which are probably usually linked as shown in Schema 1. Both can go on in many tissues, but deamination is only known to go on rapidly in liver and kidney. In transamination the amino-group is exchanged for the keto-group of an α-ketoacid, with a transaminase (or aminotransferase) which is probably specific for each ketoacid; in deamination there is an oxidation which removes the nitrogen as ammonia, while hydrogen is transferred to a hydrogen acceptor (section M 4.12). This scheme seems to account for most of the ammonia that is produced, but some aminoacids go through other pathways, such as direct deamination by L-aminoacid oxidase or by specific deaminases,

pyridoxal phosphate
L-glutamate transaminase

$$\underset{\text{aminoacid}}{R\cdot CH\cdot NH_2\cdot COOH} + \underset{\alpha\text{-ketoglutaric acid}}{COOH\cdot CO\cdot CH_2\cdot CH_2\cdot COOH} \rightleftharpoons \underset{\substack{\alpha\text{-ketoacid}\\ \text{(e.g. pyruvic)}}}{R\cdot CO\cdot COOH} + \underset{\text{glutamic acid}}{COOH\cdot CH\cdot NH_2\cdot CH_2\cdot CH_2\cdot COOH}$$

NAD or NADP
L-glutamate dehydrogenase

$$\underset{\text{glutamic acid}}{COOH\cdot CH\cdot NH_2\cdot CH_2\cdot CH_2\cdot COOH} \rightleftharpoons \underset{\text{iminoacid}}{COOH\cdot C{:}NH\cdot CH_2\cdot CH_2\cdot COOH} + 2H$$

(probably spontaneous)

$$\underset{\text{iminoacid}}{COOH\cdot C{:}NH\cdot CH_2\cdot CH_2\cdot COOH} + H_2O \longrightarrow \underset{\alpha\text{-ketoglutaric acid}}{COOH\cdot CO\cdot CH_2\cdot CH_2\cdot COOH} + NH_3$$

SCHEMA 1. Transamination and deamination, sometimes called transdeamination. (Italics indicate enzymes.)

or transamination in which acids other than α-ketoglutaric take part. It seems that lysine is not deaminated at all.

The pyruvic acid and α-ketoglutaric acids produced by these reactions are substances which, as we shall see in the next section, are found also in the metabolism of carbohydrates. We shall deal with their subsequent fate under that heading. It is obvious that, by the production of these compounds, proteins in the food can give rise to the same end products as carbohydrates. Some aminoacids give acetoacetic or oxaloacetic acid. These also appear in the carbohydrate pathways, and the same comment applies. The residues from all the non-essential aminoacids and from a few essential ones can give rise to glucose, and are called glucogenic; they are presumably of great importance in carnivores. The residues from most of the essential aminoacids either are not synthesized, or give rise to fatty substances, when they are called ketogenic.

Most of the ammonia liberated from aminoacids is a waste product, but before being excreted it is converted to urea in the liver.

First, with the help of enzymes from mitochondria, and energy and phosphate from the high-energy compound adenosine triphosphate, it joins with carbon dioxide from the blood to give carbamyl phosphate,

$$NH_3+CO_2+ATP \rightarrow NH_2COO \cdot H_2PO_3+ADP.$$

The carbamyl phosphate then reacts with arginine to form carbamyl

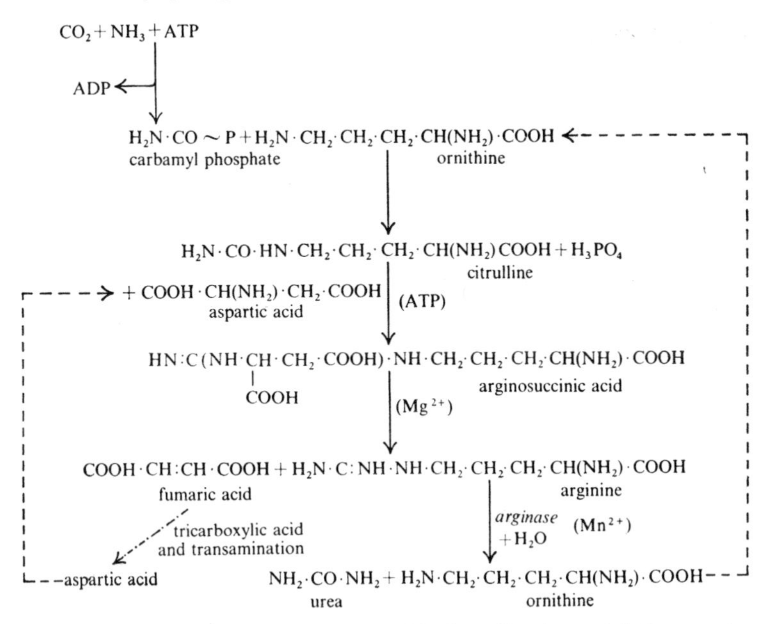

SCHEMA 2. The ornithine cycle. Arrows with chain line (·—·—·) indicate series of reactions

aspartic acid, which in its turn reacts with the aminoacid ornithine to start a cycle of reactions that ends in the formation of urea and the restoration of ornithine. This ornithine cycle is shown in Schema 2. It will be seen that the formation of one molecule of urea needs the breaking of three of the high-energy bonds of adenosine triphosphate (two in the initial formation of carbamyl phosphate, one at the citrulline-arginosuccinic transformation). The adenosine triphosphate has to be restored by energy provided by the breakdown of other substances, so that the formation of urea is a relatively costly type of metabolism.

The bases produced by digestion of nucleoproteins are purines and pyrimidines. The common purines are adenine (6-amino-purine)

adenine

and guanine (2-amino-6-hydroxypurine)

guanine

The first of these loses ammonia by two routes. Some of it is deaminated by the enzyme adenase to hypoxanthine (6-hydroxypurine),

hypoxanthine

but most of it first combines with ribose-1-phosphate to form inosinic acid (a nucleotide) which gives inosine (the corresponding nucleoside), which then splits to form hypoxanthine and ribose-1-phosphate. Hypoxanthine is oxidized to xanthine (2-6-dihydroxypurine)

(xanthine)

by xanthine oxidase. Guanine, which is very insoluble and probably much less important than adenine, is hydrolysed directly to xanthine by guanase. Xanthine is further oxidized by xanthine oxidase to uric acid (2-6-8-trihydroxypurine) with flavine adenine dinucleotide as the hydrogen acceptor. All the purines can exist in two isomeric forms, and in uric acid the keto-form (which is not shown for the substances above) probably predominates. The change for uric acid is

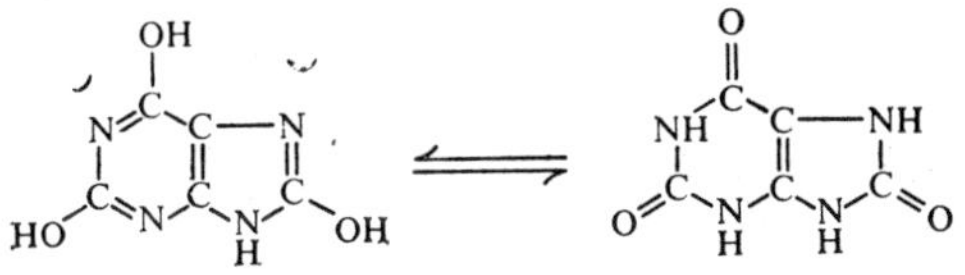

enol form ⇌ keto form

In most mammals a further enzyme, uricase or urico-oxidase, is present, which acts on the uric acid giving allantoin,

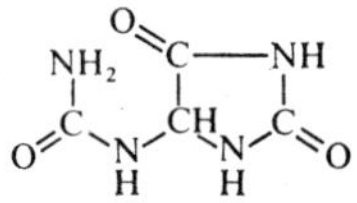

allantoin

in which the purine ring has been broken.

The distribution of the enzymes that carry out these processes is erratic, man, for example, apparently possessing no adenase. Uricase is absent from man and the higher primates, so that these animals excrete uric acid. It is not a true acid, but in its enol form it dissolves in alkalis giving monobasic acid urates of the type RHU, dibasic neutral urates, R_2U, and quadriurates, $RHU \cdot H_2O$. When it is in solution in animals it is usually in the form of acid or quadriurates, in mammals generally the latter.

Many of the reactions described in this section are reversible, and so one might expect that aminoacids could be synthesized from each other and even from ammonia. This is in general true, and if ammonium salts containing ^{15}N are provided, proteins containing this isotope can later be found in the body. The animal must therefore be able to synthesize aminoacids from ammonia, and proteins from aminoacids. Ammonia is not, however, a normal food (it is poisonous), and the synthesis of aminoacids is presumably normally through a reversal of transamination. The essential aminoacids, which cannot be synthesized, must be incapable of being formed in this way.

There is no evidence that the hydrolyses by which aminoacids are produced from proteins can be reversed *in vivo*, but before we consider how the synthesis of proteins does take place, we must look at the nucleic acids that play an important part in the process. These are polymers, consisting of from fifteen to many thousand nucleotide units. Each unit is made up of phosphoric acid, a pentose, and a base. In nuclei the pentose is generally D-2-deoxyribose, giving a deoxyribose nucleic acid (DNA), and elsewhere it is predominantly D-ribose, giving a ribonucleic acid (RNA).

The bases are generally selected from the purines adenine and guanine, and the pyrimidine bases cytosine, uracil, and thymine. The pyrimidines have a single 6-atom ring of carbon and nitrogen.

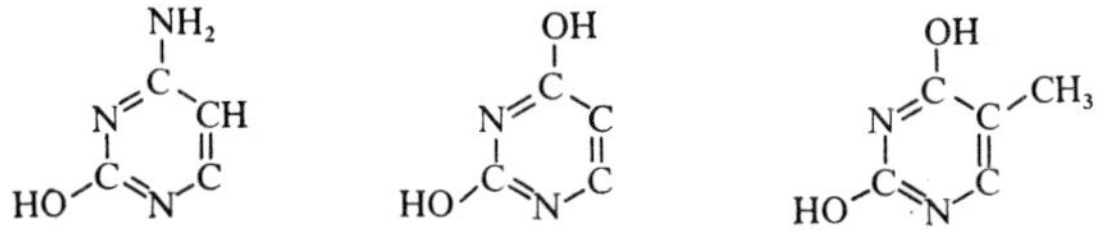

cytosine | uracil | thymine (5-methyl uracil)

Both nucleic acids generally contain adenine, guanine, and cytosine, but while in deoxyribonucleic acid the fourth base is generally thymine, in ribonucleic acid it is usually uracil.

The molecule of deoxyribonucleic acid consists of two intertwined helices, which are held together by hydrogen bonds that link the purines (adenine, A, and guanine, G) of one helix with the pyrimidines (thymine, T, and cytosine, C) of the other. Ribonucleic acid also is a helix, probably with local doubling and intramolecular hydrogen bonds joining the bases within the coil. The molecular weight of deoxyribonucleic acid runs from about 5×10^6 to 8×10^6, while ribonucleic acid is smaller, its molecular weight extending from 20 000 to $2{\cdot}5\times10^6$. Both are often combined with protein in the animal.

At this point it must be emphasized that much of our knowledge of the part played by nucleic acids in the synthesis of proteins is derived from work on microbes, and its generalization to the whole living world must be made with caution. Enough work has now been done on mammals and other organisms for us to be confident that the general principles are constant, but there may be differences in detail, and different species, for example of rodent, have different nucleic acids. What happens is in outline as follows. There are in the cytoplasm at least three sorts of ribonucleic acid, all manufactured in some way by the deoxyribonucleic acids of the nucleus. Ribosomal ribonucleic acid is combined with protein to make ribosomes. Messenger ribonucleic acid is the key to the synthesis. For every protein to be synthesized there is in the nucleus a corresponding deoxyribonucleic acid with its bases arranged in a particular order. With energy from adenosine triphosphate, and with the help of ribonucleic

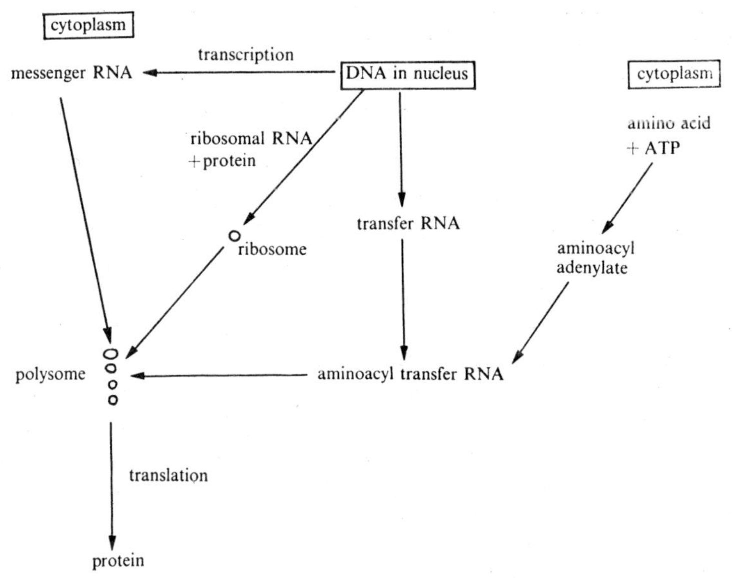

SCHEMA 3. The biosynthesis of proteins.

acid polymerase and divalent cations (Mg^{++}), this forms a corresponding messenger ribonucleic acid, which comes alongside a ribosome and forms a polysome. Transfer (formerly called soluble) ribonucleic acid is capable, with the help of adenosine triphosphate, divalent cations and enzymes, of combining with aminoacids in the cytoplasm to form aminoacyl ribonucleic acid; each aminoacid has one or more specific transfer ribonucleic acids. When the aminoacyl ribonucleic acid is brought alongside an appropriate polysome the aminoacid is released and placed in position to make part of a protein molecule. These reactions are summarized in Schema 3.

The series of bases of deoxyribonucleic acid may be read as made from a four-letter alphabet (A, G, T, C) which is accurately translated into the slightly different four-letter alphabet (U, C, A, G) of ribonucleic acid. The continuous series of letters is then read as a code to determine the order in which the twenty aminoacids (which may be compared to words 'in clear') are picked up (as aminoacyl ribonucleic acid) and joined together to make a protein. If the series is read as two-letter code words (e.g. AT TC GC AT AG CA . . .) only sixteen words are possible, so that to account for the twenty aminoacids the code words must contain at least three letters (ATT CGC ATA GCA . . .). This would give sixty-four possible words, so that either some words must be meaningless, or they mean something other than an aminoacid (or one of the hundred-odd aminoacids that do not occur in proteins), or they do not occur, or more than one word in code must give the same word in clear, that is more than one sequence of bases must determine the same aminoacid. There is evidence that the last is often true, at least *in vitro*. The triplet of bases for each aminoacid is called a codon, and the key for translating most of them has been worked out. Synthesis appears always to begin, in bacterial systems, with a triplet that picks up formylmethionine, which is often later removed enzymatically, and ends with a triplet that may be read as meaning 'stop'.

The molecules of deoxyribonucleic acid are able to duplicate themselves. Possibly the double helix becomes unwound, and then each single helix induces the formation of a new one alongside itself, adenine coming opposite to thymine and guanine to cytosine, as before. When each of the two pairs of chains has coiled up, there are two molecules of deoxyribonucleic acid where previously there was only one. However, this process is far from being understood.

There is still much unexplained. The cell is not always synthesizing nucleic acid or protein, and the initiation of synthesis, for example in developing eggs, appears to be determined by substances produced in the cytoplasm. Some of these activators (or derepressors) are not specific; for example frog cytoplasm can induce the nuclei of cells of mouse liver to synthesize deoxyribonucleic acid. Some of the reactions may be influenced by hormones; insulin and oestrogens probably increase the synthesis of ribonucleic acid, and corticotropic hormone (in mammals) and ecdysone (in insects) possibly affect the synthesis of proteins, especially enzymes. Denucleated sea urchin eggs can synthesize proteins, presumably because they contain enough stable messenger ribonucleic acid.

M 4.12. Metabolism of Carbohydrates

The chief carbohydrates absorbed from the gut are hexoses, substances with six carbon atoms in the molecule. Most of the carbon atoms—five in glucose, mannose, and galactose, four in fructose—together with an oxygen atom form a ring, and the remaining one or

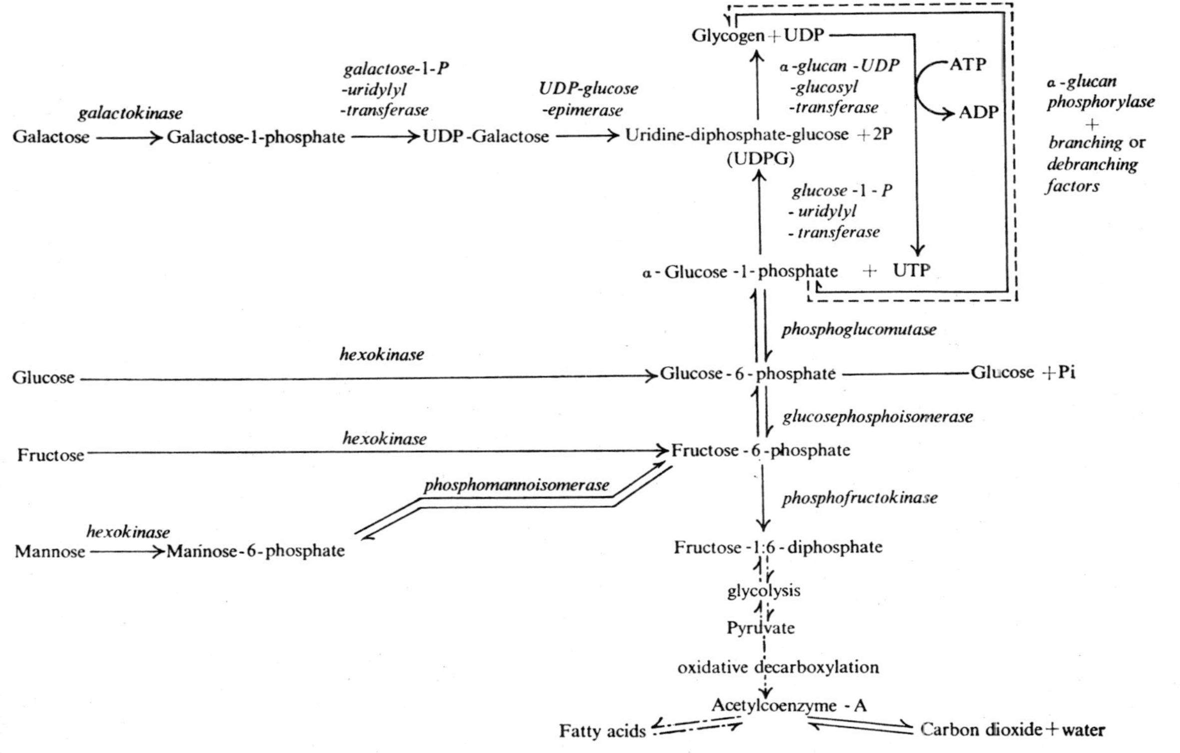

SCHEMA 4. Intermediate carbohydrate metabolism. Arrows with chain line (•—•—•) indicate series of reactions, italics indicate enzymes and co-factors. In the names of these P should be read as phospho-; P_i indicates inorganic phosphate. Adenosine triphosphate takes part in many reactions, but it is shown only in regeneration of uridine triphosphate.

two form a side-chain. All hexoses can form phosphates by reaction with phosphoric acid, and through these they are mutually interconvertible. For convenience the carbon atoms are numbered. In glucose those in the ring on each side of the oxygen atom are 1 and 5, and that in the side-chain is 6, and the numbers in other hexoses correspond.

In the liver the four common hexoses are phosphorylated, and then in general built up into glycogen by a series of reactions which are shown in Schema 4, which shows also, briefly, two alternative fates. Where phosphate is introduced into a molecule it is obtained from adenosine triphosphate which is converted to adenosine diphosphate at the same time. Uridine triphosphate (UTP), which reacts with α-glucose-1-phosphate in the synthesis, is a nucleoside similar to adenosine triphosphate but derived from the base uridine instead of adenine. Synthesis of glycogen by a reversal of the direct breakdown can take place, but is unlikely to be important in life. The branching and debranching systems, which are necessary to introduce or remove the side-chains of the glycogen molecule, are different. Most of these reactions are almost universal in living tissues, including those of plants, but they go on at different rates in different parts of the body, and the liver is the chief site of them in mammals.

The direction in which these reversible reactions move is determined by the needs and activities of the animal, and is largely under endocrine control, as described below. If energy is needed fructose-1-6-diphosphate is broken down by a series of reactions known as glycolysis (Schema 5). It involves many enzymes and coenzymes and can go on in all tissues. It begins by the splitting of the hexose molecule into two three-carbon units, or triose phosphates, and eventually each of these gives rise to a molecule of pyruvic acid. Each molecule of hexose has thus given two molecules of pyruvic acid, and at the same time it has lost four hydrogen atoms, which are taken up by substances called hydrogen acceptors, which are themselves reduced:

$$C_6H_{12}O_6 \rightleftharpoons 2CH_3.CO.COOH + 4H.$$

The most important of these acceptors is nicotinamide-adenine dinucleotide, or NAD (formerly called coenzyme I, diphosphopyridine nucleotide, or DPN). If molecular oxygen is present, the reduced hydrogen-acceptor is oxidized, so that there is a continual supply of it, but if oxygen is absent the reduced nicotinamide adenine dinucleotide is oxidized by pyruvate with the reduction of this to lactate. Some of the reactions of glycolysis are not strictly reversible, but fructose 1:6-diphosphate can be reformed from lactate or

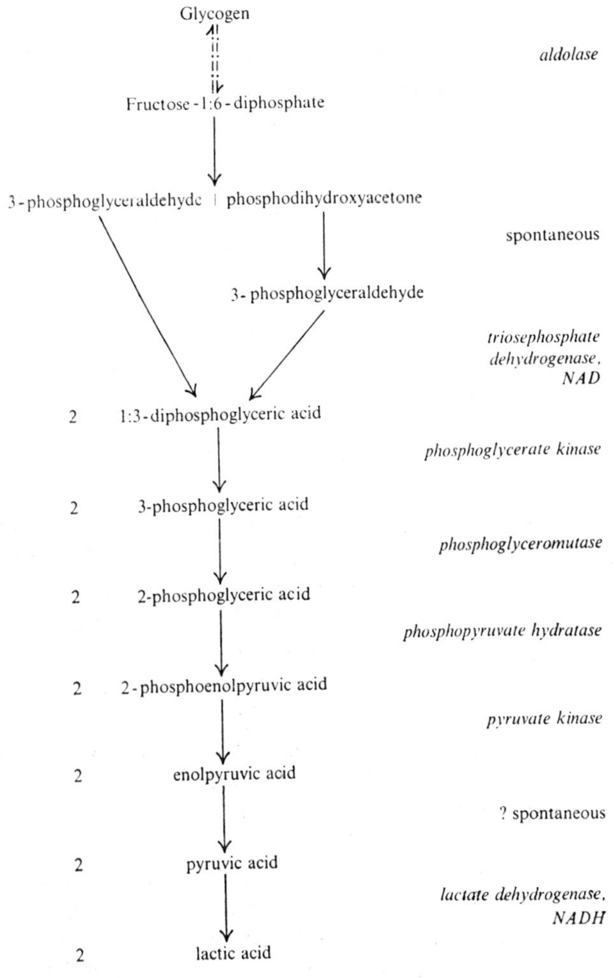

SCHEMA 5. Glycolysis. Arrows with chain line (·–·–·) indicate a series of reactions; italic indicates enzymes and co-factors.

pyruvate by comparable steps in some of which different enzymes are used.

Pyruvic acid is oxidized in two stages. In the first, it reacts with a substance called coenzyme A. This is a mercaptan, with the structure

$HS.CH_2.CH_2.$NH-pantothenic acid-diphosphate-ribose-phosphate-adenine.

It can form acyl thio-esters, and the bond by which it does so is energy-rich. Hence the acetic derivative, acetyl-CoA for short, may be written

$$CH_3.CO{\sim}S.CoA.$$

When pyruvic acid reacts with water and coenzyme A it is in effect oxidized to acetic acid, and the over-all reaction, which is called oxidative decarboxylation, may be written

$$CH_3.CO.COOH+H_2O \rightarrow CH_3COOH+CO_2+2H.$$

The hydrogen atoms are taken up by the acceptor nicotinamide-adenine dinucleotide, and the acetic acid combines with coenzyme A to form acetyl-coenzyme A, which also contains the energy liberated in the oxidation. Other co-enzymes, thiamindiphosphate (co-carboxylase) and lipoic acid, are required as coenzymes, as well as magnesium ions.

Acetyl-coenzyme A, with its high energy bond, is a very reactive substance. It can transfer its energy to adenosine diphosphate to enable it to take up phosphate and form adenosine triphosphate, in which case free acetic acid is formed, but more often it is oxidized by a series of reactions called the tricarboxylic or citric acid cycle (Schema 6), in which it reacts with oxaloacetic acid, which is itself derived from pyruvate and is ultimately regenerated.

Molecular oxygen is not used, the oxidation taking place by the addition of water followed by dehydrogenation. The hydrogen acceptors are nicotinamide-adenine dinucleotide, flavin adenine dinucleotide, and nicotinamide-adenine dinucleotide phosphate (NADP, coenzyme II). In the end two molecules of carbon dioxide are formed from each acetyl radical. In total, therefore, three have been formed from each pyruvic molecule, the carbon of which is thus completely oxidized. The oxidative decarboxylation of α-ketoglutarate is similar to that of pyruvate, but needs also the nucleoside guanidine diphosphate (GDP). All the enzymes of the citric acid cycle are present in mitochondria, but some are also present in the cytoplasm.

Since various aminoacids can give rise to pyruvate, α-ketoglutarate, and oxaloacetate, material derived from proteins can enter

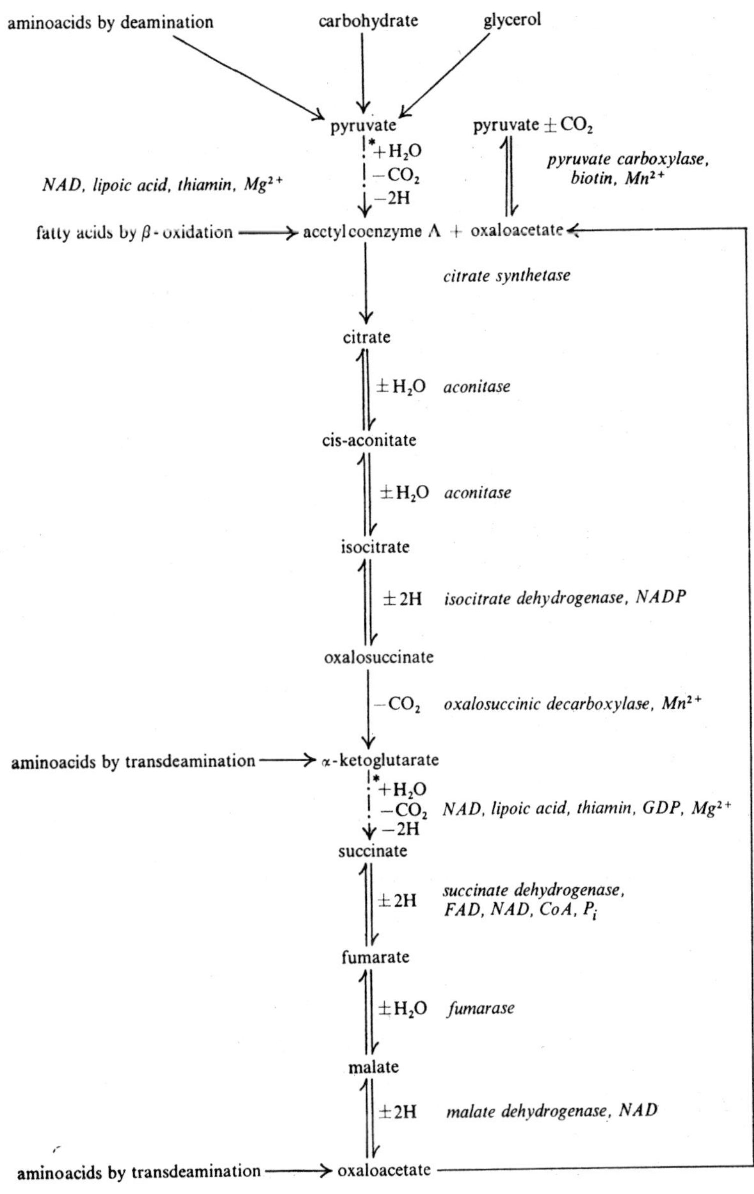

SCHEMA 6. The tricarboxylic acid cycle (citric acid cycle). Arrows with chain line (·–·–·) indicate series of reactions; italic indicates enzymes and co-factors, * = oxidative decarboxylation. Adenosine triphosphate is omitted.

the cycle at these three points. The schema shows also where derivatives of fats can enter in.

Glycolysis and the tricarboxylic acid cycle probably account for more than 90 per cent of the oxidation of glucose in mammals, but a small part goes through an alternative pathway sometimes called the shunt, in which glucose is converted to pentose with the formation of one molecule of carbon dioxide. The reactions may be summarized:

$$\begin{array}{l} \text{6 hexose phosphate} + 6O_2 \rightarrow \text{6 pentose phosphate} + 6H_2O + 6CO_2 \\ H_2O + \text{6 pentose phosphate} \rightarrow \text{5 hexose phosphate} + PO_4^{---} \\ \hline \text{hexose phosphate} + 6O_2 \rightarrow 5H_2O + 6CO_2 + PO_4 \end{array}$$

The reaction is aerobic, and the hydrogen atoms are removed by NADP. It is important in the liver and the lens of the eye.

The triose sugar glycerose (glyceraldehyde) and other sugars with 4, 5, and 7 carbon atoms are also formed in a complicated series of reactions, and it seems likely that the production of these compounds, and of reduced NADP for synthetic reactions, rather than the provision of energy, are important in the shunt. Pentoses in particular are necessary for the synthesis of nucleic acids.

Carbohydrate metabolism is regulated by several hormones, chief of which are those from the islets of Langerhans in the pancreas. Insulin, formed by the *B*-cells, causes glucose to disappear, and as it affects all four of the chief lines of change that glucose can undergo—conversion to lactate, conversion to fatty acids, oxidation to carbon dioxide, and synthesis to glycogen—it presumably acts on something that precedes or is basic to all of them. This could be the conversion of glucose to glucose-6-phosphate by adenosine triphosphate, but the most popular hypothesis at the present time is that it increases the rate of movement of glucose molecules across membranes. Since this may, as we have seen, depend on phosphorylation, the two hypotheses may be different ways of saying the same thing. Insulin is liberated into the blood whenever the concentration of glucose in the blood is raised. In the disease diabetes mellitus its production is deficient, the blood sugar rises, and is then excreted in the urine. The disease can be treated by the administration of insulin, which must be carefully regulated in amount to prevent too much glucose being removed. A deficiency of insulin also causes impaired synthesis of proteins and fatty acids, but these effects may be secondary to that on carbohydrate.

Glucagon, another hormone from the islets, probably produced by the *A*-cells, increases the blood sugar by accelerating the phosphorylation and breakdown of liver glycogen to give glucose-1-phosphate, and shifting the equilibrium to the right. It therefore assists insulin in aiding the utilization of glucose in muscle.

Adrenaline also assists the phosphorylation of liver glycogen, and as it increases also the production of lactic acid from muscle glycogen it provides more material from which liver glycogen is made. Cortin, from the adrenal cortex, has complex effects, the chief of which seem to be an increase in the formation of carbohydrates from fats and proteins, and a slowing of oxidation. Adrenocorticotropic hormone from the anterior pituitary acts by stimulating the activity of the adrenal cortex, and it increases also the production of adrenaline. Growth hormone, also from the anterior pituitary, has complex effects, most of which may be secondary to its effect in increasing protein synthesis.

The link between all the processes of metabolism is the transport carbohydrate, which is normally and chiefly glucose, though in the fetus of artiodactyls and the horse there is much fructose. Glucose enters the blood not only from the gut during digestion, but from the liver, and it is formed indirectly from muscle glycogen (via lactic acid and liver glycogen) and from protein and perhaps fat. Its concentration in the blood is kept sensibly constant in man at about 100 mg/100 ml by the action of all these hormones and by the sympathetic nervous system. Stimulation of the hepatic branch of the vagus causes the conversion of glycogen to glucose, and stimulation of the pancreatic branch causes the secretion of more insulin. The effects of the hormones are summarized in Schema 7.

Most of the carbohydrate in the food of artiodactyls is converted by the symbiotic bacteria not into hexoses but into short-chain fatty acids, especially acetic, propionic, and butyric, which are absorbed into the blood. Acetic cannot be directly converted into pyruvic acid, but it is able to react with oxalo-acetic acid and so enter the citric acid cycle. The details of its metabolism are unknown. Acetic acid by itself is poorly used, but an appropriate mixture of acetic, propionic, and butyric (which is what the animal normally gets) is 87 per cent used, compared with complete utilization of glucose. Man is able to use a little acetate, and its formation seems to be the first step in the utilization of ingested ethanol.

The function of the katabolism of carbohydrates is to supply

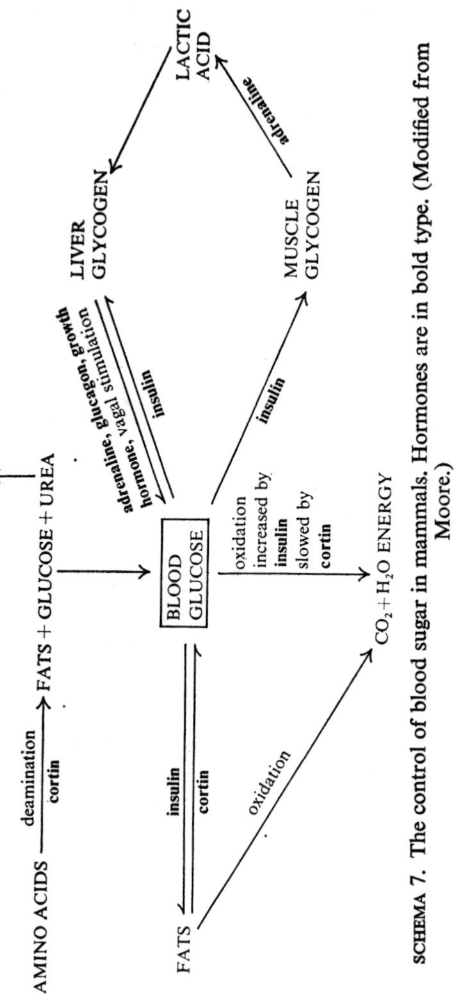

SCHEMA 7. **The control of blood sugar in mammals. Hormones are in bold type. (Modified from Moore.)**

energy, and although by the end of the citric acid cycle all the carbon from the original material has been oxidized to carbon dioxide, the hydrogen has been left associated with the hydrogen carriers. If we start with one molecule of glucose, the summary is

1. glucose $\xrightarrow{\text{glycolysis}}$ 2 pyruvate+2ATP+4H (in $2NADH_2$),

2. pyruvate $\xrightarrow[\text{decarboxylation}]{\text{oxidative}}$ $2AcCoA+2CO_2+4H$ (in $2NADH_2$),

3. AcCoA $\xrightarrow[\text{cycle}]{\text{citric acid}}$ $4CO_2+16H$ (8 in $4NADH_2$, 4 in $2NADPH_2$, 4 in $2FADH_2$).

In fact these reduced carriers do not persist, but rapidly react to give more energy by releasing their hydrogen to combine with oxygen to form water. They do this in association with substances called cytochromes, which consist of the base haem, similar to that present in haemoglobin, attached to a protein. The haem contains one atom of iron, while flavoprotein contains two.

There are several links in the chain, and at each stage there is formally a transfer of hydrogen atoms and reduction of ferric iron to the ferrous state, but since the hydrogen remains in solution as an ion, it is more correct to think of the reactions as consisting of the transfer of electrons,

$$Fe^{+++}+e^{-} \rightleftharpoons Fe^{++}.$$

At the end the hydrogen ion receives an electron from the ferrous ion (which is thus oxidized), and the atom so produced reacts with molecular oxygen to form water. The commonest pathway is that in which the hydrogen (or electrons) of nicotinamide-adenine dinucleotide are transferred to flavine adenine dinucleotide (FAD) and then to a series of cytochromes. The most likely scheme seems to be

$$\text{NAD} \rightarrow \text{FAD} \rightleftharpoons \text{cytochrome b} \rightleftharpoons \text{cytochrome } c_1 \rightleftharpoons \text{cytochrome c} \rightleftharpoons \text{cytochrome a} \rightleftharpoons \text{cytochrome } a_3$$

and finally

$$4 \text{ cytochrome } a_3Fe^{++}+4H+O_2 \rightleftharpoons 4 \text{ cytochrome } a_3Fe^{+++}+2H_2O.$$

Nicotinamide dinucleotide phosphate is used also. Cytochrome a_3 is possibly identical with cytochrome oxidase, which is necessary for the reaction with molecular oxygen. This reaction is stopped by cyanide, but oxidation of cytochrome b continues slowly in the presence of cyanide. Cytochrome c is soluble, but the others are present only in mitochondria, and are not easily detached from them.

In the scheme just given two molecules of nicotinamide-adenine dinucleotide, which hold four atoms of hydrogen, give six molecules of adenosine triphosphate when they are completely oxidized. The 24 atoms that are made available by the end of the citric acid cycle will therefore form 36 molecules of adenosine triphosphate which, with the two formed in glycolysis, give a total of 38 made by the oxidation of one molecule of glucose. We have, therefore, 38 new high-energy bonds, which are available for doing work in the body. The maximum energy that can theoretically be got from one molecule of glucose is about 3000 kJ. If each phosphate bond accounts for 50 kJ, the efficiency of these metabolic processes is therefore

$$\frac{50\times 38}{3000}\times 100 = 64 \text{ per cent.}$$

Since most of the other 38 per cent of the energy has presumably become heat, it is not entirely wasted in a warm-blooded mammal.

The reactions of the hydrogen acceptors with cytochrome are sometimes distinguished from the others that we have been discussing as tissue respiration, cellular respiration, or simply respiration. This is illogical, since they go on in life side by side with the other reactions, which cannot proceed far without them, if only because there is a limited supply of the hydrogen acceptors, which have to be continually regenerated to the oxidized form by the last step in the cytochrome reaction in which molecular oxygen is used.

If oxygen were not available, one might predict that the breakdown of carbohydrate would stop at some point back along the line, and this is indeed what happens. The block occurs at the formation of pyruvate. This does not, however, accumulate, because the hydrogen that has been produced at the same time and is held by nicotinamide-adenine dinucleotide, reduces it to lactate. The production of lactate from glycogen by a muscle working anaerobically was the first important piece of carbohydrate biochemistry to be discovered in animals (by Fletcher and Hopkins in 1907) and it has been the search for the mechanisms of its formation that has led to the elucidation of the schemes described in this section. Lactate acid is readily soluble, so that if much is produced, as in a muscle working actively and anaerobically, it is fairly rapidly carried away. In the liver it is rebuilt to glycogen; this needs energy, and one-fifth of the acid has to be oxidized to resynthesize the rest.

At the point where pyruvate is produced, one molecule of glucose has given only two molecules of adenosine triphosphate. The metabolic efficiency of anaerobic metabolism is therefore less than one-eighteenth of what it is if oxygen is available, or a little more than 3 per cent.

In fact, wherever voluntary muscle is highly active it does not get enough oxygen to run the citric acid cycle fast enough to keep pace with the pyruvate formed; so lactic acid appears, and more glycogen is broken down than would be needed if oxygen were freely available. Since, however, the glycogen is restored by the oxidation of some of the lactic acid, the muscle is said to go into oxygen debt (glycogen debt would be a better term, since it is tomorrow's glycogen that is being used). The lactic acid formed in an anaerobic muscle is the chief source of fatigue and pain after exercise.

It appears that certain tissues may be completely anaerobic. Skin can live for a week in hydrogen (though without mitosis) and can then recover. Embryonic cartilage uses oxygen at a normal rate, but that of the adult has a consumption that is very low or even zero. Glucose is broken down to lactic acid, and though in some cartilage cytochrome oxidase is present, its action is inhibited. Cartilage is without blood-vessels, so that the value of anaerobic respiration is obvious. Certain cancerous cells can grow *in vitro* completely without oxygen.

M 4.13. Metabolism of Fats

Fatty substances may be an essential constituent of protoplasm, or they may be present as formed materials that are easily removed. The structural fats are mostly complex, while the reserve or depot fats are usually in the form of neutral triglycerides, the ordinary fat of common speech. Both forms are in part built up by the animal from the fatty acids absorbed by the small intestine. There is some general connection between the fats in the depot and those taken in the food, but each tissue of each species maintains its fats fairly constant in composition, so that there must be some power to change one fatty acid to another.

Whatever its subsequent fate, most dietary fatty acid undergoes a process called β-oxidation, because the carbon atom first attacked is that next to the α-carbon, to which the carboxyl group is attached.

First, the acid joins with coenzyme A to form a high-energy derivative (acyl-coenzyme-A), the energy for this coming from adenosine triphosphate. Hydrogen is then lost to an electron-transporting flavo-protein (ETF) related to flavine adenine dinucleotide, water is added, more hydrogen is lost to nicotinamide-adenine dinucleotide, and finally another molecule of coenzyme A reacts with the fatty molecule, splitting off the end piece as acetyl coenzyme A, and forming a new acyl coenzyme A to repeat the process. These reactions may be summarized

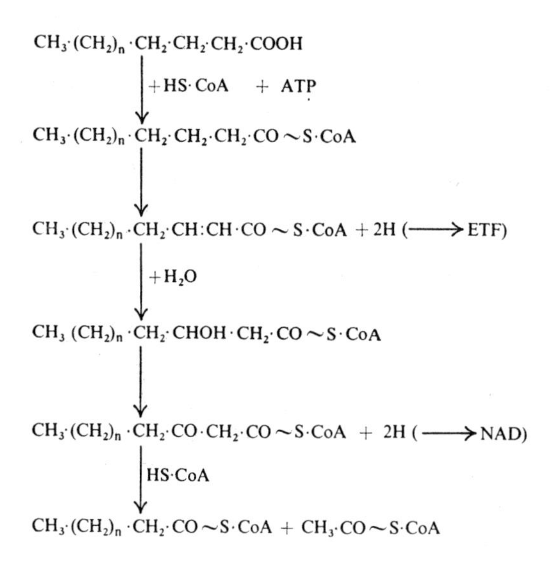

Several enzymes are needed; the first reaction goes on in the cytoplasm, the remainder in the mitochondria. The net effect of these successive oxidations is that the whole of the fatty acid is converted to acetate, which, in the form of acetyl coenzyme A, can enter the citric acid cycle and be further oxidized to provide energy.

Although the reactions of β-oxidation are individually reversible, it is unlikely that the whole process ever goes backwards in the body. Instead, acetyl coenzyme A takes up carbon dioxide to form malonyl coenzyme A. This then exchanges its coenzyme group for a similar one called acyl carrier protein, the reactive group of which is identical with that of coenzyme A. The malonyl acyl carrier protein then reacts with acetyl acyl carrier protein formed in a similar way, and by loss of carbon dioxide followed by hydrogenation and removal of water gives butyryl acyl carrier protein. These reactions are summarized in Schema 8. They go on in the cytoplasm, not in mitochondria. The butyryl acyl carrier protein can react in the same way with more

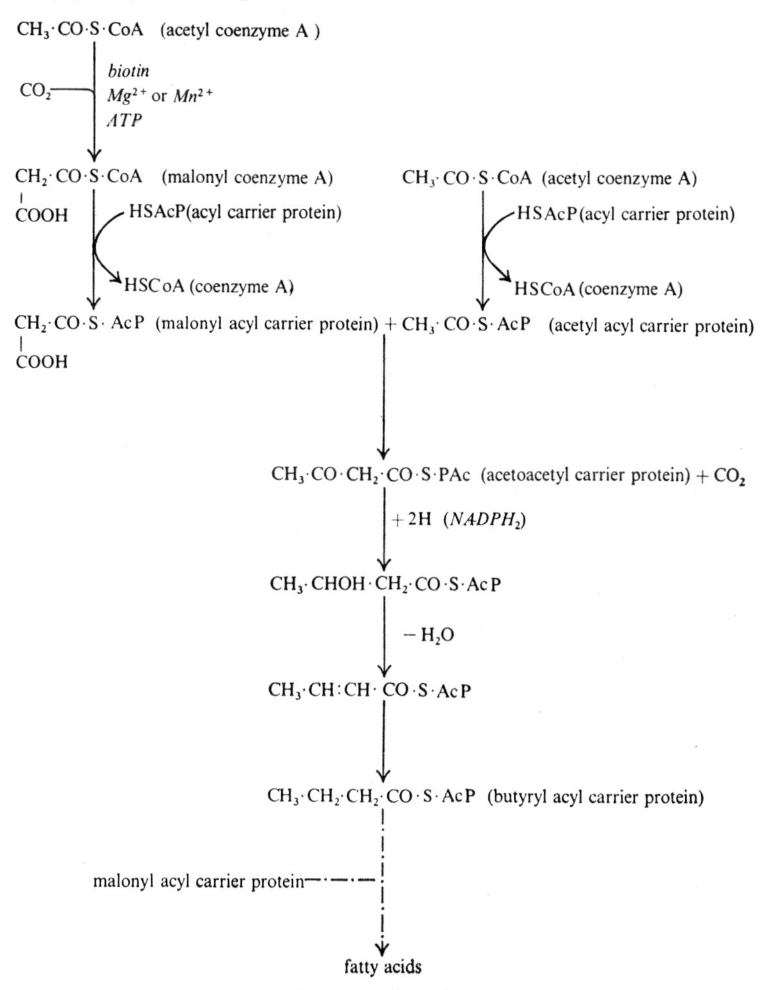

SCHEMA 8. Synthesis of fatty acids.

malonyl acyl carrier protein, and the process can be repeated several times. The result is the formation of fatty acids with successively longer chains, increasing by two carbon atoms at each step. This explains why most fats contain acids with fourteen, sixteen, or eighteen carbon atoms. The formation of neutral fats (triglycerides) takes place by reactions between the appropriate acyl coenzyme A and α-glycerol phosphate, the acidic groups being added one at a time.

Some fats probably undergo an ω-oxidation, in which the terminal methyl group is oxidized to a carboxyl, to give a dicarboxylic acid. When there is too much fat in the diet, more acetyl coenzyme A is formed than there is oxaloacetate for it to react with in the tricarboxylic acid cycle. Acetyl coenzyme A or its derivatives therefore accumulate, giving rise to a condition called ketosis.

The glycerol derived from fat is oxidized, with adenosine triphosphate as a coenzyme, to triose phosphate, which is on the path of glycolysis. The reactions are probably reversible.

M 4.14. Interrelationship of Carbohydrates, Fats, and Proteins

Schema 5 shows that all three of the chief classes of foodstuffs can enter the tricarboxylic acid cycle, and since this is the chief way in which adenosine triphosphate is formed and energy made available for the body, all three can be a source of this energy. There are obvious limitations. Excess fat leads to accumulation of ketone bodies, and although the Carnivora are able to deal with these, man is not. Excess proteins mean too much ammonia, and here again the Carnivora are better able than man to eliminate this. Nevertheless, one food can to a great extent substitute for another, and on a mixed diet the total energy content of the food can be measured by adding those of its constituent parts.

If oxygen is taken as the unit the energy available is not greatly different for the three foodstuffs. Different compounds give different values, but on the average

1 g of oxygen oxidizes carbohydrate to water plus carbon dioxide giving 15·9 kJ,

or fat to water plus carbon dioxide giving 13·4 kJ,

or protein to water plus carbon dioxide giving 13·0 kJ.

But since 1 g of oxygen will oxidize 0·94 g of glucose and only 0·34 g of tristearin, fat is evidently a much more economical form in which to store energy. One gram of fat gives about 39 kJ and 1 g of carbohydrate only 17·1 kJ so that a mammal which had its fat replaced by an equivalent amount of glycogen would be very bulky.

Fumaric acid enters into both the ornithine cycle and the tricarboxylic acid cycle. Fatty acids produce acetyl coenzyme A and proteins produce α-ketoglutarate; both of these reactions are reversible. There are therefore pathways for conversion of carbohydrates to fats and proteins (though the last needs ammonia in addition and the essential aminoacids cannot be synthesized). The change from carbohydrate to fat is a commonplace, since high feeding on an almost fatless diet leads to the laying-down of much storage fat both in man and in domestic animals.

It is not so obvious that carbohydrate could be formed from fat, since the conversion of pyruvate to acetyl coenzyme A is not reversible. Glycerol comes in above this line, so that a little of the fat (as distinct from the fatty acid) molecule could form carbohydrate, and so does some of the protein. There is some evidence from experiments with ^{13}C that glycogen can be formed from acetic acid; this is certainly important in the ruminants, which absorb chiefly acetic acid and not glucose from the gut. The biochemical details are unknown, but acetate is only well used if propionate and butyrate are available as well. In the mammary glands of ruminants the milk fat also is formed from acetate, although in rats and rabbits it comes from glucose.

V 4.1. METABOLISM IN OTHER VERTEBRATES

We have only a fragmentary knowledge of how far the above or similar reactions occur in vertebrates other than mammals. The cytochromes are very widely distributed not only in animals but in plants as well, so that the mechanisms of the final oxidation are probably broadly similar in most groups. The same is true of the synthesis of proteins through the nucleic acids.

Deamination apparently occurs in the amphibians, and presumably in other groups. The ornithine cycle has been demonstrated in the livers of elasmobranchs, *Latimeria*, *Protopterus*, amphibians, and a tortoise, but it does not occur in teleosts or birds.

There is much variation in the further fate of the nitrogen separated from the aminoacids and of the purines. Elasmobranchs, teleosts, and Amphibia possess uricase, and also allantoinase, which breaks the purine ring again and hydrolyses allantoin to allantoic acid,

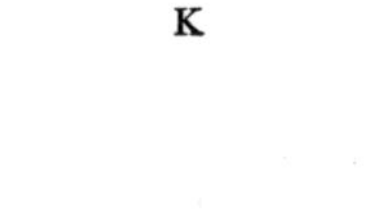

allantoic acid

They generally have also allantoicase, which hydrolyses this substance to two molecules of urea and one of glyoxylic acid, CHO—COOH, so that the nitrogen from nucleoproteins is excreted as urea, but some teleosts are without this enzyme.

The teleosts, amphibian tadpoles, and the Crocodilia do nothing further with their ammonia and excrete it as such. Elasmobranchs and adult amphibians, like mammals, convert it to urea, probably by a similar mechanism. Birds, snakes, and some lizards convert it to uric acid. In birds the reactants include also carbon dioxide, glycine, and formate, which, with the help of glutamate and ribose-1-phosphate, give hypoxanthine, which is then oxidized to uric acid by the same route as in mammals. Birds produce also a little urea, probably by the action of arginase on arginine derived directly from the proteins in the food.

The vertebrates can therefore be divided physiologically into three groups: the ammoniotelic forms, all aquatic, excreting ammonia directly; the ureotelic forms, excreting urea; and the uricotelic forms, which save water by excreting the relatively insoluble uric acid. There are some interesting special points. Aquatic chelonians excrete chiefly ammonia, or approximately equal quantities of ammonia and urea, amphibious species excrete mostly urea, and xeric species mostly uric acid with fair quantities of urea.

Dipnoi produce ammonia when they are active, urea when they are in their cocoons. The adult of the aquatic amphibian *Xenopus laevis* produces chiefly ammonia, like the tadpole. The relationship with the mode of life is obvious, and more is said of this in section G 9.11. The changeover from ammonia to urea at metamorphosis in Amphibia appears to be connected, like other changes at that time, with the activity of the thyroid, for when metamorphosis is induced in the axolotl by thyroxine, orsimilarly made to occur precociously in the frog, the ratio of urea to ammonia increases.

The carbohydrate mechanisms of all vertebrates seem to be basically similar. They all store glycogen, and glycolysis and the tricarboxylic acid cycle have been demonstrated in teleosts and amphibians. There are, however, differences in detail. Galactogen is an important energy source in some fish, and there is more fructose than glucose in the blood of the fetal sheep. While the muscles of most can work anaerobically, accumulating lactic acid and putting up an oxygen debt, the carp (*Carassius carassius*) can live in the absence of oxygen for 2 months at 5 °C without accumulating lactic acid and not repaying any oxygen debt. Young chick heart-tissue can grow *in vitro* without oxygen.

Our knowledge of the control of carbohydrate metabolism is very incomplete. The concentration of blood sugar of birds is about twice that of mammals, and this may be connected with the poverty of the pancreas in *B*-cells and its richness in *A*-cells. Removal of the pancreas from grain-eating birds causes only a mild hyperglycaemia, which disappears within a week, but a similar experiment on an owl caused extreme hyperglycaemia and death, just as it would in a mammal. Hyperglycaemia can be induced in a pancreatectomized duck by feeding it with meat. Reptiles are relatively insensitive to insulin, needing large doses to produce hypoglycaemia, but glucagon readily increases blood sugar. Amphibians respond slowly to insulin, but relatively low doses eventually cause convulsions and death. Most amphibians and reptiles that have been examined have been described as having both *A* and *B* cells, but there is some doubt about the identity of these with the mammalian cells. Lizards are said to have more *A* than *B*, amphibians more *B* than *A*. Nothing seems to be known of the hormones that they produce. Some Urodeles apparently have only *B*-cells. Urodeles are also odd in being the only vertebrates in which the hyperglycaemia caused by removal of the pancreas is not alleviated by hypophysectomy.

Removal of islet tissue from both selachians and teleosts produces hypoglycaemia, and both *A* and *B* cells have been described in the former.

If a connected story can be made from all this it is that probably fairly early in the development of vertebrates some cells from the intestinal walls, originally concerned with producing enzymes for the digestion of carbohydrates, became diverted to producing low-molecular-weight hormones that affected later stages in carbohydrate metabolism. There were two of these, insulin and glucagon, of which the former became more important in the line that led to mammals, the latter in the line that led to birds. That there has been gradual improvement in the system of control is suggested by the much greater variability in levels of blood sugar in cold-blooded vertebrates than in mammals and birds.

Little is known of the metabolism of fat. It is stored in the connective tissue of all vertebrates, though amphibians have little of it. Cold-blooded animals have a higher proportion of unsaturated fats than do mammals and birds, and marine fish have fats based on fatty acids with a longer chain, containing twenty, twenty-two, or twenty-four carbon atoms.

Fat is especially important when an animal has to undergo a long period without food. Salmon in fresh water may live for as long as a year without feeding, gradually using up their stored fats. Hibernating mammals lay up fat in autumn and use it during the winter. Some birds put on much fat, sometimes amounting to half the body weight, in spring and autumn, and burn it all up in a migratory flight of a few hours. It seems that the skeletal muscle fibres of the more active birds are very rich in fat and in lipase, so that it appears that fat is a normal fuel for flight; the advantage of this would be that more energy can be got from fat than from the same weight of carbohydrate. Whether the fat is used directly or is first converted to carbohydrate, is unknown, but there seems no reason why it should not enter the tricarboxylic acid cycle as described above.

Injected glucagon causes a great increase in the free fatty acids in the plasma of birds. According to some authors this is a primary effect, suggesting an evolution of the part played by the pancreas in controlling intermediate metabolism.

J 4.1. METABOLISM IN INVERTEBRATES

We know even less about what goes on in invertebrates. It seems that annelids, molluscs, and decapod crustaceans begin by separating ammonia from the rest of the aminoacid molecule, and as most invertebrates excrete ammonia, deamination, with or without transamination, seems likely. The chloragogen cells of earthworms carry out the deamination and produce urea. The ornithine cycle has been demonstrated in some nematodes. The insects and to some extent the gastropods, are uricotelic, in which they resemble birds, but other purines also are excreted; guanine accounts for more than 90 per cent of the nitrogen excreted by some spiders, and is important in *Helix*. The purines derived from nucleoproteins are also broken down to a varying degree, presumably depending in part on the enzymes available. Crustaceans and some bivalves can break them down completely to ammonia, while in platyhelminths and annelids there is no oxidation of the purine. Most groups produce some product in between.

The synthesis of proteins by the nucleic acids is generally assumed to occur in all living things. A free-living nematode, *Caenorhabditis briggsae*, can synthesize five of the 'essential' aminoacids.

Tsetse flies have very small amounts of carbohydrate, and use proline, which makes up 2 per cent of the haemolymph, as a circulating source of energy at the beginning of flight. Blow-flies are to some extent similar.

The general course of the metabolism of carbohydrates appears to be the same in all animals. Glycogen is a common storage material, occurring in *Pelomyxa* (a rhizopod), earthworms, molluscs, and crustaceans amongst others. In earthworms it is produced by the chloragogenous cells. The snail *Helix*, the bivalve *Anodon*, and other molluscs contain also in the gonads another polysaccharide, galactogen, which is built up of galactose instead of glucose. Where the sexes are separate it is not present in the males, but it seems to be necessary as a storage product for the eggs.

Although glucose has generally been considered to be the usual blood sugar, trehalose is the normal one in insects and nematodes; it is a disaccharide which on hydrolysis gives two molecules of glucose, from which in insects it is formed in the fat body. It is present also in the tissues of some other invertebrates, including molluscs of the three main groups.

Evidence for the presence of the glycolytic and tricarboxylic pathways has been sought in two ways: by testing for the presence of individual enzymes, and by observing the result of adding known intermediate products which, if the pathway is working, should cause predictable alterations in the rate of appearance of an end-product. By these means, parts of one or the other or both have been shown to be probably present in several invertebrates: protozoans, earthworms, molluscs, nematodes, insects. In some, however, the system seems not to be complete; for example, while succinic dehydrogenase (a necessary enzyme in the middle of the tricarboxylic acid cycle) was demonstrated in the cockroach *Periplanata americana*, it could not be found in the desert locust, *Schistocerca gregaria*. It would therefore be rash to assume that the cycle is universal. The pentose shunt has been shown in *Fasciola*, *Taenia*, and *Ascaris*, and in a few insects and some other animals. Both insects and decapod crustaceans are said to control their blood sugar by a hormone which causes hyperglycaemia, and the same claim has been made for the snail, but all the evidence is rather thin. Nicotinamide-adenine dinucleotide, flavin adenine dinucleotide, and cytochrome are common, but the last could not be found in *Ascaris*.

Many invertebrates live where the oxygen concentration is low, and some live where there is practically none; organic muds and sometimes the alimentary canal may have no detectable oxygen. In these circumstances the animals that have been examined carry out partial anaerobic oxidations comparable to those described above for vertebrate muscle. *Ascaris* produces little or no lactic acid, but much acetic and propionic acid, and various others with five carbon atoms, such as methyl butyric acid. Their treatment of carbohydrates therefore seems to be more like that of the symbiotic bacteria of ruminants than like that of mammals, and it seems unlikely that the tricarboxylic cycle is used. Both normal glycolysis to pyruvate and the shunt are said to be present. The symbiotic ciliates of ruminants also produce various acids, and so do the platyhelminths—lactic, succinic, and others in tapeworms, and propionic, with some lactic and acetic, in the liver-fluke.

The respiration of *Taenia taeniaeformis* is in part cyanide-sensitive, which suggests that cytochrome or some similar substance is used. *Schistosoma* has cytochrome enough to account for only 10 per cent of its oxygen consumption, but the whole of its oxygen uptake is abolished by cyanide. *Trichinella spiralis* does not possess glycogen and does not produce acids, so presumably the carbon dioxide that it liberates is made from some other product.

Many invertebrate tissues can put up an oxygen debt, and in the insects it may be very high, but the flight muscles are strictly aerobic.

Fat is found as a storage product in many invertebrates: in platyhelminths, in the yellow cells of earthworms, in molluscs, and in the digestive diverticula and gonads of crustaceans. The fat of insects undergoes β-oxidation and enters the citric acid cycle. Many insects, like birds, use fat as fuel. Most Diptera and Hymenoptera can use it at rest but not in flight; Orthoptera, Homoptera, some Diptera, and possibly Coleoptera use it in sustained flight; while Lepidoptera use nothing else, both at rest and in flight.

5. Excretion

In many of the chemical reactions that go on in the animal body, whether their object is to provide energy or to make some definite substance that the animal needs, by-products are formed for which there is no immediate use. Any process by which these by-products are so treated that they take no further part in metabolism is called excretion. The term is not generally applied to the voiding of material that enters the body and passes out unchanged: thus the nitrogen that leaves the lungs of tetrapods, and the undigested cellulose in the faeces of herbivorous animals, are not regarded as excretory products. The analogy with the waste products of a chemical factory is a fairly close one: the method of dealing with the unwanted material depends on its quantity and nature, and the harm that it would do in its place of origin.

G 5.01. Substances Excreted

For convenience the materials which are excreted may be put into four groups:

Water. Most animals, whether they live on land or in ponds or in the sea, take in large quantities of water with their food, and many mammals and birds drink it as well. Expulsion of this water is not strictly excretion, but some is also formed in the complete oxidation of fats and carbohydrates, and in all the class of reactions called condensations. Further, the surface of the animal cell is normally a semi-permeable membrane and the osmotic pressure of the protoplasm greater than that of fresh water; this means that freshwater animals will take up water osmotically. Unlike plants, animals seldom have rigid cell walls, and so this last process would, in the absence of special arrangements, go on indefinitely. In such animals the regulation of this osmotic water is more important than the excretion of that from the first two sources; its general physiological and ecological aspects are considered in section G 9.11. Feeding animals with deuterium oxide (heavy water) has shown that there is rapid interchange of water between the gut and the cells. It is, therefore, impossible to distinguish between molecules of water from the three sources.

Carbon Dioxide. Carbon dioxide is the normal end-product of the carbon of any material which is oxidized to give energy. The organ systems that get rid of it are generally those that bring oxygen into the animal, and so its excretion is considered under the heading of respiration; an account of the chief processes is given in section 3.42, but where the mechanism is the same as that for excretion it is dealt with in this chapter. It need only be said here that small quantities are dealt with in other ways, for example, by being converted into urea (section M 4.11) or calcium carbonate.

Nitrogen Compounds. Nitrogen compounds are so much more important than the other groups that excretion often means in effect nitrogenous excretion.

The processes of deamination and transamination, by which the nitrogen of most of the ingested protein is sooner or later separated as ammonia, and the subsequent fate of this, have been described in sections M 4.11 and V, J, 4.1.

Other Substances. There is scarcely an element that has not at some time been found amongst the excretory products of animals. Though quantitatively much less important, qualitatively they may be just as important to the animal as nitrogen. The chief are phosphorus, usually as phosphate, sulphur as sulphate derived by oxidation of aminoacid residues, and calcium. They are generally eliminated by the mechanism that deals with nitrogen.

The Felidae are peculiar in excreting fat in the urine. In tigers it appears to be secreted by the proximal convoluted tubules of the kidney.

G 5.02. Methods of Excretion

The object of excretion is to remove from the sphere of chemical action of the body end-products which may be harmful, and which in any case will, if they are allowed to accumulate, upset equilibrium by their mass-action effect. The simplest way to do this is by a method sometimes used in the treatment of poisoning, that is by converting the obnoxious substance into some insoluble derivative. In this state it can do no harm. Any part of the body that seems to be specially used for storing such materials is sometimes called a 'kidney of accumulation', though there is no suggestion that there is any similarity to the mammalian kidney other than very generally in function.

It is possible that in some animals such insoluble waste products, originally quite useless, have become of some benefit to their possessors. It is noteworthy that many skeletal materials contain elements that are often excreted. Chitin contains 4 per cent of nitrogen, the acid portion of calcium carbonate is a form of carbon dioxide, and bone is largely made of calcium phosphate, itself a common excretory product. The pigments of many insects, such as the white and yellow pterins, were probably originally excretory, although now sometimes of biological value. Uric acid appears to be regularly present, though in varying quantities, in vertebrate exoskeletons such as feathers, hair, and hooves.

Even where excretory products are made insoluble and stored they usually sooner or later escape to the exterior and are lost. Exoskeletons are shed, and insoluble material in the coelom goes out by pores. But more often there is some definite mechanism by which the unwanted material is voided. The word 'kidney' has often been applied to any organ that appears to be concerned in this, but it seems better to restrict the word to the particular type of organ that carries out nitrogenous excretion in the vertebrates. Alternative names are available in all groups, and 'excretory organ' is good enough for general purposes.

MV 5.1. THE VERTEBRATE KIDNEY

By far the best known of all excretory mechanisms is the vertebrate kidney. It eliminates water, nitrogen, and many other substances, and has shown itself capable of adaptation to very varied circumstances. In addition to dealing with waste matter it is also an important osmotic regulator; this aspect of its function is dealt with in section G 9.1. Although its structure varies to some extent with the evolutionary level of its possessor and also with the environment, it is always built on the same general plan. It consists of a mass of coelomoducts opening into a single longitudinal collecting duct. The detailed arrangement in the mammals is as follows (Fig. 5.1). The coelomoducts have lost their original openings to the general coelom, and each starts as a blind Bowman's capsule. In the concavity of this is a bundle of blood capillaries, known as the glomerulus, derived from the renal artery. Bowman's capsule and the glomerulus are together known as the Malpighian body. From Bowman's capsule the coelomoduct continues as a coiled proximal convoluted tubule, then as the loop of Henle, and finally as the distal convoluted tubule. Several distal convoluted tubules open into one collecting tubule, and this joins with others to form large ducts which finally open into the ureter. The whole part distal to Bowman's capsule is spoken of

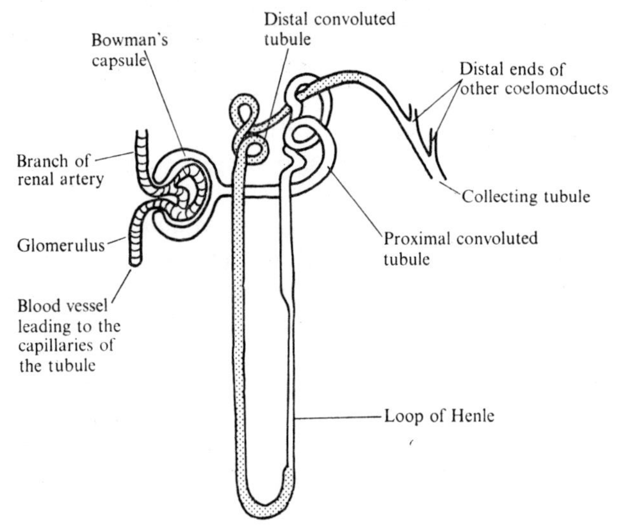

FIG. 5.1. Diagram of a single coelomoduct of a mammalian kidney.

collectively as the tubule. Blood is brought away from the glomerulus by a single vessel that breaks up into another set of capillaries surrounding the tubule. From them blood goes into the renal vein. There are some differences from this structure in other classes of vertebrates; the marine teleosts either have a much reduced glomerulus or are without it altogether, and in those animals that have a renal portal system the tubules (but not the capsules) are supplied with a second system of capillaries containing venous blood.

In the Amphibia, Reptilia, and Mammalia it has been possible by means of a micromanipulator to withdraw small samples of the fluid from a capsule for analysis. This 'glomerular filtrate' (the reason for the name will appear shortly) has been found to resemble a fluid obtained from the blood by dialysis; with respect to every constituent for which the former can be analysed they are almost identical, and their electrical conductivities are the same. There is thus little doubt that the fluid in the capsule is blood plasma without its colloids. The only assumption that need be made to explain how this comes about is that the wall of the glomerulus together with that of the capsule acts as a membrane that is impermeable to colloids but permeable to water and crystalloids; that is, as a dialyser. If this is the case the conditions in the Malpighian body are that a membrane impermeable to colloids has water, crystalloids, and colloids on its one side, and water and crystalloids on the other. The ordinary laws of osmosis apply, and in the absence of any outside forces, water and crystalloids would pass through the membrane into the blood under the influence of the osmotic pressure of the colloids. But in fact they travel in the other direction, and this can only happen if there is a hydrostatic pressure acting in the opposite direction to the osmotic pressure. This is normally always present, since the blood is maintained above atmospheric pressure by the pumping action of the heart. Filtration should stop when the hydrostatic pressure in the glomerulus falls below the osmotic pressure of the colloids. This is the case; in mammals the colloids have an osmotic pressure of about 30 mm of mercury, and secretion of urine ceases when the arterial pressure falls below about 40 mm. (It is to be expected that the pressure in the capillaries will be somewhat less than that in a large artery such as the brachial where blood pressure is usually measured.) Secretion may be stopped experimentally by ligaturing the ureter; in one experiment it ceased when the pressure in the ureter was 92 mm and that in the arteries was 133 mm. The difference, 41 mm, is obviously the maximum pressure available for forcing water and crystalloids out of the blood under these conditions, and it was just not enough to do so. The converse experiment, to increase the effective filtration pressure, has also been carried out; a certain volume of blood is removed, and replaced by Ringer-Locke fluid, which is isotonic with the blood but which contains no colloids. Under these circumstances, although the blood pressure actually falls owing to the shock, the rate of excretion greatly increases.

Since the energy for the filtration comes from the heart, one would not expect an alteration in excretory rate to affect the oxygen consumption of the kidney. This is found to be the case. Furthermore, it would be expected that low temperature or respiratory poisons, while reducing those activities of the kidney which alter the composition of the glomerular filtrate, would not greatly affect the glomerulus. The urine would therefore have a much closer resemblance to the original glomerular filtrate than it normally does. This also has been experimentally confirmed, and at 25 °C or when cyanides are added to the blood the urine (in excised and perfused dogs' kidneys) is very similar to the dialysate from plasma.

There is thus direct evidence that the Malpighian bodies act merely as ultrafilters which separate water and crystalloids from the plasma. By adding various proteins to the blood and observing if they appear in the urine, it has been found that the distinction is not strictly between crystalloids and colloids, but that the membrane is permeable to molecules of smaller molecular weight than about 70 000. Gelatin, of molecular weight 35 000, and haemoglobin, of molecular weight 67 000, are passed out, while the normal serum proteins, of molecular weight not less than 72 000, are almost completely retained.

The urine is normally more concentrated than the glomerular filtrate, and different in composition; it is richer in potassium, phosphate, urea, and other waste products, is poorer in sodium, chloride, and aminoacids, and contains no glucose, bicarbonate, or protein. Obviously, therefore, the tubules must reabsorb some substances, and they might also specifically excrete those that are most concentrated in the urine. Various experiments, including the use of tracers and in a few instances direct analysis of fluid withdrawn from various parts of the tubule, have shown that both processes go on.

Some urea—in man about two-fifths of that filtered by the glomerulus—is absorbed in both proximal and distal tubules; this is a simple osmotic effect of the concentration of the fluid, and the concentration of urea in the urine of mammals never falls below that in the plasma. Elasmobranchs, which have a high concentration of urea in the blood and reabsorb 90 per cent of that which is filtered, must have assisted transport. Most mammals reabsorb uric acid and destroy it, but the Dalmatian dog possesses a simple recessive gene that prevents absorption, so that this animal excretes uric acid in spite of the possession of uricase.

There is only a little protein in glomerular filtrate, but that is absorbed by a peculiar process called athrocytosis: the molecules are taken up by the microvilli of the cell and there flocculate and remain for some time before being passed into the capillaries.

Glucose is almost completely absorbed by the proximal tubule, and the rest is taken up by Henle's loop. Its transport is assisted, energy and active phosphate being needed.

In man, about seven-eighths of the filtered sodium ions are actively absorbed in the proximal tubule, and most of the rest in the loop and in the distal tubule. The absorption of such an important cation would lead to electrical and osmotic imbalance; the first is compensated by the secondary absorption of ions, chiefly chloride, and the second by the secondary absorption of water, so that the concentration (as distinct from the quantity) of sodium chloride reaching the distal tubule is the same as that entering the proximal tubule. More water is absorbed in the collecting ducts. In man only one-eighth of the water in the glomerular filtrate reaches the distal tubule, and fifteen-sixteenths of this is absorbed there and in the ducts, so that of 170 litres normally filtered in 24 h less than 1·5 litres reach the bladder.

Potassium ions are completely absorbed in the proximal tubule, but some are excreted again by the distal tube. Phosphate also is absorbed in the proximal tubule. So too is bicarbonate, but the relations of this ion are complicated by the production of carbon dioxide in the metabolism of the kidney cells. Bicarbonate is absorbed also by the distal tubule in such a way that the urine becomes acid.

The tubular cells form some ammonia in two ways: by the action of glutaminase on glutamine in the blood, and by the action of aminoacid oxidase on glycine and other aminoacids. It is excreted, and combines with some of the excess hydrogen ions to form the ammonium ion. This tends to neutralize the urine, and at the same time allows the excretion of anions that would otherwise need sodium to go with them.

The proximal tubule excretes many organic substances when they are added to the blood; this seems to be a safety mechanism for eliminating substances that ought not to be present. The changes in the filtrate brought about by the tubules are shown in Table 5.

There is no doubt that the kidney works in basically the same way in nearly all vertebrates, but there are some differences between classes. The loop of Henle is present only in mammals and birds, and must presumably have arisen twice. In mammals it appears to act as a countercurrent multiplier for concentrating the fluid, and has presumably evolved in connection with the need for a land animal to conserve water. There is a rough correspondence between the length of the loop and the habitat of the mammal; thus the beaver (*Castor*) lives in water, has short loops, and produces a urine of about 500–600 milliosmoles/litre; man has medium loops and produces urine of 1100–1200 milliosmoles/litre; the rabbit has mixed short and long, and urine of 1500 milliosmoles/litre; while desert animals such as the sand-rat (*Psammomys*) and jerboa (*Jaculus*) have long loops and make a very concentrated urine of 6000 or more milli-osmoles/litre.

TABLE 5. *Reabsorption of solutes by the renal tubules in Man. In 24 h 170 litres of glomerular filtrate are formed (from Samson Wright's* Applied Physiology, *Oxford University Press, London, 1965*)

Constituent	Concentration in plasma (mg/100 ml)	Total in 170 litres of plasma or glomerular filtrate	Total excreted in urine in 24 h	Total reabsorbed in tubules
Water		170 1	1·5 1	168·5 1
Glucose	100	170 g	none	170 g
HCO_3^-	150	255 g	0·1 g	255 g
Na^+	330	560 g	5 g	555 g
Cl^-	365	620 g	9 g	611 g
K^+	17	29 g	2·2 g	26·8 g
Phospha te (as P)	3	5·1 g	1·2 g	3·9 g
Ca^{2+}	10	17 g	0·2 g	16·8 g
Urea	30	51 g	30 g	21 g
HSO_4^- (as S)	2	3·4 g	2·7 g	0·7 g

In birds, the initial filtration and the subsequent absorption of glucose, sodium, chloride, water, and other substances are similar to what goes on in mammals. Although the loops of Henle have a less regular arrangement, one may guess that they have the same function. The chief nitrogenous waste is uric acid, 90 per cent of which (in the fowl) is secreted by the tubules and is precipitated either as the acid itself or, under certain conditions, as sodium urate. A similar tubular excretion occurs in the Squamata.

Amphibia, where the glomerulus is large and the renal portal vein supplies only the tubules, have been used for much of the basic work on the physiology of the kidney. Ligation of the renal artery and injection of substances into the renal portal vein have helped to show the action of the tubules. Sodium, glucose, water, and other things are absorbed in the proximal tubule, as in other vertebrates, and it can excrete injected organic substances. Urea is secreted by the tubules, especially the distal tubules, in frogs, but apparently not by the aquatic *Necturus* and *Xenopus*.

In many marine teleosts there is no glomerulus, and all the waste products dealt with by the kidneys must be excreted by the tubules.

The quantities of waste matter produced are relatively constant for an individual, but it is obvious that the kidney also eliminates water, the supply of which, in land animals, can vary very widely. Common experience shows that more urine is formed after drinking, while after sweating, as in hot climates, there is less.

There are several types of control. Quantitatively the most important is that by the antidiuretic hormone from the posterior pituitary, without which water cannot be taken up by the distal tubule and the collecting tubule. Its production is controlled by sensory endings in the hypothalamus called osmoreceptors; this is a misnomer, as they are really chemoreceptors, since they respond to the concentration of sodium and chloride in the blood but not to that of urea or glucose. Deprivation of water and excess of salt both lead to the production of more antidiuretic hormone; conversely, drinking much water stops its production.

Aldosterone from the adrenal cortex (section 6.142) is necessary for the absorption of sodium, and parathyroid hormone controls the absorption of phosphate.

Anything that increases blood pressure must, for purely mechanical reasons, raise the rate of filtration by the glomerulus; this happens under stimulation of the sympathetic nervous system (section MV 7.32) or with liberation of adrenaline as in fear or anger. There is in addition a fine hormonal control specific to the kidney. Where the arteriole enters Bowman's capsule there is a thickening in the walls called the juxtaglomerular apparatus, which is a pressure receptor and an endocrine gland. When the pressure in the arteriole falls, it is stimulated to release into the blood a hormone called renin; this is a proteolytic enzyme which converts an α-globulin in the plasma to angiotensin I; by the action of another enzyme in the plasma this becomes angiotensin II, an octopeptide which is a very powerful vasopressor substance; by causing contraction of the small blood vessels it raises the pressure in them. It leads also to a release of aldosterone from the adrenal cortex, so that as the volume of the glomerular filtrate is increased, more sodium is taken up. The kidneys contain also substances that have the opposite effects to renin. One of them, which has been called medullin, may be a prostaglandin (section MV 6.16).

Some correlation between the fine structure of the kidney of mammals and its function is possible. The walls of the capillary and of Bowman's capsule are both thin and discontinuous, with relatively large pores—up to 100 nm in diameter in the capillary, about one-third of this in the capsule; they are separated by a basement layer, secreted by the capsule, in which the pores are only about 5 nm in diameter; this must therefore be the filtering barrier. The cells of the proximal convoluted tubule have a well-developed brush-border against the lumen, and the opposite side (against the blood capillaries) is packed with mitochondria; this suggests that the chief work in reabsorption is done in passing material into the blood. The thin part of Henle's loop consists of flattened unspecialized cells; according to some authors it cannot carry out assisted transport. The cells of the distal tubule resemble those of the proximal tubule, but have fewer microvilli.

Little is known of the coordination of excretion in other vertebrates. Antidiuretic hormone has complicated and not entirely consistent effects on permeability, which are discussed in section M 6.132.

The urine passes to the bladder through the ureters, which have walls containing plain muscle, and which in man have about three waves of contraction passing down them per minute. The bladder can hold a gradually increasing volume of urine without showing any increase of pressure or of muscle tone. The act of micturition, voluntary in origin unless the bladder is over-full, is assisted by several involuntary muscles, and results in the complete emptying of the bladder.

In most other vertebrates the urine passes not directly into a bladder but into the upper part of the cloaca, from which, in amphibians and reptiles, it goes into a thin-walled bladder, homologous with the allantois, formed on the ventral surface. The bladder of amphibians varies greatly in size, being largest in desert species and very small in those that are permanently aquatic. In it there is active uptake of sodium, and other things also can be taken up. In desert forms the water can be reabsorbed when needed.

MV 5.2. OTHER VERTEBRATE EXCRETORY ORGANS

Adult vertebrates have no other specialized excretory organs than the kidneys, but water and carbon dioxide are lost in tetrapods through the lungs, and in mammals about half the magnesium and calcium and most of the iron leave through the intestine, the two former being mainly in the form of insoluble phosphate and the last chiefly sulphide. A small amount of nitrogen, about 1 g/day in man, also goes out in this way. It is largely in the form of breakdown products of haemoglobin which are present in bile. Many waste products pass out, where conditions are suitable, through the skin and the gills. The sweat contains most of the common excretory substances, but in negligible quantities, and in frogs much carbon dioxide and water go out through the skin.

In the teleosts most of the ammonia and urea that are formed are lost through the gills, the chief forms of nitrogen in the urine being creatine, creatinine, and trimethylamine oxide. The elasmobranchs are somewhat similar, but the gills are much less permeable to urea, which, with trimethylamine oxide, is largely retained in the blood (section G 8.1).

In the embryos of the amniotes the allantois is used as a deposit for renal waste that has been eliminated by the mesonephros. In those mammals where the placenta is highly permeable (e.g. rodents and man) the excretory products are apparently simply taken away by the maternal blood; the mesonephros has only a very short existence, and the allantois is a functionless stalk or (as in the mouse and rat) never develops at all.

J 5.3. EXCRETION IN INVERTEBRATES

The elimination of waste matter could take place over the whole body surface where it is thin enough, at special parts of the surface, or by a system of tubes leading from the interior to the outside world. The first two are probably adequate in the more primitive aquatic creatures, but a tubular system of some sort exists in the more highly organized invertebrates. Morphologically it may be of three main types. A nephridium is a derivative of the surface, usually ectodermal, centripetal in growth, and with an intracellular lumen bearing one or more cilia or flagella at the inner end. Primitively it is a protonephridium, ending blindly in a single flame-cell or solenocyte, with the cilia flickering in its interior, but secondarily it often opens into the coelom, and is then called a metanephridium. The external opening of a nephridium is a nephridiopore, and the internal opening of a metanephridium is a nephridiostome. Nephridia are found in Platyhelminthes, Rotifera, Nemertinea, and Annelida, and in the larvae of some Mollusca; all of these phyla can be associated together on embryological grounds, so that there is no difficulty in holding that the nephridia of all of them are homologous. Nephridia are present also in *Branchiostoma*, where they appear to be anomalous, and in some minor phyla.

A coelomoduct is the reverse of a nephridium. It is formed from the wall of the coelom and so is mesodermal, it grows centrifugally, and has an intercellular lumen, part of which may be ciliated but which is without flame-cells. The kidney tubules of vertebrates are coelomoducts, and they occur also in Annelida, Mollusca, Crustacea, Arachnida, Echinodermata, and some minor phyla. Whether they are homologous in all these is a matter of opinion. The decision will depend in part on one's views on the homologies of the coelom itself, but there would seem to be no difficulty in imagining that such a simple device could occur more than once. It seems likely that its primary purpose was always to act as a genital duct.

In some animals, particularly the polychaetes, nephridium and coelomoduct are closely associated to form a compound organ, the nephromixium.

Lastly, the gut may be excretory and there may be a system of tubes associated with it. In crustaceans, insects, and spiders such tubes make the chief excretory organ.

For many invertebrates we know only that if they are kept in a confined space certain nitrogenous compounds accumulate in the medium, and we can do no more than make intelligent guesses as to how they come to be there. For a few it has been possible to make analyses comparable to (but nowhere as detailed as) those that have been carried out on the contents of the vertebrate tubule. In this way the triple mechanism of filtration, resorption, and the addition of solutes by active excretion, has been fairly conclusively shown to be present in decapod crustaceans, gastropods, and cephalopods, so that it must have arisen three or four times in evolution. Filtration and resorption are known also in insects and lamellibranchs, and resorption, with a strong presumption of filtration, in annelids.

J 5.31. Excretion in Protozoa

There is no uniformity in the nitrogenous waste products formed; *Amoeba* forms uric acid, *Glaucoma* and *Didinium* ammonia, and *Paramecium* and *Spirostomum* urea according to some authors and ammonia according to others. Since *Colpidium* produces ammonia throughout its life-cycle but urea only when the culture is rapidly growing, there may be similar variation in *Paramecium*, or there may be specific differences. These products have been shown to accumulate in cultures of the animals named, but there is no evidence to show how they came to be outside the animal. The old story that the contractile vacuole is a specific excretory organ rests on a very insecure foundation, since attempts to show the presence in it of nitrogenous waste have failed. One author, for example, knowing the rate at which urea accumulated in a culture of *Paramecium caudatum*, and the rate of elimination of water by the vacuoles, was able to calculate that if all the urea excreted came from the vacuoles its concentration in them should be one part in from 2000 to 3000. He injected into the vacuoles Nessler's solution (for ammonia) and xanthydrol, which is sensitive to one part of urea in 12 000, but his results were negative. It is therefore probably sufficient to assume that soluble waste products diffuse out through the whole cell wall, just as it is assumed that carbon dioxide can diffuse out and oxygen can diffuse in. The contractile vacuole undoubtedly eliminates water, which probably contains waste products in solution, so that to a certain extent it assists in excretion, but it cannot be responsible for the whole or even a major part of it. Analysis of the fluid removed from the contractile vacuole of *Spirostomum* shows that it eliminates only 1 per cent of the total urea produced.

It is probable that some of the crystals found in many genera are waste products stored in an insoluble form; in *Paramecium caudatum*, for example, granules of acid calcium phosphate have been demonstrated. The shells of Protozoa, which may be of calcium carbonate, strontium sulphate, silica, or a nitrogenous material, may be regarded as excretory; this view seems particularly likely to be correct where, as in *Polystomella*, most of the protoplasm is outside the shell, which can therefore hardly be protective.

J 5.32. Excretion in Coelenterata

Most chemical investigations have been negative; uric acid has been found in *Anemonia sulcata* and urea in other sea anemones, but the chief product in actinians is ammonia. There are no specialized excretory organs, and it must be assumed that waste products escape as they do in the Protozoa. In actinians it has been shown that injected substances, such as carmine, accumulate in certain regions, and this may mean that these places are specially active in excretion. Skeletons are common, both of calcium carbonate, as in the corals, and of nitrogenous organic material, as in the perisarc of Hydrozoa and in the gorgonians (Actinozoa). One function of the symbiotic green Algae in corals seems to be to remove end-products, such as carbon dioxide.

J 5.33. Excretion in Unsegmented Worms

The flame-cell system of the Platyhelminthes, which is a protonephridium, is usually regarded as the excretory system. In *Schistosoma* there are about eighty cilia in each cell, with the usual structure of fibrils (section MVJ 7.1). The cells certainly expel a fluid, but the chemical investigations of this are few and unsatisfactory. There is some evidence that the nephridia are also osmotic regulators (e.g. their absence from the Turbellaria Acoela, which are all marine; variation in the rate of pulsation of the terminal vesicle in Cercariae with the external osmotic pressure). It has, however, been shown that this is not the case in *Gunda ulvae.* The chief nitrogenous product of planarians, *Fasciola hepatica*, and *Taenia pisiformis* is ammonia, while *Trichinella spiralis* loses one-third of its nitrogen as ammonia and most of the rest as aminoacids and peptides.

The nematodes have no cilia and no flame-cells. There are two types of tubular system which expel a fluid to the exterior, sometimes with pulsation. In many free-living species there are one or two ventral gland cells each with a terminal ampulla. In parasitic forms there is a canal formed from a single cell, usually shaped like a capital H, and opening ventrally. In some, the one lateral link is lost, and in others, such as *Ascaris,* the anterior portion is missing, giving a system like a capital *Π*. A few genera, such as *Rhabditis,* have the full H pattern as well as two gland cells.

Ammonia is the chief nitrogenous excretory product, but fair quantities of urea are produced, and some species produce aminoacids and amines. The part played by the tubules is obscure. They have been shown to produce a liquid immiscible with water, and to be able to eliminate injected dyes, so that they are potentially excretory. Nearly all the excretory nitrogen of *Ascaris* comes from the anus, and the fluid from the excretory pore is said to contain only 0·02 per cent of urea.

J 5.34. Excretion in Annelida

Ammonia is the chief form of nitrogenous excretory product in the polychaete *Aphrodite* and in the medicinal leech, both of which are fully aquatic. Various species of *Lumbricus* and other earthworms produce both ammonia and urea, but authors differ as to the relative amounts. It seems that if the animal is well fed ammonia is the more important. Other substances, such as aminoacids, creatinine, and small quantities of purines are also produced.

The chloragogen cells of the gut wall were for long said to be excretory. It seems that they form urea, but how this finds its way to the nephridia, which are the main excretory organs, is not clear.

The nephridium of the earthworm has been shown to be a true nephridium without any mesodermal elements. Its physiology is unfortunately not so well known, since observers are not in complete agreement, but certain points are fairly clear. In the first place, coelomic fluid and the finest solid particles that it contains go down the nephridium to the exterior; in lumbricids chloragogen particles are much too large, but this does not apply to the whole phylum. Secondly, cells of the middle tube and the ampulla extract particles from the coelomic fluid and store them, although they gradually disappear. Their pigment is a haemochromogen, which must presumably have come from the blood pigment, but their chemical composition is otherwise unknown. Experimentally, these cells have been shown to take up particles of Chinese ink and coloured fluids which pass the nephridiostome. The osmotic pressure of the urine of earthworms is less than half that of the coelomic fluid, which is itself hypotonic to the blood. The nephridia thus appear to be osmotic regulators. The urine is poorer than the other two liquids in almost every constituent, nitrogenous and otherwise, for which it has been analysed, so that there must be fairly general absorption of solutes. In particular, it contains much less protein than either the coelomic fluid or the blood, and agrees with the former in containing no glucose or fats, both of which are present in fair concentration in the blood. It is impossible, on the bases of these analyses, to determine whether all the urine comes from the coelomic fluid, or whether there is also filtration from the blood. The concentration of protein in the blood of *Pheretima* is about half that in human blood, so that, if the protein molecules are of about the same size, the blood pressure necessary for filtration on the principle of the vertebrate kidney would be about 15 mm of mercury. This seems unlikely to be reached, and the function of the nephridial capillaries may be only to supply oxygen. According to some authors, there is intracellular excretion of granules, through the wall of the terminal reservoir to its cavity.

The nephridium of earthworms empties at intervals of not less than 3 days, but that of *Arenicola* every 2 or 10 minutes.

There are probably also subsidiary methods of excretion. In many species of earthworm there are masses called brown bodies loose in

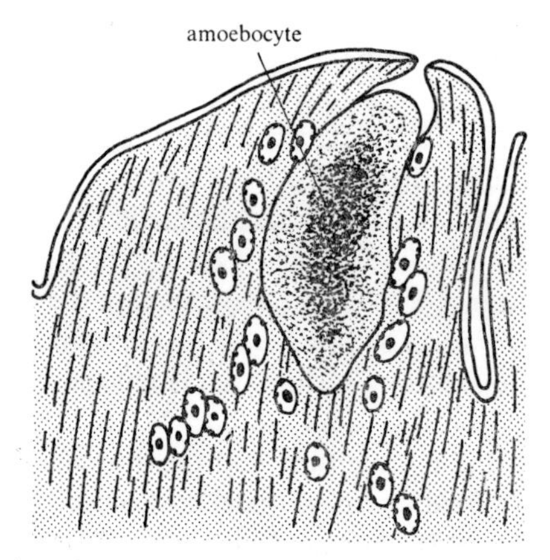

FIG. 5.2. Section of the oesophageal wall of *Allolobophora foetida*, showing a yellow cell or excretory amoebocyte in the act of traversing it. The lumen of the oesophagus is above. ×500. From Keilin, *Q. Jl microsc. Sci.* **65** (1920).

the coelom of the posterior segments. They may be up to 5 mm long, and probably eventually escape through the dorsal pores. They consist largely of cysts of nematodes and of *Monocystis* and other gregarines, and of shed chaetae, but they contain also dead amoebocytes that have ingested particles that are said to be of uric acid. Phagocytes in the blood also take up particles and pass to the gut wall, where they remain for a time as yellow cells, after which they fall into the lumen and escape with the faeces (Fig. 5.2).

J 5.35. Excretion in Arthropoda

Very little work has been done on the Crustacea except in decapods, where the chief nitrogenous excretory product is ammonia, with sometimes much aminoacid. The antennary gland of Malacostraca and the maxillary gland of Entomostraca, both of which are coelomoducts, are usually described as the excretory organs, but the liquid extracted from the green gland of decapods often contains less non-protein nitrogen than does the blood. It has, therefore, been suggested that in freshwater forms it acts as an osmotic regulator, but since it is also present in the marine species, that cannot be its only function. In *Cancer* the hydrostatic pressure of the body fluid is greater than the colloid osmotic pressure of the blood, and there is filtration of water and crystalloids followed by resorption of some ions and excretion of others, as in the vertebrate kidney. Other forms are similar, and the resorption is specially important in the freshwater crayfish and the estuarine shore crab. The liver was shown long ago to be excretory, for its cells can extract foreign matter from the blood and eliminate it through the gut. In *Cancer* and some other genera it normally deals with purines in this way, but these could not be found in the crayfish. The exoskeleton, of chitin and calcium carbonate, may be in part excretory, and chromatophores of *Eriocheir* contain uric acid.

A number of amphipods and isopods have been shown to produce chiefly ammonia; in the marine *Ligia oceanica* this makes 70 to 80 per cent of the total nitrogen, while in the terrestrial wood-lice *Oniscus asellus*, *Porcellio scaber*, and *Armadillidium vulgare* it is 50 to 60 per cent, so that these animals appear to be only partially adapted to terrestrial life. Their total excretory nitrogen is however, only one-tenth of that of the aquatic species. In the wood-lice, and also in the freshwater *Asellus aquaticus*, 5 to 10 per cent of the non-protein nitrogen is uric acid. The chief excretory organs are the segmental glands of the second maxilla.

In insects uric acid is the normal nitrogenous excretory product, but traces of urea are excreted by many, and ammonia, urea, and amino-compounds are found in the clothes moths. Ammonia is the chief substance in many aquatic larvae, including those of Odonata, Trichoptera, *Sialis*, *Notonecta*, and *Dytiscus*, and in aphids and in dipteran maggots. These last also break down uric acid to allantoin, which they excrete, as do the adults. Dytiscid and carabid beetles and some Orthoptera go further, and hydrolyse this to allantoic acid, which they excrete. Guanine is unknown.

The Malphigian tubules are the most important excretory organs in most insects. Their morphological arrangement and their histological structure vary, but they consist of from two to more than 100 tubes opening into the beginning of the rectum, and usually having cells with a brush border and many mitochondria. They excrete both urates and carbon dioxide, which they receive from the blood in solution. The mechanism bears certain similarities to that in vertebrates, particularly birds; in *Rhodnius prolixa* (Rhynchota), for example, potassium or sodium urate is secreted in the upper part of the tubule, and water and base are resorbed in the lower part, leading to the precipitation of free uric acid as solid spheres. The same water and base are circulated and used repeatedly. This leads to conservation of water, which is very important in insects, where the method of respiration leads to much evaporation. An analogous process leads to the precipitation of calcium carbonate. More water and sodium are absorbed by the rectum. There is control by a hormone released into the haemolymph by the abdominal nerves, and after a meal of blood there is a strong diuresis.

Other insects are similar in principle but different in detail. In many, especially Orthoptera, the rectum is the chief or only site for the absorption of water, which it carries out so effectively that pellets of dry solid uric acid are produced. There is normally no special mechanism for the expulsion of the contents of the tubes, but in Diptera they are muscular and peristalsis takes place.

There are subsidiary organs, of which the chief is the fat body, the parietal layer of which stores uric acid. The fate of this is different in different orders of insects. In the Collembola, which have no Malpighian tubules, it is stored throughout life, and the fat body is therefore the chief excretory organ, although the tubular glands opening by a common duct at the base of the labium in these insects and in the Thysanura are also said to be excretory. Uric acid is also stored throughout life in the Dermoptera and many Orthoptera, where the Malpighian tubules contain little or no uric acid. In the Hymenoptera it is transferred to the hind gut during early imaginal life, and is voided when the insect emerges from the cocoon. A similar process occurs in Lepidoptera and muscid Diptera; here there is a period in metamorphosis when the tubules are completely broken down, and the fat body is therefore very important. Of less importance are the nephrocytes, chains of cells along the heart or oesophagus, which store nitrogenous matter that may later be taken away by phagocytes.

There are in many insects peculiar types of excretion, as, for example, in the celery fly, *Acidia heraclei* (Diptera), where the calcium carbonate formed in the Malpighian tubules is deposited as large calcosphaerites, is dissolved during metamorphosis, and is laid down on the last larval skin and so eliminated—a process called ecdysial excretion. In some Dipteran larvae calcosphaerites formed in the fat body are eliminated in the same way. As in the Crustacea, chitin may be excretory, and the white, yellow, or red pigment of the wings of many butterflies is a pterin (closely related to uric acid) which is also excreted through the gut.

In spiders the Malpighian tubules excrete uric acid, probably as a sodium salt, and guanine. It is probable that the coxal glands eliminate urates. Uric acid is excreted by the Malpighian tubules of the Chilopoda.

J 5.36. Excretion in Mollusca

In lamellibranchs, ammonia and amino-compounds are the chief forms of excretory nitrogen, but urea and purines are often present in fair quantities, and traces of uric acid have been found. Sixty per cent of the waste nitrogen of *Anodonta cygnea* is eliminated as ammonia. The concretions in the organ of Bojanus consist mostly of magnesium and phosphate, and uric acid is not present, but there is some nitrogenous material. Keber's organ acts as a kidney of accumulation, but in what form nitrogen is stored is not known. Of the gastropods, *Helix pomatia* eliminates ammonia, amino-compounds, urea, uric acid, and other purines, all in considerable quantity, some at least of them being got rid of by the kidney. During winter, uric acid and other purines are stored in the kidney. The slug *Limax agrestis* excretes chiefly urea, and the marine *Aplysia limacina* ammonia, amino-compounds, purines, and urea, in that order of quantity. The urine of cephalopods contains much ammonia, trimethylamine oxide, and amino-compounds, and smaller quantities of urea and purines. In *Octopus* more than twenty compounds have been found by chemical analysis, some in considerable quantities, but the identity of most of them is unknown. The molluscs illustrate very well the connection between the habitat and the chief excretory products that was pointed out in section V 4.1.

The excretory organs are coelomoducts opening from the pericardium to the exterior. The cells of the inner portions generally have a brush border, while the distal parts are ciliated.

In some of the lamellibranchs, snails, and cephalopods that have been examined it is fairly certain that filtration, resorption, and secretion all occur. The sites vary. In cephalopods, filtration from a structure called the branchial heart appendage passes fluid into the pericardium, but in snails the site of filtration is probably the kidney sac. More generally and primitively, blood probably passes direct through the walls of the heart to the pericardium. Glucose, and the other substances that are absorbed, are taken up by various parts of the kidney sac or of the duct.

The cilia help to expel the fluid, and in lamellibranchs there are also contractions of the organ which drive out its contents. In pulmonates general contractions of the body are needed, which occur in terrestrial forms every two or three weeks.

J 5.37. Excretion in Echinodermata

The fluid of the water-vascular system contains small quantities of ammonia, amino-compounds, urea, and uric acid. Since it is in communication with the surrounding sea water, nitrogen is presumably lost from it.

J 5.38. Excretion in Hemichordata

The ascidians produce mostly ammonia but some species store much uric acid or other purines.

6. Coordination of Function

G 6.0. INTRODUCTION

Even in the smallest and simplest animal some coordination of the activities of the parts is necessary. When an amoeba is moving it is useless pseudopodia being formed at one end if protoplasm is not also withdrawn at the other; the cilia of a paramecium must beat in a definite rhythm and pattern if the necessary forward motion and rotation are to be maintained. To some extent such coordination may be purely mechanical. When sperms are allowed to come together and clump, the beating of their tails may become synchronized, which suggests that the rhythm of the strongest is imposed on the others in what the physicist calls forced vibrations; something similar might account in part for the regular beating of cilia. Equally, the withdrawal of the protoplasm of an amoeba at the opposite end to its pseudopodia might be brought about by stretching of a gelated protoplasm. Such simple explanations are, however, seldom if ever enough. Purely mechanical interference with cilia, as with a microdissection needle, does interfere with the rhythm of cilia, but so also do damage to the underlying cytoplasm and various chemicals; an amoeba, touched lightly at its forward end, very soon forms pseudopodia to one side of the point of contact. It seems therefore that there is some sort of conduction or coordination through the cytoplasm. It may be called neuroid, but that is no more than a name for our ignorance.

The cells of a sponge, which has no specialized tissues, may react to a stimulus received as far as a centimetre away. Here also we may speak of neuroid transmission, but since the cells are not in protoplasmic continuity it seems likely that coordination is by the secretion of chemical substances.

Even where there is a well-developed coordinating system, mechanical impulses may remain important. This is especially true in the locomotion of segmented animals, from worms to mammals, in which the stretching of one part by the contraction of another part is an important link in the maintenance of the rhythm. If an arm of a starfish is merely sewn on to a disc, its tube feet point in the right direction, so that it seems that mechanical conduction is all that is needed.

The more complex an animal is, that is, the more distinct parts it has, the more necessary becomes some method of coordinating their action, and the more complex must be the system that carries out this coordination. Higher animals have developed two types of system: that which depends only on chemical substances, or hormones; and the nervous system, in which the use of chemical substances is combined with a form of electrical conduction. It is possible that both are developments from the same fundamental neuroid system, which might have been either chemical or electrical. Since electrical conduction depends on the movement of ions the distinction is, in the limit, a fine one. Purely electrical conduction, as by currents in metallic wires or by electromagnetic radiation, is not known to occur in organisms.

G 6.1. HORMONES

Almost everything that an animal does is initiated, or has its rate determined, by substances produced within the body. The ability of a muscle to contract repeatedly depends on an adequate supply of glucose, digestion depends on enzymes, and the rate of breathing is affected by the concentration of carbon dioxide in the blood. These and many similar instances have been known for a long time, and they are not usually thought of as examples of coordination. The substances concerned may be put in four groups: the substrates of metabolism, such as glucose; excretory products, such as carbon dioxide; enzymes and coenzymes carrying out metabolic processes; and secretions passed by a duct to the gut or to the exterior. In addition to these we now know of many substances whose only function is that of initiating or controlling some process in the body; their action is thought of as chemical coordination, and they are called hormones from a Greek word for 'excite'. Their delimitation from enzymes, other than those passed to the gut or the exterior, is difficult and in part arbitrary. In part also it depends on the history of their discovery, for while enzymes were discovered as substances taking part in chemical reactions, hormones were discovered as substances that affect the response of a muscle or other effector, and only a few of them are known to act by taking part in a reaction as an enzyme does. The fact that hormones are produced in the body distinguishes them, just as it does enzymes, from vitamins, but it does not separate them. When ascorbic acid is produced by the tissues it is a hormone; when it is obtained in the food it is a vitamin. At least some of the effect of young grass in stimulating milk-production in cows is due to oestrogenic substances that it contains.

Some hormones act close to the place where they are formed, as do the evocating substances produced in vertebrate embryos and the neurohumours liberated at the end of motor nerves, but the typical hormones are made in glands that are remote from the point of application. They are carried in the blood-stream, and have been certainly proved to exist in all the main phyla in which a vascular system is present, that is in annelids, arthropods, molluscs, and chordates. The earliest known example, that of the stimulation of the pancreas by secretin, was described by Bayliss and Starling in 1902, and since our knowledge of the physiology of invertebrates other than insects is still very small, it is not impossible that hormones may yet be found to be present, in some form, in other phyla.

The glands that make hormones may have no other function, as the adrenal (though one gland may make more than one hormone), or they may also produce secretions that pass down a duct in the ordinary way; there is then generally separation of the tissues within the gland, as with the islet tissue of the pancreas which produces insulin and glucagon. A gland that produces a hormone is called an endocrine gland, and hormones are sometimes called internal secretions and autacoids.

The normal way of testing for endocrine action is, first, to remove the suspected organ from an animal and observe the effects. Extracts of the organ are then injected into the blood; if they restore the animal to normal, endocrine action may be taken as proved. Where removal of the organ is impossible other tests may be used; extracts may produce overdose effects, it may be possible to demonstrate an increased concentration of an active principle in the blood leaving the organ, the concentration of some substance in the blood may depend on the condition of an organ, or effects may be produced by the grafting of an organ from another animal. That an effector is under endocrine control may be assumed when it continues to act and to respond to stimuli when all its nervous connections have been cut. Endocrine glands are themselves under control, usually of the nervous system, and there is thus a double link between the stimulus and its response. The stimulus activates a receptor, which initiates an impulse that travels via the central nervous system to the gland, which is caused to produce or release a hormone into the blood-stream, by which it travels to the effector.

Many hormones are polypeptides, and others are steroids, but beyond this no generalization about their chemical nature is possible. It is, however, possible to say what properties they must possess in order to be effective. They must be soluble in water, or they could not be readily carried by the blood or pass into the tissues, although a few, which have been distinguished as lipohumours, may be fat-soluble and of purely local effect. They must diffuse readily through protoplasm, or they would not be able to escape fast enough from the blood-vessels or reach all the cells on which they act; this means that they must be of relatively low molecular weight. If they were not fairly rapidly removed, their effects would be permanent, and so their value as coordinating agents would be lost; their removal might be by excretion, or by oxidation, or by digestion, and examples of all of these are known. There might also be specific inhibitors, or anti-hormones, but although these are certainly formed in some circumstances it is not certain that they have any natural function.

The effects of hormones range from determining the fate of cells once for all, as in embryos, through general influences on the rate of a process that never stops, as the thyroid secretion raises basal metabolic rate, to brief action on individual effectors, as in the control of amphibian melanophores by the pituitary. It is possible that the first is the most primitive, and that it developed from the general chemical control that the nucleus has over the cytoplasm. Genes act by passing nucleic acid into the cytoplasm, and since examples of modification of a nucleus by its surrounding cytoplasm are known it seems likely that the cytoplasm also produces chemical substances that act at a distance. From this general power of protoplasm to produce chemical substances would seem to have developed the various types of hormone now known. Their primitive and fundamental nature is shown by their lack of specificity. In general they act on a tissue or cell-type rather than on an organ. Adrenalin, glucagon, vasopressin, and parathormone all activate adenyl cyclase, which converts adenosine triphosphate to cyclic 3′ 5′ adenosine monophosphate, but they possibly act in different types of cell. Each hormone seems to be chemically very similar throughout the vertebrate classes, and there are some hormones that have similar effects in invertebrates and vertebrates. Nevertheless, the same hormone, or a series of chemically closely related ones, may have different effects in the different vertebrate classes, so that there has been an evolution of the response of the target organs alongside the evolution of the hormones themselves. Sometimes there are differences in response even within a class.

MV 6.1. HORMONES IN VERTEBRATES

MV 6.11. The Gut and Pancreas

The action of the mucosa of the gut and of the pancreas in co-ordinating the secretion of digestive enzymes and the metabolism of carbohydrates is considered in sections M 2.315, V 2.31, M 4.12, and V 4.1. Their hormones are polypeptides, and are perhaps derived from the enzymes produced by adjacent cells. Gastrin is a 17-peptide. Most insulins consist of two linked chains, one of twenty-one and the other of thirty aminoacids. There are specific differences, which seem to be of no taxonomic significance; there are differences in three acids between those of the pig and the sheep, but only in two between those of the cow and the cod.

MV 6.12. The Pharynx

The pharynx of vertebrates is a very versatile organ. It has produced ciliary and mucous feeding organs, gills, swim bladder, and lungs, and several endocrine glands, which are often developed in the gill pouches. As these have serial homology the homology of the glands derived from them is difficult; they often come to lie in parts of the body more or less distant from their origin, and morphological relationships between the different classes are not always obvious. Perhaps they should all be regarded as homologous, for calcitonin, best known as a product of the parathyroids (section MV 6.122) is produced in some mammals by the thyroid and the ultimobranchial bodies.

MV 6.121. The Thyroid. The mammalian thyroid consists of two masses lying one on each side of the trachea just behind the thyroid cartilage. The division into two parts is secondary, and in the rabbit and many other animals the two halves are joined by a ventral strand across the trachea. The gland starts as a single diverticulum from the pharynx, and its position and mode of development show that it is probably homologous with the wall of the endostyle of lampreys and perhaps of *Branchiostoma*. It is made up of closed vesicles with a wall of cubical epithelium, the cells of which have a brush border. In birds the thyroid consists of two masses of tissue, one lying in the region of each carotid; it is paired also in some reptiles and in amphibians, but single in the lizard. In the frog the thyroid consists of a small reddish body on each external jugular vein, and in the dogfish of a pear-shaped structure just below the fork of the ventral aorta, while in adult cyclostomes and many teleosts it consists of groups of follicles scattered about the region of the ventral aorta.

The effects of thyroidectomy can be counteracted by a number of substances, both natural and artificial, most of which contain iodine although some bromine compounds are effective, and it seems that the important part of the molecule is a substituted benzene ring, thus,

I
—O—
I

The gland takes up iodine or iodide from the food, oxidizes it, and incorporates it in a protein, thyroglobulin, of molecular weight about 700 000. This is stored in the vesicles of the gland, and then split by a proteolytic enzyme to form thyroxine

I I
HO— —O— —$CH_2\cdot CH\cdot COOH$
NH_2
I I

and small quantities of triiodothyronine, which has only one iodine atom in the left-hand ring. Both diffuse into the blood and are there attached to a globulin. The release of thyroxine from thyroglobulin is brought about by a hormone from the anterior pituitary. Thyroxine is unique among hormones in that it is not easily digested, and so can be successfully administered by the mouth.

Over-activity of the gland produces the disease known as exophthalmic goitre, and under-activity myxoedema in adults and cretinism in children, but an account of these belongs to pathology rather than physiology. Their symptoms, however, suggest that the chief normal function of the thyroid is probably the regulation of the growth and activity of the animal. In particular it controls basal metabolism (section G 2.12), which in myxoedema may be 25 per cent below normal and in goitre 100 per cent up. The rise of temperature and increased metabolism in fevers are possibly caused by over-secretion by the thyroid, and it may be that the increased metabolism produced by low temperatures is brought about in the same way. The latter is the only case where it is likely that the thyroid normally varies in its activity during life so as to coordinate any other processes in the body. For the rest, it maintains a general control over certain aspects of metabolism, and is particularly important during the time when growth is going on; for example, the bones of cretins cease growing at an early age, and their mental powers never develop.

Thyroxine was early shown to increase the oxygen consumption of many tissues, and more recently it has been shown to accelerate several enzyme reactions, including some of those involved in protein breakdown and phospholipid synthesis and other metabolic processes, mostly oxidative. Which, if any, of these is fundamental and which are side-effects, is unknown, but many of them involve cytochrome. Thyroxine also increases the number of mitochondria in a cell, and has also been claimed to stimulate spermatogenesis and sexual activity generally. It is necessary for the maintenance of lactation, and for the conversion of β-carotene to vitamin A.

The effects on birds, so far as they have been studied, seem to be generally similar to those on mammals, but in addition the thyroid seems to be in some way responsible for moulting, since this can be induced by thyroxine; however, thyroidectomy does not prevent moulting in ducks nor does thiouracil (which prevents the gland from utilizing iodine) in hens, so that there must be some other control. The 'silky' state of feathers, in which the barbules are absent, can be induced by thyroidectomy or thiouracil, and a nearly normal feather structure can be induced in genotypically silky birds by thyroxine.

The functions of the thyroid seem to be somewhat different in cold-blooded vertebrates. The most striking is that it controls the metamorphosis of Amphibia, for if it be extirpated in tadpoles they grow far beyond their normal size and never change into frogs or newts as they should. Thyroxine, or even elemental iodine, will make such thyroidectomized tadpoles develop normally. Conversely, feeding normal tadpoles with thyroid makes them metamorphose at an early age—at 6 weeks instead of 3 years in the extreme case of *Rana catesbiana*—and the axolotl larva of *Ambystoma mexicanum*, which normally never metamorphoses, can be induced to do so.

The deficiency in this species lies not in the thyroid itself, but in the pituitary, for if an anterior pituitary gland from the related *A. tigrinum*, which does metamorphose, is grafted on to it, metamorphosis occurs. By contrast, the perennibranchiate amphibians form thyroxin only slowly, and have normal pituitaries, but their

8541112 M

tissues are unresponsive. There is a little evidence that the thyroid may control the metamorphosis of a few teleosts, notably the change from parr to smolt in the rainbow trout of North America. It appears to control the sloughing of the skin in reptiles and amphibians.

Most experiments suggest that thyroid hormones affect neither growth nor oxygen consumption in cold-blooded vertebrates. Thiourea (another substance which prevents the formation of thyroglobulin) reduces the onset of spermatogenesis and external sexual characters in the male minnow, and has comparable but less striking effects in the female. The influence of the pituitary on the thyroid of teleosts seems to be of the same form as in mammals.

MV 6.122. The Parathyroids. The parathyroids lie on the surface of, or embedded in, the thyroid, and consist in most mammals of two pairs of bodies derived from the third and fourth branchial pouches; in rats one pair atrophies. Comparable glands are derived from the same gill pouches in birds and amphibians, while in the lizard development begins in pouches 3, 4, and 5, but the tissue from the last two atrophies. No parathyroids are known in fish.

The parathyroids of mammals produce a hormone called parathormone, a polypeptide of molecular weight about 9500. A fall in blood calcium leads directly to an increased output of parathormone, and there follow increased absorption of calcium from the intestine, increased reabsorption by the renal tubules, and release of calcium from bone. Injection of the hormone has corresponding effects—a raised level of blood calcium and increased secretion of it; the level of phosphate in the blood is lowered. There are secondary effects of the alteration in the balance of the two ions. Another polypeptide, called calcitonin, with a molecular weight of not less than 3600 and opposite effects to those of parathormone, has been extracted from parathyroids, but little is yet known of its action.

Parathormone has comparable effects on the calcium and phosphate metabolism of birds and amphibians. In all these classes there are great variations in the response to parathyroidectomy, some species showing an immediate fall in blood calcium and often early death, while others are insensitive.

MV 6.123. Other Pharyngeal Derivatives. Behind the last gill pouches, and so possibly serially homologous with them, are a pair of outpushings called ultimobranchial bodies. Their tissue may become associated with the thyroid in mammals. There is a little evidence that they may be associated with calcium metabolism in teleosts; if so, they could be regarded as the forerunners of, or as homologous with, the parathyroids.

The thymus is derived from the epithelium of the gill clefts, and is largest in young animals. Its function is the production of lymphocytes. Claims have been made from time to time that extracts of it have effects on growth rate and on the onset of sexual maturity, but its status as an endocrine gland, if any, is obscure.

The pseudobranch, the rudimentary gill in the spiracle or first embryonic gill slit of fishes, persists in teleosts as a gland-like tissue. Its removal causes dispersion of melanophores, so that it possibly produces a hormone that mediates this.

MV 6.13. The Pituitary

The pituitary gland or body has a double origin, from the roof of the mouth and from the floor of the thalamencephalon; the two outgrowths meet and form a more or less unified structure, but the hormones produced by the two parts are distinct. The buccal portion, which is ectodermal, and possibly homologous with the pre-oral pit of *Branchiostoma*, is called the adenohypophysis, and soon loses its connection with the mouth. The nervous part, called the neurohypophysis, remains attached to the brain. The relationship of the two, together with the names of their parts and some names now seldom used, is shown in Fig. 6.1. The double origin of the gland can be traced in all vertebrate groups from the cyclostomes up, but the disposition of the parts varies. There are also striking and unexplained differences between the proportions of the parts in related species. Thus while in rodents the intermediate part is relatively longest in desert species, the reverse is true in anurans, *Xenopus* having a larger intermediate part than *Rana*, and this larger than *Bufo*. The nervous part is absent from fishes except Dipnoi.

The gland has a blood supply from the branches of the internal carotid artery, and in tetrapods there is also a portal system formed from vessels which come from the venous drainage of the median eminence and break up into capillaries in the distal part. Portal systems in somewhat different positions have been described in fishes.

MV 6.131. The Adenohypophysis. The distal part, formerly called the anterior pituitary, produces at least six hormones in mammals;

all are proteins, with molecular weights from 20 000 to 100 000. It seems likely that each is secreted by a different type of cell. Both in mammals and in birds the production of some or all of these is controlled by exteroceptive factors, especially light, acting through the

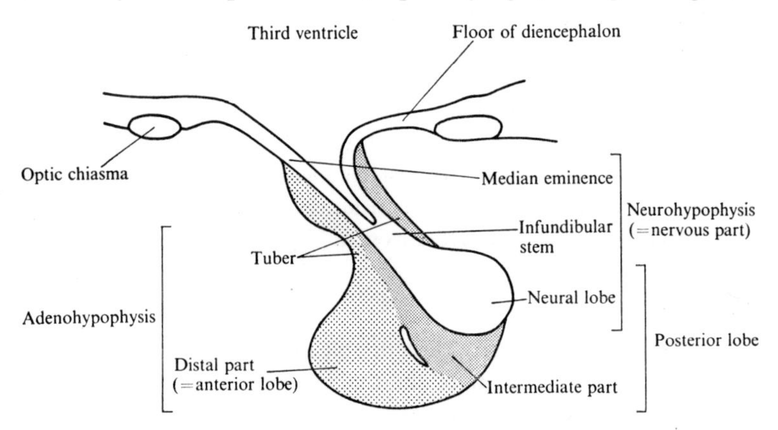

FIG. 6.1. Diagram of a sagittal section of the pituitary gland of a mammal, to show the nomenclature of the parts. From Yapp, *Vertebrates; their structure and life*, Oxford University Press, New York (1965).

nervous system and causing the hypothalamus to secrete a hormone that diffuses along the hypophyseal portal vessels. Further reference is made to this in sections M 8.23 and V 8.23. There is less certain evidence for a similar control in amphibians; the portal system is well developed in anurans, but not in urodeles. Since light has effects on the gonads and thyroids of teleosts, here too there may be a similar control. The nature of the hypothalamic hormone is not certainly known but since vasopressin (section M 6.132) and related synthetic polypeptides can substitute for it in both amphibians and mammals, it too is probably a protein.

Hormones from the mammalian adenohypophysis generally have effects in other vertebrates, but they are not uniform, and there is probably a family of related substances.

Growth Hormone. It has long been known that hypertrophy of the pituitary leads to gigantism and the disease called acromegaly, in which hands, feet, and lower jaw are of unusual size, and that deficiency produces dwarfing. The former effects can be produced by a relatively pure substance called growth hormone (GH) or somatotropic hormone or somatotropin (STH). It has also effects on the metabolism of carbohydrates (section M 4.12), protein, and fat; and probably the 'diabetogenic hormone' and 'ketogenic hormone' that have been described were simply impure preparations of it. Both in man and the rat it is not necessary for growth in early life, but it becomes important in man at the age of 2–4 years. It plays a part in lactation (section M 8.23). Injection of it increases growth in all classes of vertebrates, but it seems not to be essential in some elasmobranchs and amphibians, since they can live and grow normally after hypophysectomy, provided that they get enough food.

Thyrotropin. The action of the pituitary in stimulating the thyroid has been described in section MV 6.121. The hormone concerned is called thyrotropin, or thyroid-stimulating hormone (TSH). Its action has been demonstrated from the cyclostomes up.

Corticotropin. Two substances that stimulate the adrenal cortex to produce its hormones have been extracted from the gland, but it is simplest to assume that they are derivatives of a single hormone, which is called corticotropin, or adrenocorticotropic hormone (ACTH). It is probable that its production is increased by hormones from the hypothalamus in the portal system and by adrenaline in the blood, and decreased by circulating cortical steroids (section M 6.142), so that there is some automatic control of the latter. Corticotropin raises the oxygen consumption of the cortex, and has what may be side-effects in lowering blood sugar and mobilizing depot-fat. It, or more probably impurities associated with it, has the same effects on amphibian melanophores as does intermedin (see later in this section). Its presence has been demonstrated from the elasmobranchs up, but some species do not respond to it.

Gonadotropins. The action of follicle-stimulating hormone (FSH), luteinizing hormone (LH), and prolactin or luteotropic hormone (LTH), collectively known as gonadotropins, in controlling the sexual cycle will be described in section 8.23. Luteinizing hormone is identical with the interstitial cell stimulating hormone (ICSH) of males. There is evidence for the existence of both luteinizing hormone and follicle-stimulating hormone in all tetrapods; the latter has not yet been certainly demonstrated in fishes, but its presence seems likely. Experimentally, prolactin has various effects on amphibians and fish, including inducing the larvae of newts to move towards water, but the part, if any, played by these reactions in the natural life of the animals is unknown. Gland extracts with the same effects have been obtained from elasmobranchs, teleosts, and amphibians.

Pars tuberalis. The only hormone known certainly to be produced by the tuber (pars tuberalis) is the W-substance that concentrates the pigment of amphibian melanophores and of those of some fish (section V 7.81).

H_2N · glycine · leucine (1.) · proline · cystine (S) · aspargine · glutamine (2.) · isoleucine (3.) · tyrosine · cystine (S)

(S—S bridge between the two cystines)

	1.	2.	3.
arginine vasopressin	arginine		phenylalanine
elasmobranch oxytocin	isoleucine		serine
ichthyotocin	isoleucine	serine	
lysine vasopressin	lysine		phenylalanine
vasotocin (probably = natriferin)	arginine		

FIG. 6.2. The structure of the polypeptides of the neurohypophysis. The structure shown in the top line is that of bovine oxytocin; the *single* molecule of cystine completes the ring through its sulphur atoms: other hormones have different aminoacids at positions 1, 2, and 3 as indicated in the table.

Pars intermedia. Likewise, the intermediate part is known to be active only in poikilotherms. It produces intermedin which causes the dispersion of pigment in the chromatophores of Amphibia and of some fish and reptiles, and possibly induces the synthesis of more melanin. Intermedin can be extracted from the pituitary of birds and mammals, but has no known function in them.

M 6.132. The Neurohypophysis of Mammals. Two hormones are present in the nervous part of the mammalian pituitary, oxytocin (= pitocin) and vasopressin (= pitressin). Both are polypeptides of molecular weight about 1000, and consist of eight aminoacids of which five form a ring (Fig. 6.2). Oxytocin is the same in all the mammals so far investigated, but two vasopressins are known. Both differ from oxytocin in having phenylalanine instead of isoleucine as one of the aminoacids in the ring, but while in one the second acid in the side-chain is arginine (in place of the leucine present at this position in oxytocin), the other has lysine. Arginine vasopressin has been found in about nine species, lysine vasopressin in only one or two, but with these small numbers generalization is dangerous. If the reported presence of arginine vasopressin in *Echidna* and a marsupial is confirmed, it would seem likely to be the more primitive form.

Oxytocin causes contraction of smooth muscle in the uterus and in the mammary gland, and possibly in other places. There seems little doubt that its chief natural function is to cause the ejection of milk from the mammary gland into its ducts. The stimulus of sucking causes the release of oxytocin into the blood system, and the milk is thereby made more easily available to the young animal. Probably also oxytocin is released during labour.

Vasopressin causes a rise in blood pressure, but its chief effect is increased tubular reabsorption by the kidneys, so that it decreases the production of urine (section MV 5.1). Its fundamental effect seems to be an influence on permeability, which drives potassium out of the cells and sodium and water in. The stimulus for its liberation is an increase in the concentration of chloride and other solutes in the blood flowing through the hypothalamus. Both hormones are produced continuously and their liberation is probably mediated by acetylcholine. Each is associated with a carrier protein called a neurophysin, of higher molecular weight.

Histological observations suggest that the hormones are formed in the nervous tissue of the hypothalamus and pass down the nerve fibres, in which more of the hormones is formed, into the nervous part of the pituitary, which is thus not an endocrine gland but a storage organ for hormones. There is here a remarkable parallel with what occurs in arthropods (sections J 6.18 and J 6.19).

V 6.132. Neurohypophysis of other Vertebrates. In view of their close chemical similarity, it is not surprising that oxytocin has to a slight degree the properties of vasopressin, and vasopressin those of oxytocin. Moreover, synthetic analogues also have similar properties. Some analogues have been found in other vertebrates, but their functions in them are not entirely clear.

Oxytocin is present in birds, but its mammalian functions are not appropriate to these animals; it stimulates contraction of the oviduct, and may therefore be used in egg-laying.

Other polypeptides extracted from the neurohypophysis of cyclostomes, elasmobranchs, teleosts, dipnoans, amphibians, reptiles, and birds all seem to agree in having effects on water balance, though the means by which this is achieved vary. In crocodiles there is decreased filtration by the kidney, in birds and frogs both decreased filtration and increased absorption. One hormone, sometimes called natriferin, greatly affects the transport of water, sodium, and chloride across amphibian skin, and the others have similar but smaller effects. There is specific variation in the response as well as in the polypeptides naturally present, so that evolution seems to have gone on in two ways: by variation in the chemistry of the molecule and by variation in the sensitivity of the organs affected. There is some ecological adaptation—the more terrestrial amphibians are more responsive than the aquatic ones—but beyond this no reasonable story can be constructed. The part played by the vasopressin-like substances of fish is still obscure.

V 6.133. The Urophysis. In the tails of elasmobranchs, teleosts, and chondrosteans there are cells in the nerve cord which, from their staining properties, appear to be secretory, and in the teleosts they are associated with blood vessels to form a neurohaemal organ or urophysis or urohypophysis similar to the hypophysis. They produce a secretion which is a protein, and there is some evidence that it takes part in osmoregulation, especially of sodium.

MV 6.140. The Suprarenals or Adrenals

The suprarenals or adrenals consist in mammals of two small bodies, one in front of each kidney. Each has two parts, an inner medulla and an outer cortex. The former, the suprarenal proper, is derived from the neural crest, that is from the same set of cells as the sympathetic system, while the cortex has its origin in that part of the mesoblast which gives rise to the mesonephros. The functions of the two are quite distinct, and their association is probably an accident, for in some fishes they are quite separate. In the dogfish, for instance, the interrenal, representing the cortex, lies between the kidneys, and the suprarenals, representing the medulla, consist of small masses on the sympathetic chains. The adrenals of the frog lie on the ventral surface of the kidneys and are yellowish in colour. In some animals patches of tissue which is histologically similar to medulla or cortex are found detached from them. Since the medullary tissue in mammals is readily stained by chromic acid, it is sometimes called chromaffin tissue, and the recognition of comparable cells in other animals depends on their similar staining properties.

M 6.141. The Adrenal Medulla of Mammals. The peculiar chromaffin staining reaction depends on the presence in them of one or other or both of the substances adrenaline and noradrenaline, which are catechol amines, differing only in the substitution of a methyl group for a hydrogen in the former:

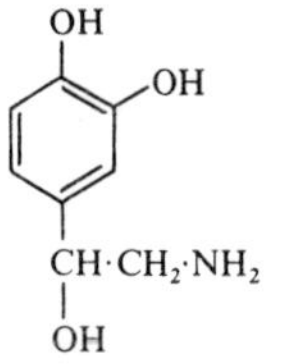

$$\text{C}_6\text{H}_3(\text{OH})_2\text{-CH(OH)}\cdot\text{CH}_2\cdot\text{NH}_2$$

Noradrenaline

$$\text{C}_6\text{H}_3(\text{OH})_2\text{-CH(OH)}\cdot\text{CH}_2\cdot\text{NH}\cdot\text{CH}_3$$

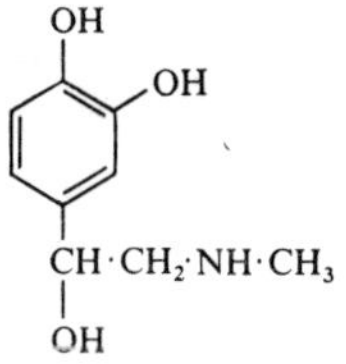

Adrenaline

Probably all the cells can form noradrenaline in stages from tyrosine, but only some of them can methylate this to adrenaline.

The natural compounds are laevorotatory; the dextrorotatory forms, when prepared synthetically, are found to be relatively inactive. In mammals about 98 per cent of the amine present in the medulla is adrenaline. When adrenaline is injected it causes effects very similar to stimulation of the sympathetic nervous system: an increase in heart rate and a rise in blood pressure by arterial constriction; contraction or relaxation of smooth muscle according to its origin and previous condition, dilatation of the pupil and erection of hairs; an increase in metabolism and a rise in blood sugar and lactic acid by breakdown of glycogen; and contraction of the capillaries of the skin. Noradrenaline has similar effects, but there are some differences; it is the substance liberated at sympathetic nerve endings (section MV 6.32).

The function of adrenaline in the living animal is obscure, and it seems not to be normally produced in amounts great enough to have any effect. Its production is increased by low temperatures or lack of oxygen, and by the emotional states of fear and anger. Some of its effects would seem to be appropriate to these stimuli, but others, such as an increased metabolism with lack of oxygen, are not.

V 6.141. Chromaffin Tissue of other Vertebrates. Adrenaline and noradrenaline have been found in many vertebrates from the elasmobranchs onwards, but generally there is a much higher proportion of the latter than in mammals—50 per cent of the total amines in anurans, 60–80 per cent in elasmobranchs and birds. The effect of the substances when injected seem to be the same as in mammals, but in addition adrenaline causes concentration of the chromatophores of most amphibians and some reptiles (section V 7.81). Its natural function is even less clear in the lower vertebrates than in mammals.

M 6.142. The Adrenal Cortex of Mammals. Removal of the adrenal cortex lowers the concentration of sugar and sodium in the blood and leads to death within a few weeks. These effects can be avoided by the administration of an extract of the gland. The active substances are certainly steroids, with the general structure

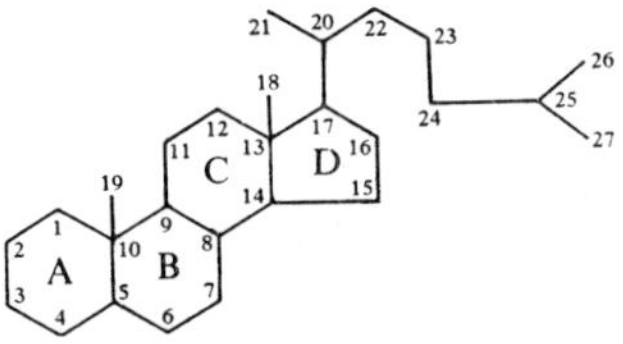

A score and a half of such compounds have now been isolated from various mammals, but it is still not clear which are the fundamental hormones. Qualitatively their effects are very similar, but quantitatively they vary widely. Some have a strong effect on carbohydrate metabolism (section M 4.12), raising blood sugar and slowing its oxidation, and are called glucocorticoids, while others act chiefly on water and sodium balance and are called mineralocorticoids. There are differences between species, but in many mammals the chief glucocorticoids are corticosterone and cortisol, while in most the chief mineralocorticoid is aldosterone (section MV 5.1).

The cortocosteroids are liberated into the blood by the influence of corticotropin from the adenohypophysis (section MV 6.131), and their function is probably to deal with stress, for the exteroceptive factors that induce their production are cold, heavy exercise, and infection.

V 6.142. The Adrenal Cortex of other Vertebrates. Steroids of some sort are produced by all vertebrate classes; corticosterone and cortisol have been found in all the living vertebrate classes and aldosterone in all except the cyclostomes. It seems likely, from the small amount of information available, that their general effects are the same in all. Cortisol injected into birds increases the secretion of sodium by the nasal gland (section G 9.12).

MV 6.15. The Pineal

In the earliest vertebrates there seems to have been a pair of dorsal eyes, and traces of these, in which a lens-like structure can sometimes be discerned, remain as the pineal and parapineal bodies; it is rare for both to be well developed. There is slight evidence that they have become endocrine glands. There is some clinical evidence for a connection with sexual development, and a substance called melatonin, closely related to tryptophan, which has been isolated from mammalian pineals, has a very strong concentrating action on the chromatophores of frogs (but not of urodeles). There are indications that the pineal is concerned with colour-change in lampreys and salmon. Its cells contain vesicles comparable to those of secretory nerves.

MV 6.16. The Gonads and Related Structures

The hormones of these are discussed in section 8.23. In addition the mammalian prostate and seminal vesicles, and probably other organs, produce long-chain organic acids called prostaglandins. Experimentally they have various effects on metabolism and muscular contraction, but their natural function is unknown.

J 6.1. HORMONES IN INVERTEBRATES

Control of reactions by diffusing chemical substances is probably universal in living tissues, but a fully developed and rapidly acting hormonal system is possible only where there is a vascular system. Hormones have therefore been searched for, and found, chiefly in the annelids, arthropods, and molluscs. The simplest, but least conclusive, line of attack, is to stain possible endocrine tissues and see whether their cells contain substances that look like secreted granules that might later be liberated. In this way cells called neurosecretory have been found in the nervous system of polychaetes, oligochaetes, scaphopods, opisthobranchs, prosobranchs, cephalopods, branchiopods, decapods, myriapods, and insects, as well as of polyclads. In some of these, transplantation and other experiments have shown the presence of hormones, concerned chiefly with growth and reproduction.

J 6.11. Hormones in Annelids and Molluscs

If the posterior segments of *Nereis* (or a few other polychaetes) are amputated, they are normally regenerated, but if the supra-oesophageal ganglion is extirpated before the amputation, or within 3 days after it, they are not. Later extirpation has no effect. In some experiments, injection of crushed cells from another worm amputated 3 days before, caused growth of a new tail in a worm from which the ganglion had been removed. The explanation seems to be that the shock of cutting through the body starts nervous impulses that cause the ganglion to produce a hormone that passes along the axons and is liberated into the coelom (in *Nereis*) or into the dorsal blood vessel (in *Nephthys*).

The supraoesophageal ganglion produces also hormones that influence the reproductive system. In *Nereis* and *Nephthys* one of these is a juvenile hormone, since removal of the ganglion causes precocious sexual development, including the development of the heteronereis (section J 8.23). Another hormone, found only in the posterior part of the ganglion, is needed for ripening of the eggs and spawning in *Arenicola*, and in *Nereis* also full maturation of the eggs needs a hormone. The total number and the nature of the hormones in polychaetes remain unknown.

There is histological evidence for comparable neurosecretion in leeches and oligochaetes, and some slight experimental evidence for the action of hormones in regeneration and the sexual cycle. Removal of the brain from the leech *Theromyzon rude* prevents maturation both of eggs and of sperms.

Some early experiments suggest that the movement of the chromatophores of cephalopods, though primarily under nervous control, is affected also by hormones, and that they control the development of the genital ducts of gastropods.

J 6.12. Hormones in Crustacea

The control of decapod chromatophores (section J 7.82) was shown in 1928 to be by a hormone, the first clear case of endocrine coordination in an invertebrate. The hormone is formed chiefly in nervous tissue of the eye-stalk, but also in other ganglia, and passes along axons that end in club-shaped swellings in the sinus gland. This is formed from the neurilemma, which encloses the ganglia of the eye-stalk, and is a storage organ for hormones, not a gland; it has been shown to contain several other hormones besides the one that controls colour. They control the movements of the pigments of the retina, respiratory metabolism and blood sugar, regeneration of limbs, and the moult. For the last, three hormones are probably necessary; one inhibits the onset of the preparatory changes of the pre-moult period, and another accelerates and controls the pre-moult once it has begun; these two together repress the formation of a third, called ecdysone, which is produced in a structure associated with the suboesophageal ganglion and is necessary for the moult itself. The relative production of the moult-inhibiting hormone and of ecdysone depend in part on external factors. The total number of sinus-gland hormones, and their nature, are unknown, but the chromatophore hormone appears to be protein.

Nervous tissue in the wall of the pericardium of decapods, called the pericardial organ, contains a substance that accelerates the heart-beat. It is possibly related to tryptophan; a derivative of this called serotonin, which is present in mammalian and other tissue, has similar effects when injected.

The sex-determining hormones are discussed in section G 8.24.

J 6.13. Hormones in Insects

The best-known hormones in insects are those that control growth and moulting. A part of the brain called the pars intercerebralis produces a hormone, called ecdysiotropin or prothoracotropin, which travels along the axons to the cardiac body, where it is liberated into the blood and goes to the ecdysial or thoracic gland. This structure takes various forms in different orders, and is known as prothoracic, peritracheal, pericardial, and so on, according to its position. Ecdysiotropin causes it to produce its own hormone, ecdysone (or PGH, or growth and differentiation hormone). This stimulates the growth and mitosis of epidermal cells and the secretion of a new cuticle. It therefore induces moulting. It is not specific, is non-nitrogenous, but is normally carried by a protein. The thoracic gland atrophies in the adult and so moulting does not occur. Where, as in the Thysanura, the gland persists, moulting continues two or three times a year throughout life, and moulting has been induced in adults of the bug *Rhodnius* by grafting them on to nymphs and by the implantation of thoracic glands.

In young stages (up to the fourth nymphal instar in *Rhodnius*) a juvenile hormone, which has been called neotenin, is also present. It is formed in another gland in the head, called the corpus allatum, and actively promotes juvenile characters and suppresses imaginal ones. There is some development of adult characters, or metamorphosis, at each moult, but their full development waits on the reduction of juvenile hormone. Whether this is gradual, or in two well-marked stages, determines the absence or presence of a pupa.

In some insects the corpora allata of the adult female produce a hormone which is necessary for yolk formation; it may be identical with the juvenile hormone. The chromatophores of insects are controlled by hormones, and the gonads may produce sex hormones. In some insects, and perhaps in all, the arrest of growth known as diapause, which occurs at various stages from the egg onwards, is under hormonal control. In the silkworm (*Bombyx mori*) diapause occurs in the embryo if the mother's suboesophageal ganglion produces the appropriate hormone. The synthesis, or perhaps the release, of this is determined by nervous stimuli from the brain, so that diapause is facultative. In the giant silk moth (*Hyalophora* (= *Platysamia*) *cecropia*) diapause occurs in the pupa because of a failure of the pars intercerebralis–prothoracic gland system that also controls metamorphosis. A period of low temperature is necessary to stimulate the pars intercerebralis, and when this is activated development continues and leads to metamorphosis.

There is histological evidence that a substance produced in the brain of the cockroach *Leucophaea* passes along axons into the corpus cardiacum. Extracts of this body in *Periplaneta* are widely active physiologically; they induce or accelerate contraction of the hind gut, Malpighian tubules, and heart, and they affect the chromatophores of insects and crustaceans. Chromatographic analysis suggests that the principle that is active on the heart is an orthodiphenol, similar to, but not identical with, adrenaline. It is therefore not surprising that the insect heart is affected by adrenaline (and acetylcholine). There is some evidence that the effect of extracts of corpus cardiacum on chromatophores is due to a different substance. Extracts of the corpus cardiacum of *Rhodnius* and of mealworm larvae (*Tenebrio*) have similar but weaker effects. It seems likely that while the chromatophore principle may be produced in the brain and transferred to the corpus cardiacum, the heart-stimulating principle is made in the gland itself.

There are obvious analogies between the neurosecretory systems of crustaceans, insects, and vertebrates. In all, cells closely associated with or derived from the central nervous system produce hormones that travel as droplets along their axons, and after liberation into the blood cause the production of further hormones that affect various processes. One may see in all, the special development of a fundamental property of nervous tissue which appears also in the secretion of the transmitter substances of synapses and nerve endings (sections MV, J 5.32). Beyond that it is unwise to go. There is no homology between the hypophysis of vertebrates and the eye-stalk–sinus gland system of crustaceans, nor does it seem likely that there is any between the latter and the pars intercerebralis-thoracic gland system of insects. It seems probable that the two were evolved separately; if, as there are strong grounds for holding, the group Arthropoda is an unnatural assemblage, this is what one would expect.

G 6.2. ECTOHORMONES

The distinction between a hormone, which by definition is produced and transmitted and acts within one individual, and the same or a similar substance that is supplied from outside, is, as has been said above, a fine one. There is a class of substances that are produced by one individual, are shed into the outside medium, and are then received by other individuals of the same species, which are thereby caused to behave or develop in certain ways. They are called ectohormones, or pheromones.

The best-known examples are strong-smelling substances that cause definite behaviour in animals that perceive them. Many male animals produce such scents, either in the urine or in special glands, and use them for marking their territory, where they presumably serve to warn off other males. Similar scents attract the opposite sex (even in man, where, however, their place is now usually taken by synthetic, or at least borrowed, substances). More definite in its action is the androgen secreted by the male bitterling in the breeding season, which is not only a hormone for him, but, excreted into the surrounding water, causes the growth of the ovipositor of his mate. When the skin of many species of teleost (Ostariophysi) is damaged, substances are liberated that cause a fright reaction in the same or related species. Thus an attack on one individual in a school causes the others to scatter, and the liberation of the alarm substance may even cause a cannibalistic fish to give up its attacks on its smaller brethren.

Ectohormones are probably best developed in insects. Male moths and other insects are attracted to females by very powerful scents, and can travel long distances to them. Three different unrelated substances have been extracted from two moths and a bee. Smells laid on the ground by ants, and usually lasting for only a few minutes, enable other ants of the same species to follow the trail. Honey bees indicate that they have found food by exposing a scent gland on the dorsal surface of the abdomen.

Some ectohormones probably need to be absorbed by the recipient in order to be effective. The queen honey bee produces in her mandibular gland an unsaturated fatty acid and an unidentified component that together make queen substance. It is licked from her body by the workers and distributed throughout the hive. It has two effects, inhibiting the development of ovaries in young workers, and preventing the workers from constructing queen cells. Hence, the removal of the queen from the hive causes the preparation of the hive for a new queen and allows some workers to develop ovaries and eggs that give rise to more workers. A comparable substance in the faeces of the king and queen of a termite colony similarly prevents the sexual development of the pseudoworkers.

Earthworms whose giant fibres have been stimulated (section G 6.42) produce a slime that causes escape reactions in other worms.

G 6.3. THE NERVOUS SYSTEM

G 6.30. General Properties of a Nervous System

Whatever may be the case with the endocrine system, a nervous system is easily shown to be present in all groups of the Metazoa. All nerve cells have common staining properties (though in the limit, as in sponges and even in the coelenterates, there has been disagreement between histologists as to whether certain cells are nerve cells or not), but more important are the physiological properties of the system. Simply, these are that a stimulus at one point on the body causes a response, which is almost immediate, at some other point, and that this response is abolished by cutting through the nervous tissue, and only the nervous tissue, connecting the two points. There must therefore be conduction of some sort of impulse along the cells.

The early work on the physiology of the nervous system consisted largely of histological investigation and of relatively simple experiments on the conduction. These produced some important conclusions, which formed the basis for the more detailed studies, with the electron microscope and other sophisticated apparatus, of recent years. It is probably true to say that these have refined but in no case contradicted the early conclusions.

One of the most striking histological observations is that all nerve cells have one or more processes, usually branched, and that some of these, the axons, may be very long indeed. Another is that these processes come into contact with other cells, either other nerve cells or cells of a different sort such as muscle fibres. Such a contact between two nerve cells is called a synapse, and although contacts between nerve cells and other types are usually given special names, such as nerve-muscle junction, some workers now call these synapses also. This usage is not adopted in this book. Although some histologists at first believed that there was protoplasmic continuity between the two cells at a synapse, there was general agreement by the 1930s that there was merely apposition of two processes.

In the coelenterates, and to some extent elsewhere, nerve cells are arranged so that each has contact with many others, forming a network. Although not fundamentally different from other systems, the nerve net has some special properties and will be dealt with later (section G 6.34). More often the axons of nerve cells are packed together in parallel, so that they form nerves. Synapses are concentrated in special places, which generally contain also the bodies of the cells, with their nuclei, and so are wider than the nerves; they are called ganglia. There is a special concentration of synapses and cell bodies, usually running more or less the length of the animal, called the central nervous system; in invertebrates this generally consists of ganglia joined by connectives.

Selective cutting shows that some nerves, or sometimes parts of nerves, normally send impulses towards the central nervous system, while others only send them outwards to an effector organ such as a muscle or gland. Finer investigation shows that this directional difference is a property of the individual cells. We therefore have the concept of the reflex arc. In cellular terms this consists at its simplest of a sensory cell, which picks up a stimulus and as a result conducts an impulse along its afferent or sensory nerve fibre to the central nervous system; a synapse to another nerve cell, called internuncial; another synapse to a motor nerve cell, and the motor or efferent fibre of this leading to the muscle or other effector. There are two variants on this pattern; sometimes the sensory cell is distinct from that which has the afferent fibre, so that there is an extra synapse (or occasionally two or three, with intercalated cells), and usually more than one internuncial neurone lies between the afferent and efferent cells. The theoretical possibility that the afferent and efferent cells should be directly connected is probably seldom or never realized. There are always complications in that in the central nervous system the internuncial neurones do not simply connect one afferent to one efferent cell, but have synapses with several of each sort; by these means it seems likely that, at least in vertebrates, every sensory cell in the body can be connected with every motor cell. There is also branching of both motor and sensory fibres, so that one nerve fibre, for example, supplies several muscle fibres (making what is called a motor unit), and several sense cells may, as in the eye, converge on a single afferent fibre. The physiology of the reflex agrees with these histological observations. If two sensory nerves are stimulated, the appropriate muscle does not give as strong a contraction as the sum of the contractions got by stimulating the two nerves separately. There is overlap between the fibres innervated by the two.

In physiological terms, the reflex arc consists of the changes initiated in the sense cell by the stimulus; the propagation of the resulting impulse; its transmission across one or more synapses to start the motor impulse; and the initiation of the response in the effector. The first and fourth of these we shall leave until later (sections G 6.40, MV 7.31), and will deal here with the propagation of the impulse and the properties of a synapse.

The explanation of both of these is in part electrical and in part chemical, and students sometimes ask which is fundamental. This is as sensible a question as to ask whether the changes in a Leclanché cell or an accumulator are chemical or electrical. The two are aspects of the same thing. Since all valency bonds depend on electrical forces, all chemical change is electrical; since electric currents depend on the movements of ions or electrons, all electrical change is chemical.

While the propagation of the nervous impulse seems to be the same, in all but the finest details, in all animals, there are several different types of synapse.

MV 6.31. Propagation of the Nervous Impulse in Vertebrates

Simple experiments are usually carried out by applying electrical stimulation to an isolated nerve, but it must be remembered that this is not how nervous impulses are normally started; there is, however, good evidence that they are the same whatever their cause (section G 6.40). It is easy to show certain properties of the impulse.

(1) A certain strength of stimulus is necessary to start it, so that there is a threshold. This is roughly constant for a given type of nerve fibre, but is higher in small fibres than in large.

(2) After one impulse has passed there is a refractory period during which a second stimulus, however large, is ineffective; in frog's sciatic nerve it is about 1 ms at 20 °C, and longer at lower temperatures. If an inadequate stimulus is followed by another one

within 0·5 ms there may be a response, so that there can be summation of two inadequate stimuli.

(3) The absolute refractory period is followed by a relatively refractory one, during which the threshold is increased.

(4) The impulse takes a certain time to travel along the nerve. The larger the fibres the more rapidly they conduct, and medullated fibres conduct more rapidly than non-medullated. In mammals impulses in the fastest fibres, both motor and sensory, have a velocity at body temperature of 100 $m\ s^{-1}$, but there are others that range down to 0·3 $m\ s^{-1}$. A high value for the frog is 40 $m\ s^{-1}$. Above the minimum temperature of 5 °C at which mammalian nerve ceases to conduct there is a temperature coefficient of about 1·7 for a rise of 10 °C. Frog nerve is active at lower temperatures.

(5) The impulse can pass equally well in both directions. That it does not normally do so is merely because the nerve is normally stimulated only at one end.

More information about the nervous impulse only became possible when, in about 1930, the cathode-ray oscillograph was introduced to physiologists, and it became possible to record the electrical changes that accompany the impulse in a single fibre. The availability of this instrument stimulated technical methods, especially the use of intracellular electrodes, so that the properties of single fibres, which were previously mainly inferred, could be observed directly. The electrical facts are as follows.

At rest, a nerve fibre shows the electrical imbalance characteristic of living cells (section G 1.12), but it is much larger than usual. Electrodes placed one on the outer surface and one inside the fibre show a potential difference of 60–80 mV, the outside being positive to the inside. When a nervous impulse passes there is a very rapid reversal of this potential difference, until the outside becomes 30–50 mV negative, and then a return through the normal value to a slightly increased positive potential, and finally a return to normal (Fig. 6.3 (*a*)).

These electrical facts can be largely explained in terms of the chemical composition of the cell and its surroundings. The cytoplasm has a concentration of potassium twenty to sixty times that of the interstitial fluid, while that of sodium is only one-tenth to one-third of that outside. The concentration of chloride also is less inside the cell than outside, but other anions are present inside but not outside. This distribution, which could only be maintained if the cell wall were at least to some degree impermeable, would be expected to lead to a concentration of sodium ions on its outer surface and of chloride on its inner surface, which would account for the resting potential. No membrane is completely impermeable, so that it is to be expected

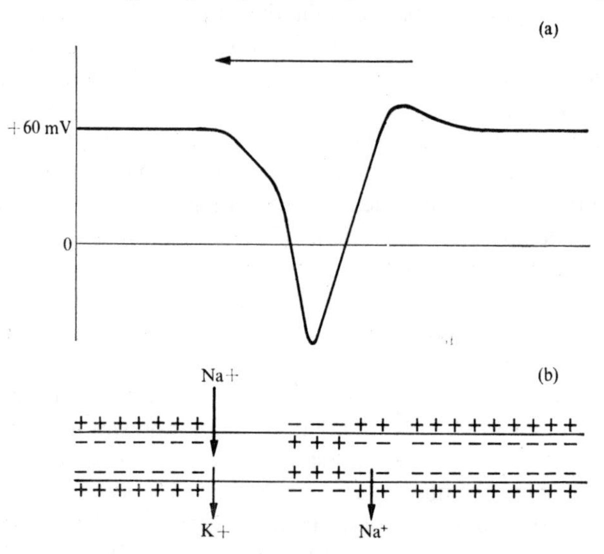

FIG. 6.3. Diagram of a nervous impulse travelling from right to left. (*a*) The electrical potential difference between the outside and the inside of the nerve fibre. (*b*) The passage of ions; the size of the arrows indicates approximately the magnitude of the flow. From Yapp, *Vertebrates; their structure and life*, Oxford University Press, New York (1965).

that there would be some leakage, and the state would be unstable. Resting nerve consumes oxygen, and is presumably doing work to restore the ions that leak through; this is known as the sodium pump, since, although the membrane is rather more permeable to potassium and chloride, sodium is the ion that is quantitatively most important. If the pump is stopped by lack of oxygen, by inhibitors of catabolism, or by repeated stimulation, sodium slowly leaks in and in time the ions come to be uniformly distributed.

When the fibre is stimulated, there is a rapid and complex change in its permeability. At first, there is a great increase in the ease with which sodium ions can pass through, so that they rush in down the concentration gradient and cause the reversal of potential difference. The impermeability to sodium is rapidly restored, but there follows a longer-lasting increased permeability to potassium ions, during which they pass out and the polarization is restored. These movements are illustrated diagrammatically in Fig. 6.3 (*b*).

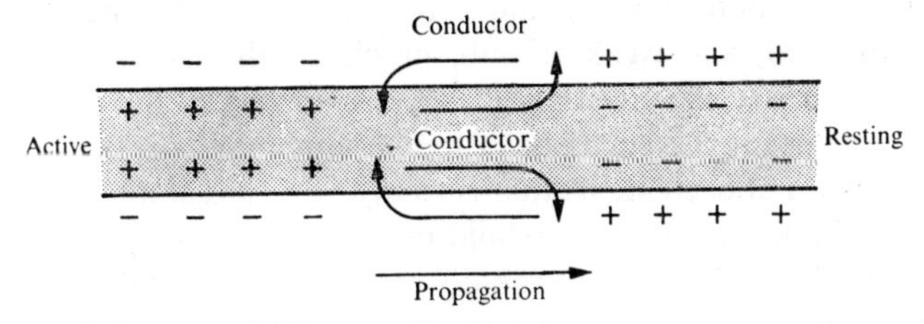

FIG. 6.4. Local circuits in nerve. For explanation see text.

These changes, while they would appear as the usual form of pointed wave or 'spike' of voltage in the common recording apparatus, do not in themselves make a propagated wave, or impulse. This can be explained in terms of local electrical circuits (Fig. 6.4). The adjacent patches of active and resting cell surface represent, because of their charges, two generators in series. Since both the cytoplasm and the external fluid are conductors, a current will flow, and its direction is such that sodium ions move in through the active membrane. In so doing they may be supposed to de-polarize it, causing the sequence of events that give rise to the spike, which is thus propagated along the fibre. If the fibre is stimulated in the middle one would expect the wave to travel in both directions, and this, as we have seen, it does. If two waves are started at opposite ends of a fibre, they would be expected to extinguish each other where they meet, since the conditions for a local circuit are not then present; this is found to be so. The local circuits can presumably not move backwards, or be reflected from the end of a nerve, because the membrane in the final phase of the activity cycle cannot be made more permeable—it is refractory. By the time it has regained its excitable state the active region is too far away for its potential difference to be enough to start the local circuit.

A slightly different system of conduction holds in myelinated fibres. Here a relatively thick layer of fatty material surrounds the fibre, which is exposed to the extracellular fluid only at intervals, the nodes of Ranvier. Hence the local circuit can run only in wide loops, from one node to the next, and the changes in permeability take place only at the nodes; conduction is said to be saltatory. It is, as would be expected, much faster than in ordinary fibres, and another advantage is that since ions can be lost only at the nodes, transmission is much more economical in usage both of them and of energy to restore the resting condition.

The restriction of entry and exit of ions to the nodes has been demonstrated, and weak stimuli, or chemicals that interfere with stimulation, are effective only if they are applied to the nodes.

An important characteristic of the nervous impulse is that it is conducted without decrement; that is to say, if it runs at all it maintains its magnitude. Once the threshold of stimulus has been reached the full size of electrical change rapidly appears, and the refractory state that follows ensures that it cannot be increased either in duration or in voltage. The nervous impuse is therefore 'all-or-none'. An exception to this rule is that if a length of fibre is narcotized, the impulse may pass through it, but the action potential is lowered, rising again to the normal value when the impulse emerges into the un-narcotized fibre.

It is possible to conceive of another type of electrical change, in which there would be an increasing change in potential with increase in stimulus, and which would die away gradually from the point of origin without being propagated. Such changes, known as graded responses or generator potentials, are characteristic of parts of the nerve cell other than the axon. They occur in sensory cells (section G 6.40), in the cell body, and in dendrons. Observations from the insertion of microelectrodes into cells suggest that the all-or-none impulse in the axon originates near its base, and follows only on the integration of more than one graded potential in other parts of the neurone.

The upper limit to the number of impulses that can pass in a second is determined by the refractory period, and experimentally, rates as high as 450 or more per second have been obtained. In life, rates between 5 and 50 per second are more usual, but the electric nerves of fish can carry 1000 per second. If there is continuous stimulation there is a continual succession of impulses.

It is not yet possible to give a complete account of what happens when an impulse is initiated and travels along the nerve. The membrane of the axon has the usual structure of a fatty layer between two protein layers (section G 1.22). It may be supposed that this is pierced by pores through which small ions can pass. It is possible that in the resting state most of these pores are blocked by calcium ions, and that on stimulation they are removed; sodium and potassium ions can then go through, and there will necessarily be discrimination between them because the smaller sodium ion can go through more easily. But the discrepancy between their rates is too great to be explained wholly in this way, and it may be that the ions move not by themselves but as complexes with organic compounds of unknown nature; the fact that some chemicals will stop passage of one ion but not the other suggests the same conclusion. The large organic anions within the cell (e.g. proteins), which are too large to pass through the membrane, are important in determining the equilibrium of the other ions. The removal of the calcium ions is presumably chemical, and may be connected with the liberation of acetylcholine or some comparable substance (section MV 6.32).

The resting nerve uses glucose and oxygen, and produces heat and carbon dioxide. Frog's nerve at rest in oxygen has a metabolism of $300\,\mu W\,g^{-1}$ (7×10^{-5} cal $g^{-1}\,s^{-1}$); when an impulse passes there is an extra heat production of $4{\cdot}2\,\mu J\,g^{-1}$ (10^{-6} cal g^{-1}). Both appear to be connected with carbohydrate chemistry of the normal type (section M 4.12). Active phosphate is necessary, and if the nerve is deprived of oxygen lactic acid accumulates but can be removed by re-admission of oxygen. Without oxygen a nerve gradually loses its irritability, the more rapidly the more often its is stimulated.

There can be no doubt that the resting metabolism is that of the sodium pump, and the recovery heat is connected with the provision of energy for restoring the resting distribution of ions; in other words, a stepping-up of the pump. The resting nerve, like a stretched spring, contains stored potential energy which is liberated on stimulation. At high frequencies (above 200 per second in the frog) the total heat per impulse falls off, suggesting that the metabolism cannot keep up with the rate at which the stored energy is being used.

J 6.31. Propagation of the Nervous Impulse in Invertebrates

The nervous impulse in cephalopods and crustaceans seems to be basically similar to that of vertebrates; indeed, many of the facts of ionic movement were first demonstrated in the giant nerve fibres of the squid *Loligo*, which, since they are about 1 mm in diameter, are relatively easy subjects for investigation by micro-electrodes and chemical analysis. The assumption that all nervous impulses are the same, though commonly made, is, however, dangerous. As we shall see in section J 6.32, there are important chemical differences between the synapses and nerve-muscle junctions of different groups, and the same may be true of the nerve fibres. Some insects have only traces of sodium in the body fluids, and many of them have much potassium in their haemolymph. In these at least the ionic relations must be different.

Cephalopods do not have medullated nerves with regular nodes of Ranvier, and saltatory conduction has not been demonstrated. They have giant nerve fibres mostly of diameter from 100–700 μm, but sometimes outside these limits, which may be enlarged single fibres or may be syncytia formed by the fusion of several axons. At the lower limit they overlap in size with the large medullated fibres of vertebrates. As would be expected, they conduct faster than the ordinary narrow fibres, but not so fast as would a nodal vertebrate fibre of the same diameter, if such existed. A large cephalopod fibre of diameter 700 μm conducts at only 20 m s^{-1} at 20° C, the same velocity as a frog nerve of only one-seventieth the diameter. The vertebrates appear, therefore, to have achieved a great evolutionary advantage by the invention of the medullated and nodal fibre. In general, the velocity of conduction in a non-medullated fibre is approximately proportional to the square root of its diameter, that in a medullated fibre to the diameter itself. At a diameter of about 1–2 μm the velocities are equal, so that below this a non-medullated fibre is the faster and above this the medullated fibre is faster. This is about the point at which all vertebrate fibres become medullated, so there appears to have been selection for speed of conduction.

The crustaceans have done better than the cephalopods; the conduction velocity in a prawn's giant fibres of diameter 35 μm is 20 m s^{-1} at 17 °C. These fibres have a myelin sheath, and are constricted at intervals; it looks as if here, too, there is saltatory conduction.

The giant fibres of earthworms also are myelinated, and have a conduction velocity comparable with that of the prawn's fibres, but the giant fibres of the polychaete *Myxicola*, with a diameter of 1 mm, conduct at only 20 m s^{-1}.

All these giant fibres seem to have been evolved where there was an evolutionary advantage in the rapid and simultaneous contraction of a group of muscles, such as is needed for the withdrawal of an earthworm into its burrow or a polychaete into its tube, the jet propulsion of the squids, or the rapid flick of the telson of the prawns. Even in the anemone *Calliactis* there are through conducting systems with a velocity of 1 m s^{-1} which are used for the quick withdrawal of the body.

MV 6.32. The Synapse in Vertebrates

Although the nervous impulse can travel in either direction along a nerve fibre, stimulation of a motor fibre does not initiate an impulse in any of the sensory fibres with which it is connected. The synapse must therefore act as a valve, and its action must be at least in some ways different from that of the axon. Further evidence for this comes from the time taken for a response. The finite period between the application of a stimulus and the response of an effector may be broken up into parts; the latent period between the application of the stimulus and the initiation of the impulse in the afferent nerve, the time taken for this impulse to travel to the central nervous system, the time before an impulse emerges in the motor nerve, the time for this to travel to the effector, and the latent period of the effector. All of these can be easily directly measured, or calculated from known conduction velocities, except for the time in the central nervous system, so that this can be got by subtraction of the others from the total. When the leg of a frog was stimulated the difference for the contraction of the gastrocnemius muscle of the same side was 8 ms, while for the contraction of the corresponding muscle of the opposite side it was 12 ms. The simplest explanation is that the delay in the central nervous system (the distances being negligible) is due to the time necessary for a new impulse to be started at a synapse, and that each synapse imposes a delay of 4 ms, so that for the ipsilateral reflex there is one internuncial neurone and two synapses, and an extra neurone and synapse for the contralateral reflex. This type of experiment can only give the minimum number of synapses and maximum synapse-time, but many observations suggest that in ganglia the delay is usually 2–4 ms, and in the brain about a quarter of this.

The central nervous system has such complex connections that every sensory cell could be put into communication with every motor unit, but this does not normally happen. If a single sensory nerve is stimulated it does not produce as great a contraction in the appropriate muscle as can be got by stimulating the motor nerve directly; presumably the sensory nerve is only put into communication with a limited number of the fibres in the muscle, and this must mean that the synapses are selective, or have different thresholds. When an animal is poisoned with strychnine a light touch on the skin causes vigorous contraction of every muscle in the body; presumably the resistances of the synapses have been lowered, so that every possible circuit functions.

Frequent passage of impulses along a path leads to fatigue, but a less frequent use of the pathway makes it easier for subsequent impulses to travel along it, since a response may now be elicited by a smaller stimulus than before. This is known as facilitation, and has the effect that a strong stimulus, which often acts like many repeated stimuli and sends a volley of impulses along the nerve, may affect more distant organs and cause a more vigorous reaction than does a weak one. Each successive impulse weakens the synaptic resistance, and makes it easier for its successor to follow.

If two stimuli, which by themselves would produce two different reflex actions, are applied simultaneously, in general the response for only one of them is obtained; this phenomenon is known as inhibition. It may be explained if one impulse, arriving at a synapse slightly ahead of the other (for the two are never likely to be strictly simultaneous) increases the resistance of nearby synapses, so that the pathway for the second impulse is closed.

The electron microscope shows that there are many different forms and arrangements of synapses, but they nearly all have the same basic structure, which is illustrated, very diagrammatically, in Fig. 6.5. The most important feature is that the membranes of the two cells are separated by a space, the synaptic cleft, about 10–50 nm wide. It sometimes shows traces of some contained material, and in some synapses from the brain it is crossed by filaments. The presynaptic cell membrane on the side of the cleft at which the nervous impulse arrives, and the post-synaptic or subsynaptic membrane on the side where the new impulse starts, have the usual triple structure. The electron micrographs regularly show a polarization, in that in the cytoplasm near the presynaptic membrane there is an accumulation of spherical or ovoid objects called synaptic vesicles. Their diameter is often equal to the width of the cleft, but they are sometimes smaller.

Explanation of the action of synapses was originally based largely on analogy with the nerve-muscle junction, and antedated our knowledge of their fine structure obtained with the electron microscope, but there is now little doubt that it was correct. The receipt of an impulse at the presynaptic side causes the liberation of acetylcholine into the cleft, and this acts on the postsynaptic membrane, just as it is known to be able to do on any nerve fibre, to increase its permeability and so start a new impulse. The acetylcholine is presumably contained in the synaptic vesicles, and their absence from the post-

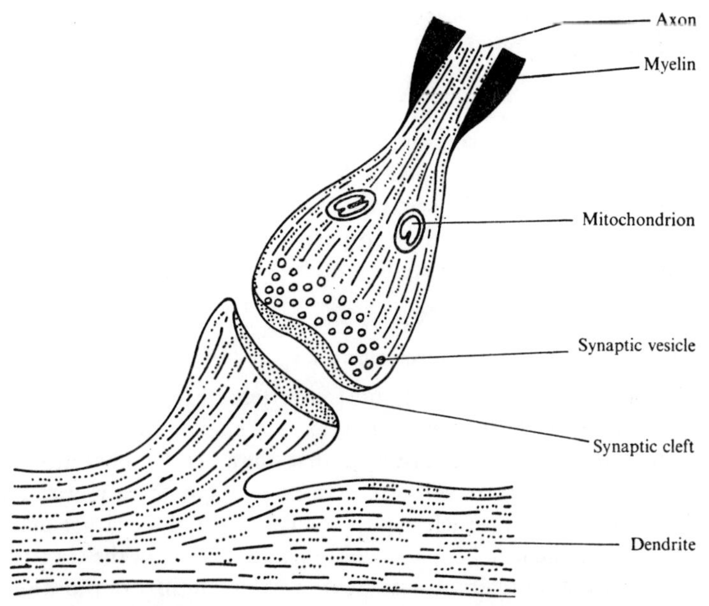

FIG. 6.5. Generalized diagram of a longitudinal section through a synapse, as seen in electronmicrographs. From Yapp, *Vertebrates; their structure and life*, Oxford University Press, New York (1965).

synaptic side accounts for the valve-like action of the synapse. Facilitation presumably depends on the greater ease of escape of the acetylcholine when the presynaptic membrane is weakened by successive impulses. Inhibition must occur either because acetylcholine blocks the escape of the synaptic vesicles from neighbouring cells, or because some other substance that does the same thing is liberated at the same time. After liberation the acetylcholine is rapidly destroyed by the enzyme cholinesterase.

The analysis has been taken further by means of microelectrodes inserted into motor neurones. Stimulation, and so presumably liberation of acetylcholine into the cleft, usually causes a general increase in permeability, quantitatively a large movement of sodium ions, and a postsynaptic potential. If this is large enough, it starts a new impulse by local circuit action. But it may not be—its intensity varies with the intensity of the stimulus. If two sub-threshold stimuli are applied within 15 ms their effects are added together, and an impulse may start; presumably the total acetylcholine liberated is enough. Similarly, there may be spatial summation, two sub-threshold stimuli to nearby presynaptic cells that touch the same postsynaptic cell causing a propagated impulse. In inhibition there is also a postsynaptic potential, but the increased permeability is to chloride and potassium only, not sodium.

It is probable that other substances are used as transmitters. Noradrenaline and 5-hydroxytryptamine (serotonin) are found in relatively high concentration in parts of the central nervous system, and it is likely that they are liberated at synapses.

J 6.32. The Synapse in Invertebrates

The general physiology of most invertebrate synapses, and their appearance in electron micrographs, are similar to those of vertebrates, and it may be assumed that they act by liberation of a chemical transmitting substance. In crustaceans there is a range of size of the synaptic vesicles up to about 3 μm, and the larger ones are bounded by the usual triple membrane. The nature of the transmitter substances of invertebrates is not known, but all three of those known to act in vertebrates have been found in chemical analyses of various invertebrates, and acetylcholine has been shown to be present in the nerve cord of the cockroach.

In a few invertebrate synapses, such as those of the giant fibres in the abdominal ganglia of the crayfish, the membranes are in close contact, or at least the synaptic cleft is very narrow, and the transmission appears to be electrical. Currents pass across with only a very brief delay, but since they go in one direction only the membranes must act as a rectifier. The junctions between the giant fibres of earthworms can conduct in both directions. That they do not do so in life depends on the conditions of the sensory cells; those in the part of the body in front of the clitellum are connected to the medial fibre, which therefore conducts backwards, while the posterior sense cells are connected to the lateral fibres, which therefore conduct forward.

G 6.33. The Central Nervous System

Anything that can be called a central nervous system consists of a large number of internuncial neurones, and so of a large number of synapses; since each neurone is connected to many others the total number of synapses is much greater than the number of cells.

Central nervous systems have been little studied except in vertebrates, and especially mammals. Most of their special properties can be understood, at least in outline, in terms of chemical transmission at the synapses. In some parts of the brain the rate of production of acetylcholine is high, but in others it has not been found, and other transmitters must be used; but the principle is presumably the same.

The resting oxygen consumption of grey matter is about five times as great as that of peripheral nerves, and when it falls below a certain value the system ceases to function normally. Grey matter needs also a continual supply of organic food—glucose or substances easily converted to this—while nerve fibres can live actively for some time in plain Ringer. This implies that the central nervous system requires more energy than it can get by anaerobic respiration, and that it carries no reserves of substrate. The continual manufacture of the transmitter substance is presumably an expensive process, which needs both organic material and a relatively large amount of energy, whereas the sodium pump is simple, and can be carried on for a time by anaerobic use of active phosphate.

The neurones of the cerebral cortex are continually active, even in the absence of stimulation. The receipt of a nervous impulse therefore does not so much start a new circuit as alter the pattern of the resting activity.

Defined types of behaviour can be induced both in mammals and in birds by electrical stimulation or by the application of chemicals to particular parts of the brain, but the matter is extremely complicated. Thus noradrenaline applied to one site in the brain of rats causes eating, while acetylcholine at the same site causes drinking, but acetylcholine at the corresponding site in cats induces the reactions of anger and fear, or a sleep-like trance. In spite of the difficulties, the explanation of memory, instincts and conditioned reflexes must be sought, not in terms of indefinables such as action specific energy and innate releasing mechanisms, but in the building up and variation of a complex pattern of circulating nervous impulses which cease only on death.

The central nervous system is probably important as a co-ordinating and dominating mechanism in all animals that possess it, and in many there is some degree of concentration of the central tissue into a brain. Removal of the cerebral ganglia from turbellarians interferes with their locomotion and other activities, and the effect is greater in polyclads than in triclads. Earthworms, with their receptors scattered over the surface of the body, suffer relatively little from the removal of the supraoesophageal ganglia; they can crawl normally, eat, and burrow slowly. *Nereis*, on the other hand, which has its chemoreceptors effectively confined to the palps and tentacles and has eyes instead of scattered photoreceptors, is strongly affected by the removal; it is over-active and neither feeds nor burrows. More complex animals show an increasing degree of dependence on the brain, but even many insects, with their high degree of cephalization, can carry out activities such as copulation when headless. This is perhaps connected with the presence of chemoreceptors on the legs.

G 6.34. The Nerve Net

The bare nature of a nerve net is adequately described by its name; it is a nervous system in which one cell connects with many others, so that localization seems impossible. It is the only type found in the coelenterates, is conspicuous, though not the dominant system of control, in the echinoderms, and is found in some other places such as the small intestine of vertebrates and on the foot and labial palps of mussels. The central nervous system of vertebrates is a network of nerve cells, but differs from the others in that it consists almost entirely of association neurones that have no direct connection with any effector or sense organ. The cells of the nerve net in the usual sense of the word supply both effectors and receptors, so that there are no motor or sensory nerves and no necessary similarity to the central nervous system. In coelenterates no central nervous system is present at all.

Most of the more recent work has been done on sea anemones, and it is to these that the following account chiefly applies, although Scyphomedusae are basically similar. The nervous impulse travels with a finite velocity, but one which is much less than that for vertebrate nerve; for the general nerve net of *Calliactis* it is from 0·1 to 0·2 $m\,s^{-1}$ against a minimum of 0·3 in the frog. But the values for the frog are those for the velocity in a single nerve fibre, and, as has been pointed out above, a finite time appears to be taken for an impulse to pass each synapse. The nerve net contains very many synapses, and it is to be expected that the velocity of conduction in it would be slow. In some parts of the sea anemone, such as the mesenteries, where there are specialized conducting tracts, it is over 1 m s^{-1}. In *Calliactis* the refractory period is from 40 to 60 ms.

In most cases the conduction is diffuse, with a complete absence of polarity. A stimulus applied at one point starts an impulse which spreads equally in all directions, and reaches any other given point in the animal by the shortest possible path. Considerable cuts may be made in the tissue, and provided the nerve net still has any continuity the impulse can reach any part of the body. A jellyfish may be cut into a spiral strip, and the nervous impulse will travel right along it. It does not, however, necessarily cause contraction of the tissue through which it passes, since a stimulus can produce contraction of distant tentacles without affecting the intervening parts. Physiologists have claimed that this means that there must be two separate networks, and in the hydrozoan *Velella* two different networks have been demonstrated histologically. The one with larger fibres, 1–5 μm in diameter, appears to have no synapses and to be syncytial, and so would be suitable for fast through-conduction. If there were only a single network, the same result could be achieved if the muscle cells in different places had different conditions for response. The rhythmical stepwise contractions of *Hydra*, which take place every 5–10 min when the animal is undisturbed, depend on a pacemaker in the hypostome, and are presumably dependent on the nerve net. The electrical impulses that accompany them travel at about 0·15 m s^{-1}.

The impulse apparently behaves like an electric current in a network, a result that is to be expected from the structure of the nerve net. There is no distinction of motor and sensory nerves, and impulses can be transmitted in both directions across the nerve-cell junctions. In some places, for example on the tentacles of anemones, there is a certain amount of polarity, which appears to be of two sorts. From the tentacle into the mesentery, conduction is always much easier than anywhere else, which suggests that there is some anatomical peculiarity of the cells allowing this; from mesentery to tentacle, the polarity is physiological, for when repeated stimuli are given facilitation occurs much more readily in this direction than in any other. This presumably means that facilitation is easier in one direction across a nerve-junction than in the other—perhaps a foreshadowing of the coelomate synapse which conducts in one way only·

In two important points the nerve net appears to be greatly different from a localized nervous system, but they can be satisfactorily explained in terms of the properties of all nervous tissue. As an impulse spreads out from its point of origin its response becomes weaker and weaker, so that there appears to be conduction with a decrement, which is unknown in ordinary nerve fibres. Secondly, there is graded response; a light touch on a tentacle of *Hydra* will cause that tentacle alone to contract, a stronger will cause other tentacles to do so, and a very strong one will cause the whole animal to respond. These two phenomena are basically the same, and both can be explained in terms of facilitation. The usual stimulus, whether mechanical or not, is really compound, and causes a succession of impulses presumably relayed to the nervous system by the sense organs. A weak stimulus means a low frequency of impulses, a stronger one a higher frequency. The greater the frequency, up to a limit set by the refractory period, the easier will be facilitation. In most cases in anemones the important junction where facilitation takes place is that between nerve and muscle, but in some that between the nerve-cells is involved. In both types a stronger stimulus activates more muscles or its influence travels farther than a weak one because it causes a more rapid succession of unitary impulses.

Coelenterate synapses have a narrow cleft about 2 nm wide, and have vesicles on both sides, so that they probably conduct in the ordinary way by a chemical transmitter. Facilitation presumably depends on the accumulation of this; its effect lasts three times as long for a fall of 10 °C, which is compatible with the destruction of the transmitter by chemical action.

The more active movements of the arms and tube feet of starfish are controlled through tracts of nerve fibres that behave much like those of vertebrates, but the local responses of the pedicellariae and spines show decremental conduction and facilitation like those of the anemones, and are probably controlled by the nerve net of the dorsal integument. It is probable also that strong stimuli may affect the tube feet through the net.

6.4. SENSE ORGANS

G 6.40. General Properties of Sense Organs

It is a truism that an animal cannot respond to a stimulus unless the latter is first picked up by a receptor or, in more general language, a sense organ. On the subject of sensation, physiology becomes even more intimately connected with psychology than it does on behaviour, for strictly speaking all sensation is subjective. Nevertheless, something is known about the means by which an event outside the animal becomes the starting-point for an event inside it, and it is with this that the present section is concerned.

The traditional statement that there are five senses is inadequate biologically for two reasons: first, even in man there are certainly more—the temperature sense is an obvious one left out of the usual list; secondly, and more important, the list is compiled not from observed reactions, but from subjective feelings which are not applicable to animals other than man. An amoeba reacts to light, but it would be over-bold to conclude that it sees or has the sensation of sight. The rational way of dealing with sense organs is to classify them according to the type of stimulus to which they normally or chiefly react. An isolated nerve can be stimulated in a number of ways, and the same applies to sense organs, the essential parts of which are in all higher animals made of specialized nervous tissue. But in spite of this, sense organs are usually specialized so that they react particularly well to one type of stimulus, which is said to be adequate for them, and they may be so shielded that other sorts do not normally reach them. For instance, the nerve endings of the ear are so situated in the interior of the skull that only influences of vibration in a material substance and of acceleration can be transmitted to them.

In spite of all this it is difficult to get away from the subjective aspect of sensation, for it is found that the same stimulus, when received in man by different sense organs, produces different sensations. A striking example of this is given by a tuning-fork or piano wire of low frequency. The sensation it gives through the ear is that of a note appropriate to its rate of vibration, but if it be touched only a shaking feeling is obtained. The converse of this is also generally true, that however a sense organ is stimulated the same sensation is obtained, a generalization first made by Johannes Müller in 1824. This is easily demonstrated by a simple experiment. The right eye is shut, and turned as far to the left as possible, and the eyeball is then pressed sharply with the finger at the outer (temporal) corner of the lids. A flash of light is seen, apparently just above the nose. A little consideration will show that an object in this position would produce an image on the retina at about the spot where pressure has been applied, and the obvious explanation is that pressure and light both have the same effect on the retina, but that the brain is accustomed to appreciate all stimuli received there as light. In other words sensation is a central phenomenon, and all impulses received by a particular set of cells in the brain are interpreted in the same general way. Normally such cells receive impulses only from the fibres of their proper sense organs, but direct electrical stimulation, for example of the part of the human cortex concerned with vision, shows that it is the brain cells, not the afferent fibres, which determine the sensation. The independence of stimulus and sensation provides further justification for discarding the traditional list of senses in dealing with animals; we do not know what sensations an animal feels, but we can find out to what stimuli it reacts.

This dependence of sensation on the sense organ that is stimulated can be explained in terms of the properties of the nervous impulse. It is, as explained in section MV 6.31, all-or-none, does not vary in magnitude, and is of the same form whatever stimulus excites it. With a stronger stimulus, more impulses are started than with a weak one, so that intensity of stimulus can be judged in the brain by the frequency and number of the impulses that arrive in a given fibre (and, since a large stimulus will in life generally activate more neurones, by the number of fibres carrying impulses). One impulse following another has its speed reduced, so that a train of closely but unequally spaced impulses arrives at the end of the nerve fibre more equally spaced, but apart from this tendency to obliterate any differences between the spaces between pairs of impulses in a train, there is no theoretical reason why different stimuli should not produce different patterns of impulses and so give information in this way. Histological evidence suggests that sensations of touch and temperature in human skin are mediated by the same nerve-endings. Here, if anywhere, are exceptions to Müller's generalization likely to be found.

It is often obvious without much experiment that an animal reacts to a stimulus, which must therefore have been picked up by a sense organ; when a paramecium meets some strange chemical in water and gives the avoiding reaction, or when a dog lifts its leg and scratches its side where it is violently rubbed, there is clearly response. On the other hand, where a stimulus causes no sensible reaction, it cannot be assumed that the animal is incapable of reacting to it or that there is no change in the nervous system. As long ago as 1749 attempts were made to use the method of the conditioned reflex to find out if fish can hear, but it was only after Pavlov had rationalized the study of this type of behaviour that the method was widely used. It is obvious that a normally neutral or ineffective stimulus could not take the place of an effective one in eliciting a response unless the neutral stimulus were first received by some sense organ. By this means it has been found that vertebrates can pick up many stimuli which are normally without apparent effect on them, but unfortunately the method has been little used with other animals. This procedure also enables one to test the power of an animal to discriminate between two similar stimuli. A dog, for instance, is trained to salivate to the sound of a tuning-fork or the sight of a circle, and is then tested with a fork of different frequency or with an ellipse.

More recently, precise information on the effect of stimuli on sense organs has been obtained by electrical recording, particularly with the cathode-ray oscillograph. There are dangers in this type of approach which physiologists have not always avoided. It does not follow, for three reasons, that a stimulus that can induce action currents in a sensory nerve has any natural function in connection with the organ. In the first place, the receptor may be so placed that the stimulus cannot normally affect it. Secondly, the stimulus may not normally be available in the environment in which the animal lives; this is obvious if the stimulus is electrical, but is true in other cases, such as for temperature in a deep-sea fish or light in an animal living in the dark. Thirdly, the impulse may produce no central or behavioural effect; it is a common experience that a man who is working intently may not hear the clock strike, although we can be confident that the sound waves started action potentials in his auditory nerve. For these reasons, electrophysiological experiments need to be followed up by observations on the behaviour of the living animal, preferably in its natural environment.

All sense organs have a threshold, and if the stimulus is below this magnitude they do not respond. The threshold is not constant; not only may it vary with temperature, and with the physiological condition and state of health of the organism, but in most sense organs there is adaptation, so that if the same type of stimulus is applied continuously or continually the threshold rises.

The generalization known as Weber's law, or the Weber–Fechner law, may be regarded as dealing with adaptation. It states that the smallest perceptible increment of stimulus is, for a given sense organ, a constant fraction of the total intensity. Experiments show that this is not true, and various mathematical relationships have been proposed in its stead. Since different workers do not agree on what these should be, even when they have been working with the same type of stimulus, it is safer to say merely that, the larger the total stimulus, the greater must be any increase in it for it to be appreciated.

Sometimes the terminal sensory element, so far as can be seen, is a naked end of an axon, in which the nervous impulse is both initiated and carried to the central nervous system. Often there is another neurone, sometimes misleadingly called the secondary sense cell, which first reacts to the stimulus and then, through a synapse, starts the impulse in the afferent fibre. Often, perhaps usually, the sensory cell is invested by another non-nervous accessory cell, derived for example from the epithelium. The function of the accessory cells is generally to mediate the stimulus (see, for example, section MV 6.42). The sense cell responds to this by developing an electric potential that is graded according to the size of the stimulus; if it is large enough, the all-or-none spike potential is formed and the impulse is initiated. All this may be interpreted by changes in permeability and movements of ions comparable to those we have described as happening in the subsynaptic membrane. The graded potential has been called a generator potential if it appears at the end of a sensory neurone in which the impulse will later flow, a receptor potential if it occurs in another sense cell and itself initiates the generator potential. But these terms have been confused and reversed, and since the two types of potential do not differ from each other, or indeed from the graded potential at a synapse, the single term generator potential will be used in this book.

We have spoken so far as if the action of a stimulus were always Thisto initiate a change in an otherwise inactive organ. Sometimes, however, this is not what happens. Many receptors are continuously active, and send a steady stream of impulses to the central nervous system. The effect of a stimulus is then to increase or decrease the frequency of these impulses, or sometimes to stop them altogether.

No completely rational classification of sense organs is possible. One scheme would divide them according to the human senses; this is incomplete, for the human body certainly reacts to stimuli that give rise to no sensation. Another uses the histology of the receptor cells; it can be argued that this must be fundamental, at least in an evolutionary sense, but unfortunately our ability to associate structure with function is so poor that little use can be made of it. Thirdly, a division can be made by the types of stimulus that normally affect the receptors. This has some similarity to a classification by the senses, but leads to difficulties, especially in human physiology. Thus the term mechanoreceptor, which is widely used for cells or organs that respond to touch, vibration, position, acceleration, tension, or pressure, and may mediate the sensation of hearing, covers many unrelated types of receptor and will not be used in this book. Lastly, it is sometimes convenient to make a broad division into exteroceptors, which receive stimuli from the outer world and in man usually give rise to well-defined sensations; interoceptors, which deal with similar stimuli arising within the body; and proprioceptors, which react to internal stimuli but give rise to no definite sensation. Some authors make the proprioceptors a subdivision of interoceptors. In this book we shall base our classification on the human senses, but modify it by the other schemes where they help.

MV 6.41. Proprioceptors in Vertebrates

Stimuli that originate within the animal's own body cooperate in the coordination of the position of the limbs. The sense with which they are concerned is called kinaesthetic. In man they permit the voluntary placing of a hand or foot in any desired position. It is possible with the eyes shut to touch almost any part of the body with fair accuracy; the ability to do this has presumably been acquired partly by practice, with unconscious memory of the exact tensions and movements required to place the finger in the right spot, but it is also possible to make the two index fingers meet, or nearly so, at any point out of sight behind the head that can be reached, even though the experiment has never been done before, so that some complicated unconscious calculations in solid geometry must occur also. The proprioceptors are concerned with recording these tensions and displacements of the muscles and joints. The same mechanism is used to a much higher degree in all cases of manipulative skill where the movement is one that has been learnt but is now carried out unconsciously. Examples are found in piano-playing and skating. The statement, surprising or incredible to a beginner, that one should be able to drive a golf ball blindfold, is correct because the movements of the body and limbs should be the same on every occasion, and with adequate training of proprioceptors this can be attained. From this point of view success at musical performance or games-playing or typewriting depends on a good proprioceptive mechanism. In learning all these things the eyes are used, but unless they can very early be put in a subordinate position progress is impossible.

In the appreciation of limb movement by the proprioceptors two things are important, the angular displacement and the angular velocity; the smaller the first, the greater must be the second for the movement to be felt. If the joints of the arm be considered in order from the shoulder down to the finger-tips, it is found that the sensitivity gradually decreases. This is of adaptive value in that a small angular displacement of a proximal joint of a limb means a big spatial movement of the distal parts.

In addition to making it possible for acts that were once voluntary to be carried out unconsciously, the proprioceptors assist in many reflexes that have never been consciously learnt. The most important of these are those by which the erect posture is maintained. The human body is a very unstable structure, and it is only by continual adjustments of the muscles of the legs and trunk that man can stand upright. The eyes and ears help in this, but the proprioceptors are also important, particularly those that record the pressure on the soles of the feet. The ability to balance the body when one is skating or crossing a stream in the Lake District is acquired by modifying the normal reflexes through practice.

Sense organs of various types, which presumably take part in proprioception, have been described as occurring in muscles, tendons, joints, mesenteries, and various other places, but little is known about the function of most of them. Anaesthesia of those in the joints abolishes conscious perception of position but has little effect on the control of movement. The distinction can be correlated with the presence of one histological type of receptor in the ligaments, whose impulse-discharge varies with position but is relatively insensitive to movement, and with other types, in both ligaments and capsule, which are very sensitive to movement.

Two of the larger proprioceptors have been much studied; one, the Pacinian corpuscle, occurs also in the skin and will be dealt with in the next section. The other, the muscle spindle, is described in section MV 7.31. It is the chief receptor concerned not only with muscle tone, as described there, but with the maintenance of posture and of the rhythmical movements of the limbs in walking and of the respiratory muscles in breathing. All these depend on the stretch reflex, by which tension on the spindle causes contraction of muscles which relieves that tension but causes a new one elsewhere, so that there is a continual dynamic equilibrium of antagonistic muscles.

Proprioceptors are widely distributed in the vertebrates. Reference to their importance in breathing is made in sections M 3.21 and V 3.22.

J. 6.41. Proprioceptors in Invertebrates

Outside the vertebrates proprioceptors have been little studied. It seems likely that active muscular movement would be impossible without stretch receptors of some sort, and they have been found in earthworms, in decapod crustaceans, and in many orders of insects. In crustaceans the dorsal abdominal muscles have a pair of receptor cells in each segment, with the branches of their dendrites twining round ordinary muscle fibres. In the violent backward swimming that the crayfish uses to escape from strong stimuli, these receptors are put out of action through the release of α-aminobutyric acid by impulses carried in the giant fibres, which carry also the necessary motor impulses. More complicated organs in the joints of the limbs respond to movement or to position, and presumably act in a similar way to those in the joints of vertebrates.

Insects possess stretch receptors of more than one type. Chordotonal organs, which are fibrous neurones stretching from one part of the body wall to another, are universal; some at least of them respond to stretching. Simple nerve-endings in the cuticle, covered with a more or less dome-shaped cap and called campaniform sensilla, probably respond to deformations of the exoskelton. Tufts of sensory setae, which are primarily receptors for touch, are situated on the membranes of the joints, and are stimulated when the body bends.

MV 6.42. The Senses of the Skin: Touch, Temperature, and Pain in Vertebrates

Many different sensory end-organs have been described in skin, but it now appears that most of these are artefacts. Three types can be made out with certainty. An arborization of fine, naked, axoplasmic filaments, less than 1 μm in diameter and probably ending freely, is found in all types of skin, both in the dermis and the epidermis. In hairy skin there are also special naked axoplasmic filaments surrounding the follicles, and these appear in an outer and an inner series. In glabrous skin and mucous membranes there are special capsules of cells containing arborizations of naked nerve-endings. Similar capsules are present in other places, such as the mesenteries, the pancreas, the bladder, and the thyroid; they are generally called Pacinian corpuscles, although many other names have been used and attempts to subdivide them have been made. It seems likely that, at least in mammals, there is only one basic type. The constant feature is that a non-myeinlated nerve fibre is surrounded by non-nervous cells, often arranged in layers more or less like an onion. The capsule is usually ovoid, and may be 1 mm in diameter.

Since what are usually considered as the four primary skin-sensations of touch, heat, cold, and pain can be appreciated by glabrous skin such as that of the sexual parts as well as by the general hairy skin of the body, no allocation of these sensations to particular nerve-endings is possible. It is certain that the same nerve-ending responds to more than one type of stimulus. How the stimuli are then distinguished is not clear, but one possibility is suggested by electrical recordings from fibres in the skin of both toad and monkey, which show that while slow pressure with a fine probe causes a single spike, quicker pressure causes two or three, and reduces the threshold. Comparable differences could distinguish temperature from touch. Another possibility is that there are groups of receptors responding to one, two, or three types of stimulus. The total of impulses received from these for a single stimulus would then be unique. But this seems a wasteful way of conveying information.

Shaving reduces, but does not abolish, sensitivity to touch, and it is likely that hairs act mechanically as levers in magnifying the deformation of the sense organ caused by a light touch. The sensitivity of the skin may be investigated by exploring the surface with glass fibres of varying thickness. Each fibre bends at a particular pressure, which can be observed, and so can be made to press on the skin with its own constant pressure. By this means it has been shown that the sensitivity of the tongue and nose is twenty-four times that of the loins, and other parts of the body come between these extremes. Separate stimuli up to as many as 600 per second can be perceived as discrete.

Subjectively, man can distinguish pain from touch, though from the point of view of the stimulus, pain, as caused for instance by pressure from the point of a pin, is merely an increased touch. But pain can also be provoked by excessive stimulation of other sorts, high temperature for example, and, quite apart from the subjective side, it can be shown that different sense organs from those of touch are being used. Some regions of the body, such as the cornea, and internal organs like the intestine, are deficient in ordinary touch sense, but are very sensitive to deformation—the slightest depression of the cornea causes pain. In the condition of analgesia that occurs in certain diseases the sense of pain is abolished but not that of touch.

The impulses that cause pain seem to be carried in special fibres. First pain, such as that caused by a pinprick, follows the passage of impulses at 15–20 m s^{-1}, along large myelinated fibres, and second or lasting pain, such as occurs after burning, follows slow impulses, travelling at less than 2 m s^{-1} in the smallest fibres of all.

The physical difference between a pressure on an end-organ that causes pain and one that merely causes a feeling of touch is unknown, so that the distinction remains a subjective one that cannot be applied to animals other than man. It has been claimed that the first effect of a pain-stimulus on the skin is to cause the formation of histamine, and that this then acts on the nerve-endings.

Herbst corpuscles, similar to the Pacinian corpuscles of mammals but much smaller, and even smaller Grandry corpuscles, are found in various places in the skin of birds and, especially in ducks and wading birds, in the beak. Some of the Herbst corpuscles have been shown to respond to vibrations of 20–1000 Hz. It seems likely that it is through these that the ducks and waders determine the presence of prey when they probe in the mud. Those in the beak are supplied by the ophthalmic branch of the trigeminal nerve.

It has been possible to remove most of the capsular cells of a large Pacinian corpuscle of a cat's mesentery and explore the results of almost direct stimulation of the nerve ending with a pulsating probe driven by a crystal. The response is a generator potential that falls off logarithmically with the distance from the point of application. It is proportional to the magnitude of the stimulus and to the area of deformation. If two spots are stimulated simultaneously their potentials are added together, but if they are close together the total is less than the sum of the two. If the generator potential is large enough it initiates an all-or-none impulse in the fibre, which appears to begin not in the unmyelinated fibre but at the first node of Ranvier. The generator potential is almost abolished by the removal of sodium, so it is probably of the usual type. The function of the capsule seems to be to reduce the pressure applied to the nerve ending and also by its elasticity to enable the nerve to respond to cessation of pressure as well as its beginning, for while the intact organ gives an 'off' impulse as well as an 'on', the isolated nerve ending does not.

It seems inevitable that a receptor that works, as does the lateral line system, by maintaining a flow of impulses the rate of which is varied on stimulation, must respond to changes in temperature by altering its activity. It could only not do so if the chemical reactions on which its impulses were based had a temperature coefficient of one. It is therefore not surprising that the ampullae of Lorenzini of elasmobranchs and the lateral line of goldfish should respond to changes of temperature, but it has not been shown that the fish are naturally subject to variations rapid enough to produce any effects. That in man it is change of temperature, rather than absolute temperature, that is felt is suggested by the observation that, if the right hand be placed in hot water and the left in cold, and then both be put into the same tepid water, the latter feels cold to the right and hot to the left hand; warmth is felt when the skin is warming up, cold when it is cooling. But after a hot or cold object has been placed against the skin and removed, its appropriate sensation remains, so that now cold must be felt when the skin is warming up and vice versa, and it is more likely that the different sensations can be explained in terms of adaptation.

Electrical recording shows that there are receptors in the tongue which have a resting discharge the frequency of which varies with temperature. A change in temperature causes an initial change in the frequency, followed by a return to a new steady state that may be above or below the previous resting rate. In general, a rise in temperature gives a fall in rate, and a fall gives a rise. A theoretical explanation of this in terms of changes in permeability has been made.

The threshold of sensitivity for human skin is 0·15 °C when 600 mm^2 are heated. Rattlesnakes (Crotalinae) have a much more sensitive organ, which responds to changes as little as 0·002 °C. It is a small facial pit near the eye, and is a true radiation detector, sensitive to medium and long infra-red rays from 1·5 to 15 μm; it enables the animals to find their mammalian prey in absolute darkness. It would seem to be more comparable to receptors for light than to any of the ordinary dermal sense organs.

J 6.42. Touch, Temperature, and Pain in Invertebrates

Probably all animals are sensitive to touch, and there is some evidence that some of them react violently to certain forms of it; this probably corresponds to the human pain sensation, for pain has no conceivable biological value unless it be the source of protective reactions. Tree-living caterpillars are continually being touched by the leaves and shaken by the wind, and give no apparent reaction, but quite a light touch of an unusual sort will cause them to drop down on their silken threads. The ordinary movements of the earthworm are controlled by the main part of the nerve cord, but the convulsive movements which it makes when dropped into alcohol, stepped on, or otherwise strongly stimulated, are organized by the giant fibres. It is therefore likely that the two types of movement are started by different sorts of stimulus received by different types of sense organ. If violent or protective reactions to touch are said to be started by pain receptors, one must be careful to avoid any dogmatic assumptions of pain as a conscious state, for there can be no certainty of the existence of this in any animal other than man.

In earthworms there are variations in sensitivity to touch in different parts of the body, and there is fairly rapid adaptation. In insects, seta-like hollow outgrowths of chitin, found all over the body, are important as tactile organs; some of them are also chemoreceptors, and are described further in section J 6.43. Pressure on those of the foot causes tonus in the leg muscles and so assists standing, just as pressure on the soles of the feet is one of the stimuli that enables man to stand upright.

It is obvious that many animals are sensitive to temperature, and the thresholds in *Rhodnius* (a bug), *Ixodes* (a tick), and *Agriolimax* (a slug) are comparable to that in man. The first two are ectoparasites of mammals and might be expected to be more sensitive than most. The bed-bug (*Cimex*) will orientate to a tube 1 °C above the ambient temperature and 10 mm away, presumably a useful reflex for finding its warm-blooded host. The temperature-sensitive organs of insects have a high concentration on the antennae.

MV 6.43. Chemoreception in Vertebrates

Man is aware of two senses, smell and taste, which are induced by the presence of chemical substance. In terms of gross anatomy, the former is served by the first cranial nerve, the latter by the branchial nerves, and this has led to the extension of the terms smell and taste to the senses served by these two groups of nerves in all other vertebrates. But there is another and more important distinction between the two senses. Smell is a response to airborne chemicals, and is therefore a distant sense, like sight or hearing, and gives us information about the outside world. Taste does not; it is a response to water-borne substances, and can therefore be aroused in a land animal only when the chemical has entered the mouth, or at least touched the lips, and dissolved in the saliva. In aquatic vertebrates there is no such distinction. The receptors connected to both the first and seventh nerves respond to dissolved substances, and all are distance receptors, since the soluble chemicals can diffuse to the animal from the water outside it and do not have to be brought into the mouth. As we shall see, many fish have chemoreceptors on their skin. We shall use smell in its strict sense to mean appreciation of airborne chemicals.

Besides taste and smell there are at least two other types of chemoreception. There is a response to strong and damaging chemicals, perhaps analogous to pain, called the common chemical sense; and various internal organs respond to the chemicals passing through them without giving rise to any definable sensation.

MV 6.431. Taste in Vertebrates. The taste organs in man are the taste buds that are found in the stratified epithelium of the tongue and, to a lesser extent, in other parts of the buccal cavity. They are lemon-shaped bodies, made up of a few bipolar nerve cells, each with a fine process projecting on the surface. Nerve endings of the lingual branch of the fifth cranial nerve, the chorda tympani branch of the seventh, and the ninth and tenth nerves ramify between them. Similar taste buds are found in the oral cavity of other vertebrates, and in fish some are found in other parts of the body as well. The catfish *Ameiurus* and some others have them all over the body, and will turn and snap at meat or meat juice placed in contact with the side of the animal.

Four primary taste sensations, sweet, sour or acid, bitter, and salt, can certainly be distinguished, and there are possibly two others, alkaline and metallic, as well. Other tastes are made up of mixtures of these, but a good deal of the apparent taste of foods is derived from their smell, as is clear when the effects of catarrh on one's enjoyment of a meal are considered. The simple experiment of holding the nose while eating shows the same thing. For these qualities derived partly from taste but chiefly from smell the name 'flavour' has been suggested. Different end-organs are concerned with the different primary tastes, for the sensitivity of different parts of the tongue varies from one to the other, and there are drugs which suppress one or two of them without affecting the rest.

In some way taste must be connected with chemical constitution: alcohols, sugars (which are polyhydric alcohols), and α-aminoacids are nearly all sweet, and so is the beryllium ion. Bitterness is produced by the ions of ammonium, magnesium, and calcium—it is noteworthy that these two metals belong to the same group as beryllium—but chiefly by the alkaloids and some other organic substances. All compounds with three nitro-groups are bitter and so usually are those with two, but one seems to have no effect on taste. Saltness is characteristic of some anions, especially those of low molecular weight: chloride is more potent than bromide, which is more potent than iodide. Sulphate and nitrate are also salty. Sourness seems to be exclusively connected with hydrion but its intensity is not proportional to this. The metallic and alkaline tastes, if they exist, are correlated with cations of heavy metals and with hydroxyl respectively. The four tastes are so distinct that it is incorrect to speak of a single sense of taste. Why any substance should have a taste at all is unknown, but the first stage seems to be that it combines with a protein in the taste bud, each taste having a different protein. There are a few compounds, such as dulcamarin, found in bittersweet, which definitely excite two tastes.

Taste is found throughout the vertebrates. Lampreys and trout orient towards water in which the fish on which they prey have been placed for a short time. Single-fibre recording showed that the receptors on the barbels of the catfish *Parasilurus asotus* respond to the four modalities of acid, salt, sugar, and quinine, but much more to the first two. The minnow is able to distinguish water that has passed over one member of its shoal from that which has passed over another. Frogs and toads appear to be poor in taste, responding, according to some authors, only to salt, and according to others to salt and acid; but to make up for this they can taste distilled water, as, apparently, can some mammals. Birds have few taste buds.

Very little is known of the functions of the nasal organ in fish. It is responsive to some chemical substances, but not to others, such as oil of cloves, which man considers to be strong-smelling and strong-tasting. In *Ameiurus* it responds also to light touch.

MV 6.432. Smell in Vertebrates. In mammals the organs concerned with appreciating smells are nerve cells with their bodies in the epithelium of the upper part of the nasal cavity, and the nerve connected with them is the first cranial. On the surface each cell has, in the rabbit, 9–16 long stereocilia comparable to those of the labyrinth (section MV 6.44). The nasal epithelium is innervated also by the ophthalmic branch of the trigeminal nerve and this responds to smell.

Animals can only smell volatile substances, but mere traces of a compound are enough—mercaptan can be smelt at a concentration of 4×10^{-11} gl^{-1}. Eight molecules can start an impulse and 40 nerve-endings have to be stimulated for a smell to be perceived. Even so, it is well known that the sense of smell in man is very poor compared with that in dogs and many other animals. The number of molecules that must enter the nose is, however, very large—about 2×10^{11} in the example just given. Most substances (other than gases) of strong odour have large molecules, so that their rate of diffusion is comparatively slow. Electrical analysis of the discharge from the olfactory bulb in the rabbit has shown that the different parts of the nose have different sensitivities to different substances. The receptors in the anterior part respond mainly to water-soluble substances and those at the back to those that are fat-soluble.

The primates and seals have a poorly developed olfactory organ and are termed microsmatic. Most mammals are macrosmatic, and many of them are known to be very sensitive to smells. A few, such as the toothed whales, have no olfactory organ and are termed anosmatic.

Although wild-fowlers believe that some birds have a good sense of smell, physiologists have long maintained that, with the exception of the kiwi, which can follow trails like a dog, they have none. There is now reasonably convincing evidence from studies both of behaviour and of electrophysiology that some birds, especially geese and vultures, have a reasonably good sense of smell, while others, such as pigeons, can respond at least to strong smells. The olfactory lobes of the higher birds, the passerines, are very small, and smell is probably of little or no importance in them.

Reptiles and amphibians have some sense of smell, and in the former there is an accessory organ, called Jacobson's or vomeronasal, innervated by the olfactory and the terminal nerves. It consists of two pits opening into the roof of the buccal cavity, and in lizards and snakes it responds to odoriferous particles picked up by the forked tongue, the tips of which fit into the pits.

As with taste, no general correlation between smell and structure can be made. Two types of incomplete theory are current. According to one, there are seven primary smells—camphoraceous, musky, floral, peppermint, etherial, pungent, and putrid—and each of these is elicited by a molecule of a characteristic shape, which fits on to a correspondingly shaped site on the lipid cell membrane and in so doing causes depolarization and a generator potential. The other starts from the observations that the olfactory epithelium is nearly always yellow, and that most strongly smelling substances have strong Raman spectra. Each of these can be connected with a particular type of molecular structure, and the necessary depolarization is supposed to be due to an appropriate atomic vibration or other source of energy.

MV 6.433. The Common Chemical Sense in Vertebrates. In man, the mucous surfaces that are more or less exposed to the exterior—those of the nasal cavity, mouth and larynx, the anus, and genital apertures—are sensitive to many chemicals, such as the spices that cause a hot sensation in the mouth, smokes that cause crying, and mustard oil and gas that irritate the skin. No taste is perceived, but the sensation leads to such reflex actions as coughing and sneezing. This common chemical sense is active only with relatively high concentrations of substances, and leads usually to protective reflexes. The distribution of the sensitive endings is much the same in other land-living vertebrates as in man, but in Amphibia and fishes they are found all over the skin. The endings are derived from the spinal nerves and, in the head, from the facial, and are distinct from those that are sensitive to touch.

The sense of itch may be put here. It may arise in any part of the skin, and has been shown to be distinct from pain; it is carried by different fibres, and responds differently to drugs. Indeed, itch is temporarily abolished by pain, hence the relief of scratching. It can be produced artificially by injection either of histamine or of certain proteinases, and natural itch is almost certainly due to the presence of one or other of these classes of substance. It is therefore a chemical sense.

MV 6.434. Osmoreception and Internal Chemoreception in Vertebrates. We have already referred to cells in the carotid body, the arterial walls, the hypothalamus, and elsewhere, which respond to changes in the chemical composition (usually the acidity or chloride concentration) or osmotic pressure of the blood passing over them (section G 3.6). By so doing they initiate reflexes that maintain the standard state of the body, so that in function they are closely comparable to the stretch receptors and other organs that maintain muscle tone and posture. They are in fact proprioceptors, although for no good reason that term is not usually applied to them.

J 6.43. Chemoreception in Invertebrates

Most animals react by reflexes and taxes to chemical stimuli, and any animal that takes food but rejects other solid particles—a discrimination possible even to *Amoeba*—may be loosely said to have a sense of taste. Sea anemones have a different pattern of chemical sensitivity from man. They respond both by movements of the tentacles and discharge of the cnidae, especially to proteins and their derivatives. Carbohydrates, except possibly glycogen, are ineffective, and so are fats, but some lipids that can be extracted from food by alcohol but not by ether are adequate. The jellyfish *Aurelia* is somewhat similar, but responds to fats. Planarians can find food by a distinct sense of taste, and if one of the auricles is cut off they make circus movements comparable to those of an insect with one eye blackened. Earthworms (*Lumbricus* and *Allolobophora*) have cells that initiate action potentials on contact with acid or sodium chloride all over the body, and the prostomium is sensitive in addition to quinine, glycerol, and sucrose (but not glucose), and has a concentration of the acid-sensitive cells. Different species of worm will only burrow in soil between certain limits of acidity, which determine the distribution. *Limulus* is insensitive to the four standard tastes, but its chemoreceptors are stimulated by extracts of the shellfish on which it feeds.

Insects have organs of taste in various places, with often more than one site in the same species. They occur widely on the labial and maxillary palps, for example in cockroaches, beetles, and flies; within the mouth in Diptera and Hymenoptera; on the tarsi of butterflies and muscids; on the antennae of Hymenoptera; and on the ovipositors of blowflies.

Some of the receptors on the labellum of blowflies are relatively large, and can be seen to contain three neurones, each of which has a distal process running up into the projecting seta, one ending at the base and the other two going to the tip (Fig. 6.6). The organ is sensitive to touch, to various chemicals, and to small temperature changes, but if the tip is cut off the response to chemicals is abolished, so that it is presumably mediated only by the cells with long processes. Electrical recording from the organ shows that spikes of different

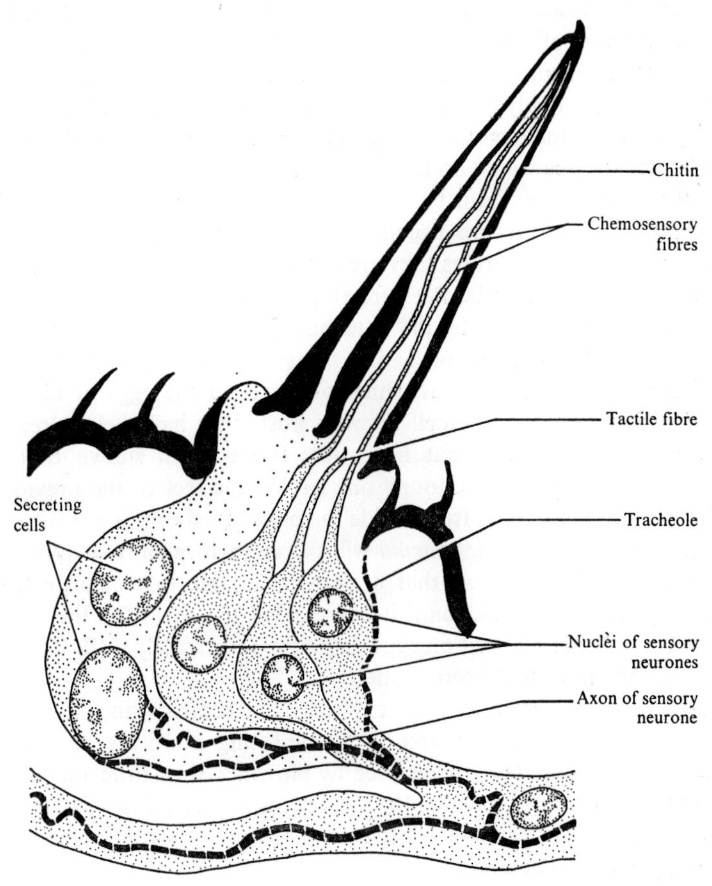

FIG. 6.6. Sensory seta of a blowfly, *Phormia*. Redrawn from Dethier, *Q. Rev. Biol.* **30** (1955).

size are generally given by sugars, by salts, acids and alcohol, and distilled water. The explanation of this that is generally given is that each spike is carried by the axon of a different neuron, but the fourth neuron has yet to be found. It is possible that the response to distilled water is really a response to cooling by evaporation, since it is given also, though with a longer latency and to a lesser degree, by solutions.

Bees and butterflies can distinguish the four standard tastes, but the list of sweet things to which they respond is not the same as for man. Their sensitivity to some sugars is about 200 times that of man.

Land snails and slugs find scented food, such as strawberries, by smell. A sense of smell is present in woodlice, which can detect and move towards an area occupied by members of the same species. Insects are probably the most skilful of all animals in detecting faint odours. Some male moths find the females from a distance of over a mile, but they can only do this if there is a breeze to bring the scent to them and allow them to orientate by flying up wind. The smell receptors of insects are usually situated in the antennae, but in Orthoptera and butterflies they are present on the palps also, and in some flies on the labella. Bees are able to some extent to distinguish between odours, again by the antennae, and their discrimination is somewhat similar to that of man. *Trichogramma evanescens* is a chalcid parasite (Hymenoptera) which oviposits in the eggs of moths, and it distinguishes by smell between hosts that have been already parasitized and those that have not. It has been shown that it recognizes at least two odours, one left by the feet of the previous chalcid and one arising from inside the parasitized egg. If parasitized eggs are washed, *Trichogramma* will pierce them with its ovipositor but will not lay eggs, so that it must be unable to recognize that the eggs are parasitized until it has pierced the shell. On the other hand, it will not oviposit on eggs that have been merely walked upon unless they have first been washed.

Many insects, such as Diptera, bees, and beetles, can find water from long distances, apparently by detecting gradients of water-vapour in the air. The organs used by blowflies are on the antennae, and this is perhaps general. In wireworms the response is more to saturation deficit than to relative humidity, so that the organs are perhaps evaporimeters.

The common chemical sense, to judge from reactions to noxious vapours and chemicals, is present in many or all animals, and in earthworms, insects, and others is spread more generally over the body than in man.

Little is known of chemical proprioception in invertebrates. Cells in the pedal ganglion of slugs increase their spontaneous activity with dilution of the medium bathing them, which agrees with the increased activity of the animals in damp conditions. One would expect that control of this sort might be widespread, though not easy to find.

MV 6.44. Receptors for Balance and Movement in Vertebrates

Man is conscious of standing upright, and when he suffers from certain diseases of the ear he is unable to do so. The ear, is not, however, his only organ of balance; the eyes also are used, and to a lesser degree the touch and pressure receptors of the soles of the feet. Not less than two of these must be active at one time if he is to maintain his position. In a steady state the eyes seem to be the most important balancing organ, for an aircraft without instruments in cloud may be flown tilted, right or left or nose up, without the pilot being aware of it. The disinclination of many birds to fly in mist suggests that the eyes may be important for them also. In this section we shall deal with that part of the inner ear, or, as it is better called, the labyrinth, known as the pars superior, whose chief function is to respond to position and acceleration, and with some associated structures.

The labyrinth is a system of membranes, canals, and swellings; in most vertebrates it is enclosed in bone, but in the elasmobranchs this is replaced by cartilage. Because this is relatively easily dissected to expose the nerve endings, elasmobranchs have been investigated more than any other animals; there is enough similarity of structure to make it safe to generalize. Although there are differences in detail, the general structure of the labyrinth is always the same, and may be illustrated by the frog (Fig. 6.7). The labyrinth contains a fluid called endolymph, and processes form its sensory portions—cristae, maculae, papillae, according to the nature of their non-cellular parts —project into this. They are connected to the eighth cranial nerve.

These receptors enable the animal to react to gravity in such a way as to right itself, or, if this is physically impossible, at least to counteract to some extent the way in which it is held or the inclination of the ground on which it is standing. A frog, for instance, placed facing downwards on a slope, stretches out its forelegs in front of it and raises its head, thus bringing itself as nearly as possible into the ordinary position. The traditional case of an animal reacting to gravity is the cat, which falls on its feet no matter in what position it is dropped. Stimulation of its labyrinth, caused by its being in any but the vertical position in the air, produces muscular movements which rotate it until its feet are below and it is in equilibrium. The physical principles on which this is based are simple: first, Newton's Third Law of Motion, that action and reaction are equal and opposite, and secondly that the moment of inertia of a body depends on its radius as well as its mass. Suppose that the cat is held supine and dropped, and that its head inclines slightly to its right. It stretches out its hind legs, draws up its front legs, and twists the front part of its body to the right; by Newton's third law the hind part of the

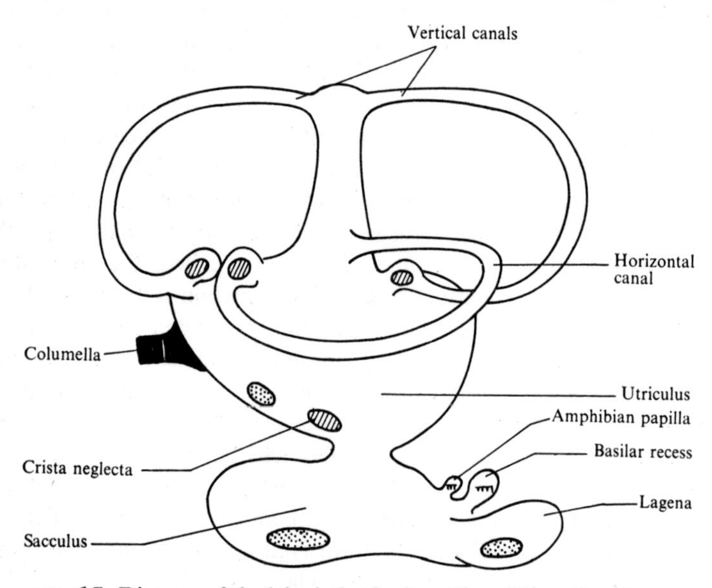

FIG. 6.7. Diagram of the labyrinth of a frog. From Yapp, *Vertebrates: their structure and life*, Oxford University Press, New York (1965).

body must also rotate in the opposite direction, but it will not go so far to the left as the front part does to the right, because the hind legs are stretched out and the moment of inertia is therefore high. The cat then draws in its hind legs, stretches out its front ones, and twists to the left; this time, for the same reasons as before, the front part moves less than the hind part, and also moves less than it did at the first turn; the final result of the two twists is that the cat's ventral side has been turned towards the ground.

The labyrinth responds also to acceleration, both linear and angular. Its function in appreciating the latter explains the phenomenon of nystagmus. Consider a man turning his head and his gaze from the straight ahead position to a direction due right. If the movements of the head and eyes are examined carefully it is seen that first the head alone moves, there being a smooth compensatory movement of the eyes to the left which maintains their direction in space unchanged. While the head continues to move, the eyes now perform a jerk to the right to catch up with the head position for an instant, and are then maintained in this new direction of fixation while the head goes on. Thus while the head moves continuously, the eyes perform a movement in space which is a series of jerks to the right with stops, while relative to the head they move slowly to the left and in jerks to the right.

The cupula, the jelly-like covering of the crista, is gelatinous and very slightly springy, and extends across the canal, interrupting the continuity of the fluid. Before the movement of the head, the endolymph in the horizontal canal will be stationary; then when the head movement starts, the angular acceleration of the canal wall will cause the endolymph to lag behind. Since the inertia of the endolymph is small, after a fraction of a second it will assume the same angular movement as the canal wall. During this fraction of a second, while there was a relative movement, the endolymph displaces the cupula, bending it to the left, and stimulating the cilia of the crista. After the relative movement the cupula remains displaced since there is no relative movement of endolymph and canal, so the cilia stay bent. By a mechanism that is not fully understood, this maintained bent position of the cilia of the crista brings about the alternating movement, already described, of the eyes relative to the head, namely rapid jerk to the right and slow drift to the left. This double movement, which is referred to as nystagmus to the right (direction of nystagmus equals direction of rapid movement) can be evoked in fish held stationary, by causing a small displacement of the endolymph in the horizontal canal.

When the head movement ceases, the endolymph continues to move for a very short time, and covers a distance that is just sufficient to restore the cupula to its equilibrium position, and the eyes are no longer stimulated. The springiness of the cupula is so weak that it plays no part in the process just described.

The effect of this nystagmus mechanism, during normal head movements, is to keep the image on the retina stationary for most of the time, and to cause rapid movements of the image during the jerks. There is some evidence that a person is blind or partially blind during these jerky movements. If the reader looks at one of his eyes in a mirror, and then changes over to look at the other, he will not be able to see the eye movement, though he can see it easily if he watches someone else perform the experiment. Further, a steady movement of the image across the retina causes distress, as may be experienced by watching a film taken by an amateur who deliberately waves the camera about. Thus the nystagmus mechanism is useful in eliminating any sensation of motion of the image across the retina.

If a steady rotation of the body and head is maintained for 10–15 s, then just as before, there is a transient movement of endolymph relative to the canal, causing a displacement of the cupula, after which endolymph and canal move together. Now, however, the very slight springiness of the cupula causes it to return to its initial position over a period of the order of 15 s. At the end of this time the crista is no longer stimulated. If a steady rotation is induced on a turn-table, it can be seen that the movements of nystagmus cease, and if the eyes are closed no sensation of rotation remains. If now the rotation ceases, the endolymph by its inertia will maintain its rotation for a fraction of a second, and the cupula will be displaced. Nystagmus will be induced, and there will be a sensation of giddiness. Let us suppose the steady rotation was to the left (anti-clockwise as viewed from above), then when the motion ceases the cupula is displaced to the left; this is the direction in which it would be displaced in a normal rotation of the head to the right. This, as shown above, induces nystagmus to the right, and it is understandable then that this is in fact the nystagmus observed when rotation to the left ceases.

The labyrinth is responsible for some other special reflexes, such as the one by which a duck ceases to breathe when it dives. Pointing the bill downwards stimulates the labyrinth so that respiratory movements cease.

Corresponding reactions to both angular and linear accelerations have been shown to occur in all vertebrate classes except reptiles. The resulting muscular movements largely consist in rotations of the eyes, but there are also movements of the head and limbs, the latter, when stimulated by a downward acceleration, sometimes moving in such a way as to prepare for landing after a jump or flight.

The distribution of these functions amongst the various parts of the labyrinth is obscure, and not necessarily the same in all the vertebrate classes, but in general the ordinary static reactions to tilting and those to rapid linear acceleration are controlled by the macula utriculi, while the semicircular canals are concerned with angular accelerations. The canals have a much smaller reaction time than the utriculus. In elasmobranchs the sacculus and lagena also respond to tilting. The sensory patches all have the same basic structure. Several supporting cells surround a bunch of nerve cells, and each of these has several processes (often miscalled hairs) projecting into the endolymph. They have almost the same structure in cross-section as cilia (section MVJ 7.1), but lack the central pair of filaments; they are

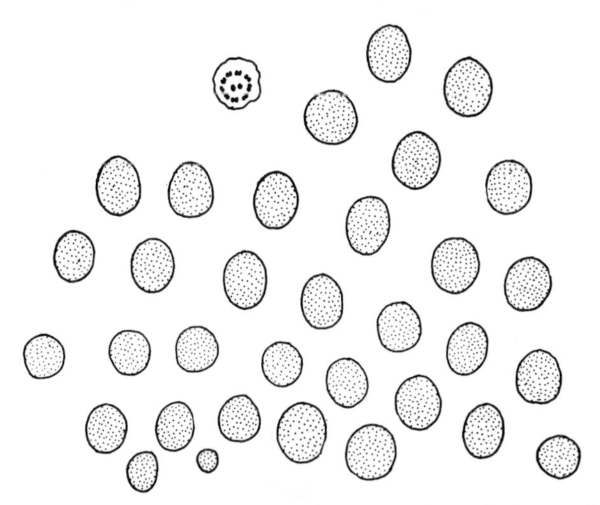

FIG. 6.8. Transverse section of a sensory process from the lagena of the skate *Raia clavata*, ×36 000, showing one kinocilium at the top, and several stereocilia. Redrawn from Lowenstein, Osborne and Wersäll, *Proc. R. Soc.* **B160** (1964).

therefore sometimes distinguished as stereocilia. There may also be a single ordinary cilium, or kinocilium, so that each tuft of cilia is asymmetrical (Fig. 6.8). The fibre running from the receptor cell is continually carrying impulses; on stimulation, the rate of these is altered, either up or down. The immediate cause of this is a bending of the cilia, which can be produced by the weight of the solid particles or otoliths in the endolymph acting on the maculae, or the movement of the fluid passing over the cristae. The asymmetry given by the single kinocilium probably accounts for the ability of the receptors to distinguish between accelerations in different directions, for a response is given only when the organ is bent in the plane that goes through the kinocilium and the centre of the group of stereocilia.

Fish, amphibian larvae, and the adults of those Amphibia that are permanently aquatic, have a series of sense organs called the lateral line, which make a pattern over the body and especially mark a line down each flank which gives the system its name. The sensory cells may be exposed on the surface, as in amphibians and some teleosts, but more often they are enclosed in tubes that are just below the surface and have occasional openings. The receptors called neuromasts,

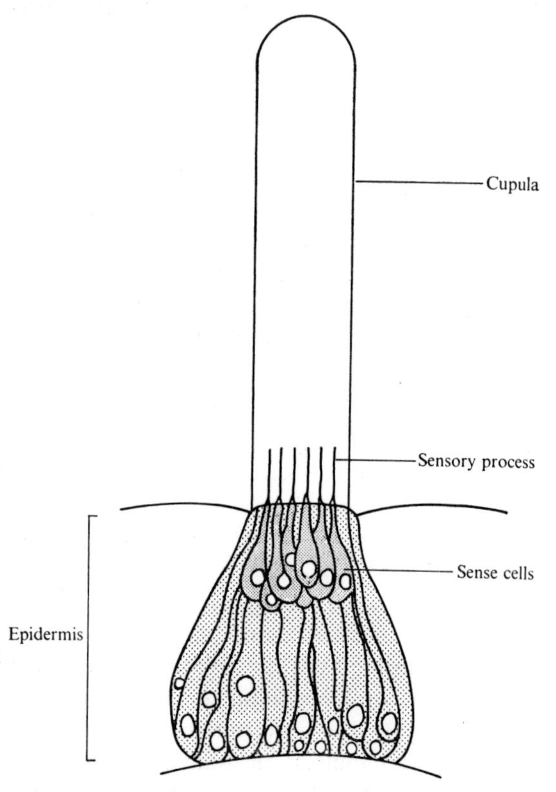

FIG. 6.9. Diagram of a neuromast. From Yapp, *Vertebrates; their structure and life*, Oxford University Press, New York (1965). Redrawn from Dijkgraaf, *Biol. Rev.* **38** (1958).

which are innervated by the seventh, ninth, and tenth cranial nerves, closely resemble the cristae of the labyrinth, for the cells have processes, each consisting of a bunch of twenty to fifty stereocilia, and one kinocilium, projecting into a jelly-like cupula (Fig. 6.9). Most zoologists believe that the labyrinth was derived from the lateral line system.

The neuromasts are continually sending impulses to the brain, and the rate at which they do so is increased by a headward movement of the water in the canal, and decreased by a tailward movement. Such movements bend the tufts of cilia through the plane of symmetry, and may be brought about by relative movement of the water surrounding the fish. By these means the fish can detect its own movement through the water, the approach of other fish, and the neighbourhood of solid objects, which reflect waves caused by its own motion.

J 6.44. Receptors for Balance and Movement in Invertebrates

Many animals possess organs called statocysts, which are hollows with receptor cells on their inner surface, and contain particles of a higher specific gravity than the medium that surrounds them, whose position can change with acceleration or the relation of the animal to the force of gravity. When these particles are in an unusual position they stimulate the receptors that they touch, and either initiate action currents or alter the frequency of those that normally flow.

Some of the Protozoa, such as *Paramecium*, react to gravity, and so presumably possess some form of acceleration receptor. Statocysts occur in flatworms and polychaetes, and are common in the coelenterates and molluscs. In the Arthropoda they are invaginations from the external surface lined with chitin, and in the Crustacea the particles they contain are grains of sand acquired from outside at ecdysis. If a moulting prawn is provided only with iron powder instead of sand it must of necessity use this to fill its statocyst, and after this has happened it can be forced to swim on its back by holding a magnet above it. In determining the movements that result in the characteristic posture, the statocyst organs act in collaboration with the proprioceptors. Each statocyst of a lobster contains about 300 fine processes, of which some respond to position and others to acceleration. Some appear to detect vibrations carried through the ground on which the animal stands, so subserving a primitive form of hearing.

Statocysts are rare in insects, but occur for instance in some dipteran larvae. *Notonecta*, a bug that swims upside down, orientates by responding to the pressure that its respiratory bubble exerts on the antennae; in darkness, with the bubble removed, it swims dorsal side up. The halteres of Diptera, which signal yawing (that is, rotation round a dorsoventral axis), work on the principle of the gyroscope, and contain their own proprioceptors. The beetle *Sisyphus*, which flies with its hind wings only, uses its middle legs in the same way.

M 6.45. Hearing in Mammals

The sacculus of the labyrinth (the pars inferior) in most vertebrates has only two small diverticula, the basilar recess and the lagena, but in mammals these are replaced by a spirally coiled cochlea, which is the organ of hearing. Vibrations in the air are transmitted to it through the external ear and the middle ear.

In most of the mammals there is a more or less trumpet-shaped pinna which collects sound waves and concentrates them to the opening of the external auditory meatus. Moreover, it can be moved by muscles and so used to locate approximately the direction from which sound is coming. In man it is probably functionless. At the inner end of the external auditory meatus is the tympanum, a stretched membrane which is set into vibration by the waves that strike it. Its movements are imparted to a chain of three ossicles, the malleus, incus, and stapes, which stretch across the middle ear. The base of the stapes is sealed by a membrane into an opening, the fenestra ovalis, in the cochlea. The ossicles are so hinged on one another that the movements of the tympanum are accurately transmitted to this membrane, but they are reduced in size because of the lever system of the ossicles. Variations in pressure in the air are therefore transmitted to the cochlea. The ossicles, under appropriate experimental conditions, can be seen to move, and in rabbits and dogs it has been shown that their movements are to some extent reflexly reduced on receipt of a loud sound, by muscles attached to the malleus and incus. There is thus a certain amount of automatic volume-control. The Eustachian canal, which leads from the cavity of the middle ear to the pharynx, prevents excessive differences of pressure on the two sides of the tympanum. Sounds may also find their way to the cochlea by conduction through the bones of the skull.

The cochlea (Fig. 6.10) consists of a tube wound helically round an axis of bone called the modiolus. In this runs the eighth (auditory) nerve, from which branches are given off to the sense cells. The tube of the cochlea is divided longitudinally into three, by two membranes running along its length; apically Reissner's, and basally the basilar. The uppermost subdivision of the tube thus formed is called the scala vestibuli, and the lowest the scala tympani. These two contain perilymph and are in communication at the inner end by a small opening, the helicotrema. At the other end the scala vestibuli opens into the vestibule, in which the fenestra ovalis is an opening. The scala tympani has an opening, the fenestra rotunda, which is separated only by membrane from the cavity of the middle ear. The third division of the cochlea, between the two membranes, is filled with endolymph and is called the scala media. On the side of the basilar membrane which is bounded by endolymph is a series of sense cells, similar to those of

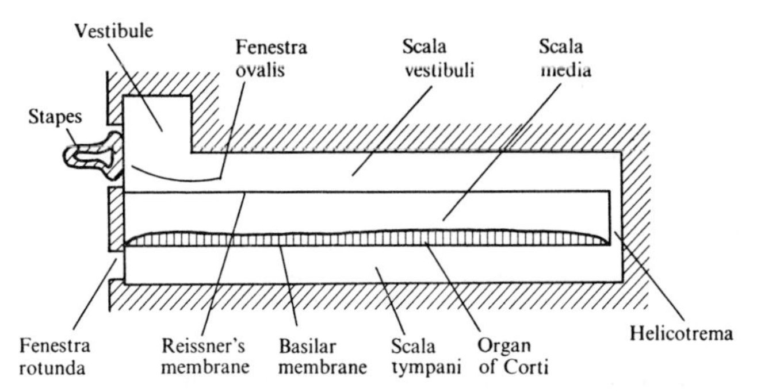

FIG. 6.10. A very diagrammatic representation of the mammalian cochlea, unwound. Diagonal shading indicates bone. The stapes is in the middle ear, and the fenestra rotunda also looks into this space.

the other parts of the labyrinth. Their processes project freely into the endolymph and are covered, but not normally touched, by the tectorial membrane. Collectively they make the organ of Corti, the essential organ of hearing, which is homologous with the papilla of the basilar recess in lower vertebrates. When the base of the stapes is pushed inwards, the resulting increase of pressure moves Reissner's and the basilar membrane downwards, and the membrane of the fenestra rotunda bulges outwards to compensate for this. Vibrations of the ear-drum are thus transmitted to the organ of Corti.

If the organ of Corti is damaged with a fine drill, the animal becomes deaf for a narrow range of frequencies, and the nearer the damage is to the oval window the higher the note for which deafness is produced. This agrees with the fact that the basilar membrane is narrow and stiff near the window, soft and wide near the helicotrema. Such observations suggest that the organ is acting as a resonator, but the mechanism is not simple. Direct observation in various mammals shows that any sound, if loud enough, causes a movement of the basilar membrane that travels along it and varies in amplitude as it does so. High notes give maximum amplitude near the oval window, and the maximal point goes away from this as the frequency falls, reaching the helicotrema at 400 Hz in the mouse, 200 in the guinea pig, and 30 in man. It seems, therefore, that by some central mechanism impulses coming from points on the basilar membrane other than those of maximum movement are suppressed. At the threshold of human hearing, for a sound of 2000 Hz the amplitude of vibration of the air particles is 10^{-8} mm, or one-tenth the diameter of a hydrogen atom, and that of the stapes is about 10^{-11} mm. At these intensities the displacement of the membrane must be effectively confined to a short length near its maximal point, so that it is in effect a resonator. It is possible that at low frequencies, where the maximum is off the scale, successive impulses in different neurones accurately reproduce in the auditory nerve the frequency of the note.

Sinusoidally fluctuating potential differences (called microphonics) can be recorded from the cochlea on receipt of a sound, and are greatest near the sense cells, so that they are presumably an expression of the generator potentials. Their energy is greater than that of the acoustic energy that causes them, so that there must be an amplifying system, presumably supplied with energy by the metabolism of the cells.

The ability of the ear to appreciate small sounds depends on their frequency; the human ear is most sensitive to notes of about 2000 Hz (roughly the third C above middle C—usually the highest but one on a piano). There is very little adaptation, so that the threshold remains more or less constant, but it is slightly increased after a loud sound. Frequencies from about 40 to 40 000 Hz can be detected, and over the middle part of this range, from 500 to 4000, people who have not been specially trained can detect differences of three parts per thousand, or about one-twentieth of a semitone. With two notes so close together as this, however, only a person with a musically trained ear can decide which is the higher.

The direction from which a continuous sound comes is determined by the phase difference between the images obtained by the right and left ears, which depends on the different distances which the sound has to travel to reach the two ear-drums. For sounds of high pitch (more than 800 Hz) and therefore of short wavelength, the same phase difference will be given with more than one angle of the incident sound (for a phase difference of $1+\alpha$ is the same as one of α) so that multiple images will be formed. Hence it is difficult to detect the direction from which a high-pitched note comes, and a very high note, such as the chirp of a cricket, seems to come from every side at once. For high notes some appreciation of direction can be got by turning the head, since this casts a sound shadow that reduces the intensity at the ear on the side away from the source. Sounds that begin and end sharply may also have their direction detected by the different times at which they arrive at the two ears; this is most noticeable with a succession of sharp clicks, but if they are not separated by about 1 ms or more, they sound like a continuous note.

In all mammals the general mechanism of hearing is the same. Dogs are most sensitive to rather higher tones than man, and can certainly detect differences of a semitone, and possibly of less, and have a memory for absolute pitch better than that of most men. Aquatic mammals possess various devices suitable to their special environment. Bats are specially sensitive to very high-pitched sounds, such as those that they make themselves. Such sounds, because of their short wavelength, travel very nearly in straight lines, so that their reflection gives an accurate sound-picture of the solid world surrounding the animal. By this echo-sounding bats flying in the dark can avoid even quite small obstacles, and can learn to distinguish falling meal worms from plastic disks of the same size. Behavioural responses suggest that small rodents can hear sounds up to 60 or 100 kHz, and their young in the nest emit sounds in this range. Shrews also emit pulses of 30–60 kHz, and use them to detect surfaces by echolocation. Some toothed whales echolocate both by clicks and by whistles.

V 6.45. Hearing in other Vertebrates

The ear of most other vertebrates is simpler. There is no pinna; the external auditory meatus is short in Sauropsida and absent from Ichthyopsida. Fish have no middle ear, and in amphibians, reptiles, and birds it contains only one bone, the columella auris. Birds alone have an outgrowth of the sacculus with the same general structure as the cochlea of mammals, and in spite of the fact that it has only a slight curve it is given the same name. It has the same parts, but the scala vestibuli is almost entirely filled up by the tegmental membrane, and its small spaces communicate with the scala tympani by two channels, one at each end. The evolutionary history of birds and mammals and the absence of any form of cochlea from reptiles show that in spite of the similarity the two structures are not homologous.

The cochleas of birds and mammals work basically in the same way, but there must be important differences. Song birds can hear much the same range of tones as mammals, but some are less sensitive to the lower notes. Some cave-living species find their way, like bats, by the echoes of high-pitched notes, and an owl finds a mouse by a highly coordinated homing on to the animal's squeak. Song birds have very good pitch discrimination, and birds in general can appreciate much more rapid changes of sound intensity or pitch than can man. The appreciation of frequency by the pigeon appears to be by a travelling wave, as in mammals.

Reptiles have a small outgrowth from the sacculus, and lizards seem fairly sensitive to sound. Chelonians are apparently unresponsive, but sounds up to about 300 Hz cause impulses to pass along the auditory nerve; the sensitivity of the ear of the alligator is very similar, and in both, the threshold for stimulation is lowest at about 100 Hz. Snakes have no tympanum and have the columella attached to the quadrate; they cannot hear airborne sounds, but they can hear those which reach the skull directly from the ground. In Amphibia the cochlea is represented only by the lagena, a small projection on the sacculus. Frogs can be shown to respond to tones of from 50 to 10 000 Hz, and there is some evidence that they can appreciate differences of frequency in sounds below 500 Hz.

Fishes have no cochlea, but various parts of the labyrinth respond to sound: in herrings the utriculus, in Ostariophysi the lagena and the macula of the sacculus, and in elasmobranchs both utriculus and sacculus. Conditioning experiments show that most teleosts respond only to frequencies from 20 Hz or thereabouts to about 1000, but a few can hear higher notes than this, and *Ameiurus* can hear notes of more than 13 000 Hz. Discrimination of pitch is generally poor, being something less than an octave, but some freshwater fish can do better than this; the minnow, *Phoxinus phoxinus*, can distinguish semitones in the range 987–1046 Hz. This increased sensitivity may be correlated with the presence of Weber's ossicles, extending from the sacculus to the air-bladder. Elasmobranchs are nearly deaf; there are no good conditioning experiments, and the electrical response from the macula of the sacculus is given only at frequencies below about 120 Hz.

J 6.45. Hearing in Invertebrates

Outside the vertebrates, well-developed auditory organs are found only in the insects. Some grasshoppers and crickets (Orthoptera) possess a quite complicated organ in the tibia of each front leg; it consists essentially of a membrane that covers a series of sense cells.

Grasshoppers respond to chirps made by other grasshoppers, and they will also respond to similar artificially produced notes of a range of frequency from 430 to more than 90 000 Hz; in general they are more sensitive in the higher ranges. A female cricket finds a male by flying in the direction from which the latter's chirp comes, and no other sense than hearing is necessary, so that she will answer a telephone call from him. If one of the female's ears is destroyed her sense of direction is impaired, but she can still find the male. Each organ is directional, and must respond not to pressure, but to something that is a vector quantity. The structure and dimensions of the tympanal organ, with two membranes enclosing a third, would provide such a quantity in the displacement of the third membrane caused by the differences in pressure that the sound wave produces on the two outer membranes. The organ is in effect measuring pressure gradient, which is a vector, whereas pressure itself is not.

Neither the behaviour of the animals nor electrical recording suggests that there is any response to pitch, but the organ is very sensitive to amplitude modulation, and the insect presumably recognizes its specific song by this means.

Somewhat similar organs are found in the abdomen of Hemiptera (*Corixa* being sensitive from about 2000 Hz up to 40 000 and cicadas up to 20 000), and on the abdomen, thorax, or labial palps of many moths (Lepidoptera). Most of the latter have maximum sensitivity at about 50 000 Hz, and one of them has a range from 3000 Hz up to 240 000; this ability to hear high-pitched sounds possibly enables the moths to hear the approach of bats. As some moths produce sounds in the same range they may indulge in jamming.

The filament of the sense cell of the locust's auditory organ has the internal structure of a kinocilium, with 9+2 filaments, in spite of the absence of ordinary cilia from insects. Some caterpillars of the Lepidoptera respond to sounds, the receptors being the hairs on the anterior part of the body. Structures called chordotonal organs, similar to parts of the ears of grasshoppers but without the tympanum, are found on the halteres of Diptera, on the legs of ants (Hymenoptera), and on the antennae of most insects, but they may be proprioceptors. Bees and gnats can hear, and the ability may be general in the class. Beetles, cockroaches, butterflies, and others can pick up ground-borne sounds through their legs. Echolocation has not been proved in invertebrates, but is possible. The whirligig beetles (*Gyrinus*) avoid each other by a sense, something between touch and hearing, which responds to the ripples that they make on the surface of the water. The cladoceran *Mixodiaptomus laciniatus* can swim a maze, and can apparently detect obstacles 1 mm away.

G 6.46. Photoreception

Light is a form of electromagnetic radiation that is conveniently dealt with, for most purposes, in terms of wave-theory. It has three qualities. Intensity, corresponding to the amplitude of the waves or the energy flow in them, is interpreted in human vision as brightness; wavelength, or its reciprocal, frequency, we appreciate as colour; and the plane in which the maximum displacement of energy takes place, or plane of polarization, man is unable to detect.

The range of wavelengths to which the visual organs of animals respond is nearly always from about 0·4 μm, which gives violet for man, to about 0·8 μm, which gives red. There seems no reason in principle why other electromagnetic radiations should not stimulate receptors, and there are a few cases known where they do. The infrared detectors of snakes have been mentioned in section MV 6.42, and a few animals, such as planarians, *Daphnia*, and many insects react to the near ultra-violet, while X-rays falling on the retina give man the sensation of light. Conversely, some bees are insensitive to the longer red rays.

Sensitivity to light is a fairly general property of protoplasm. The pseudopodia of *Amoeba* react to light, and so do the muscle cells of *Hydra* and of the sea anemone *Metridium*, and the spines and tube feet of some echinoderms. The euglenoid flagellates turn to move towards the light. The sensitive spot is probably a swelling near the base of the flagellum (an observation that may be significant in view of the structure of rods and cones), which is alternately exposed and shaded by the stigma (miscalled eye-spot) unless the animal is correctly orientated. There are also some cells or organelles that are specially sensitive; notably the chromatophores of echinoderms, crustaceans, fish, amphibians, and reptiles. Some form of general sensitivity to light is known in all the major groups except cephalopods. As would be expected, it is less common in those groups with a thick epidermis or well-developed eyes. It is widespread in the lower vertebrates, but rare in the amniotes. The hypothalamus of ducks is sensitive to light and this property is used in the stimulation of the pituitary (section MV 6.13). The photoperiodic response of the aphis also depends on a receptor on the top of the head, probably in the cranium.

For perception of form, as distinct from mere sensitivity to light, a real image must be produced. There are four known ways in which this can be done: by reflection from a concave mirror, by a pin hole, by refraction through a lens, and by diffraction. All of these except the first are used by animals, and an organ that forms an image in any way is an eye.

It seems likely that in all forms of photoreception the first effect of the light is to cause a photochemical reaction, which in the more highly developed cases causes a generator potential that in due course initiates a nervous impulse. There is a regular association of sensitivity with pigments, which are often chemically related to carotene.

M 6.46. Vision in Mammals

The eye of vertebrates (Fig. 6.11) forms an image by refraction. Its sensitive part, the retina, is surrounded and nourished by a surprisingly dense layer of blood vessels, the choroid, outside which is the 'white' or sclera. The transparent cornea protrudes in front, which the reader may feel in his own eye by moving the eye while holding the finger over the closed lid. Behind this is the iris. This contains radial muscle-fibres supplied by sympathetic nerves, and circular muscle-fibres supplied by the parasympathetic system. Its aperture decreases in diameter in bright light. Behind the iris, lying in the aqueous humour and supported by the suspensory ligament, is the lens. About two-thirds of the refraction of the light rays is due to the cornea and about one-third to the lens.

Accommodation, i.e. focusing for near objects, is brought about by making the lens thicker and its surfaces more curved. The ciliary muscle, shown only in section, is a ring running round the circumference of the iris and behind it. Its fibres have their origin on the sclera and their insertion in the ciliary body, and they run radially and towards the back of the eye. When they contract the ciliary body is pulled forwards and towards the centre of the pupil, so that the tension in the suspensory ligament that supports the lens is reduced. The lens now assumes its natural thick shape with highly curved surfaces, the change being probably brought about by the elasticity of the thin capsule that encloses it. Were the lens free, it would have its maximum curvature, and would therefore be suited for the vision of near objects, but in the eye the suspensory ligament is pulling on it all round, so that in the condition of rest it is stretched and is focused for distant objects. In accommodation the lens also moves forward slightly, but this is not important. In many small mammals and in the horse there is no accommodation. The otter has a large sphincter

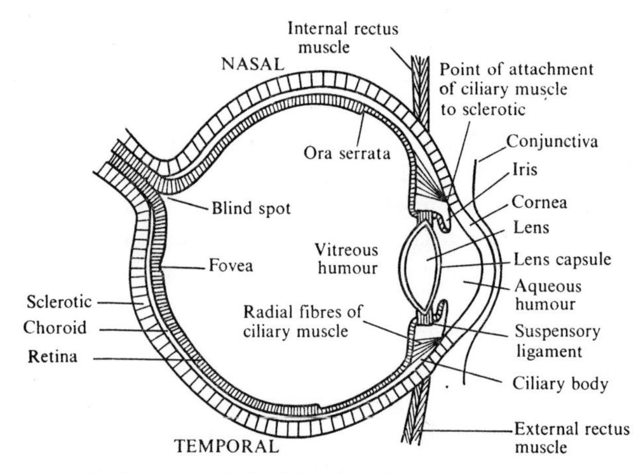

FIG. 6.11. A diagrammatic horizontal section of a mammalian eye. The thickness of the choroid and retina, and especially of the anterior continuation of the latter over the posterior surface of the ciliary body and iris, is exaggerated.

muscle which squeezes the anterior part of the lens and so gives the good accommodation necessary for seeing under water.

The binocular vision of man enables solids to be seen in the round and distances to be estimated by the convergence of the two eyeballs. Monocular estimation of distance, presumably by proprioceptors in the ciliary muscles, is much less accurate. Except in the Anthropoidea and in *Tarsius* binocular vision is very limited.

Fig. 6.12 shows a diagram of the structure of the retina. Light rays forming the optical image pass through the transparent nerve-cells of the ganglion cell layer and the bipolar layer, and only about 10 per cent are usefully absorbed by the sensitive elements, the rods and cones. The rest go on and are mostly absorbed by the layer of pigmented epithelium. Between them the rods and cones in man number about 10^8 cells. Many converge on to a single bipolar cell, so there are fewer of these, and many of these in turn converge on to a single ganglion cell, of which there are only about 10^6. Each of these sends a fibre to the brain. It will be seen that this convergence of the retinal elements would appear to reduce the visual acuity, for if all the cells converged on to one cell no form could be perceived at all. It is important also to notice certain cells in the bipolar layer which make connections laterally, for if these cells with their branches served to

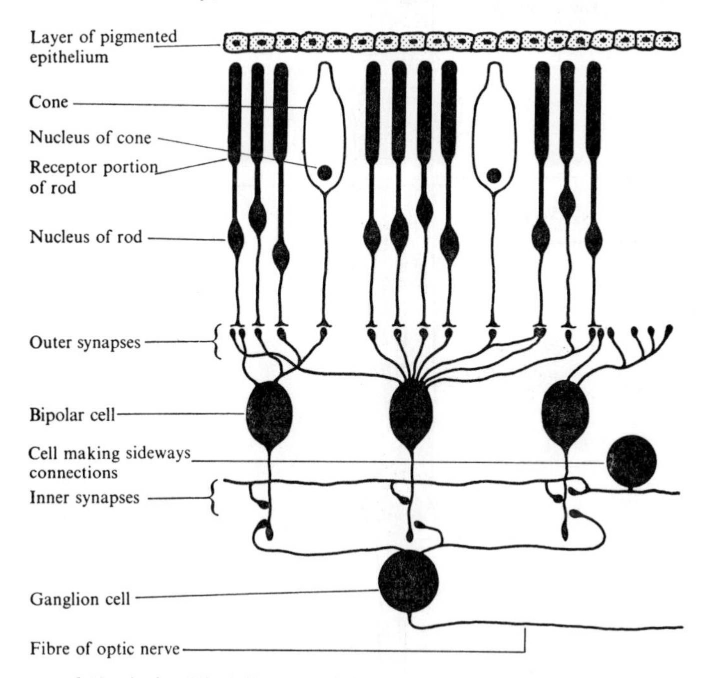

FIG. 6.12. A simplified diagram of the nervous connections in the mammalian retina.

conduct impulses starting in the rods and cones they would only muddle the whole picture. It is tempting to think that they enable one part of the retina to control adjacent parts and have something to do with simultaneous contrast.

Electron microscopy shows that each rod or cone usually has five parts, which, beginning from the outside of the eye, are (Fig. 6.13): (1) the outer segment; (2) a stalk, which in transverse section has the structure of a stereocilium; the two central fibrils are sometimes visible; (3) the inner segment, which contains oil-droplets and large mitochondria; (4) a stalk with neurofibrils, leading to (5) the synaptic peduncle. Most of the outer segment is made up of a pile of discs; their structure is difficult to make out, but each probably consists of

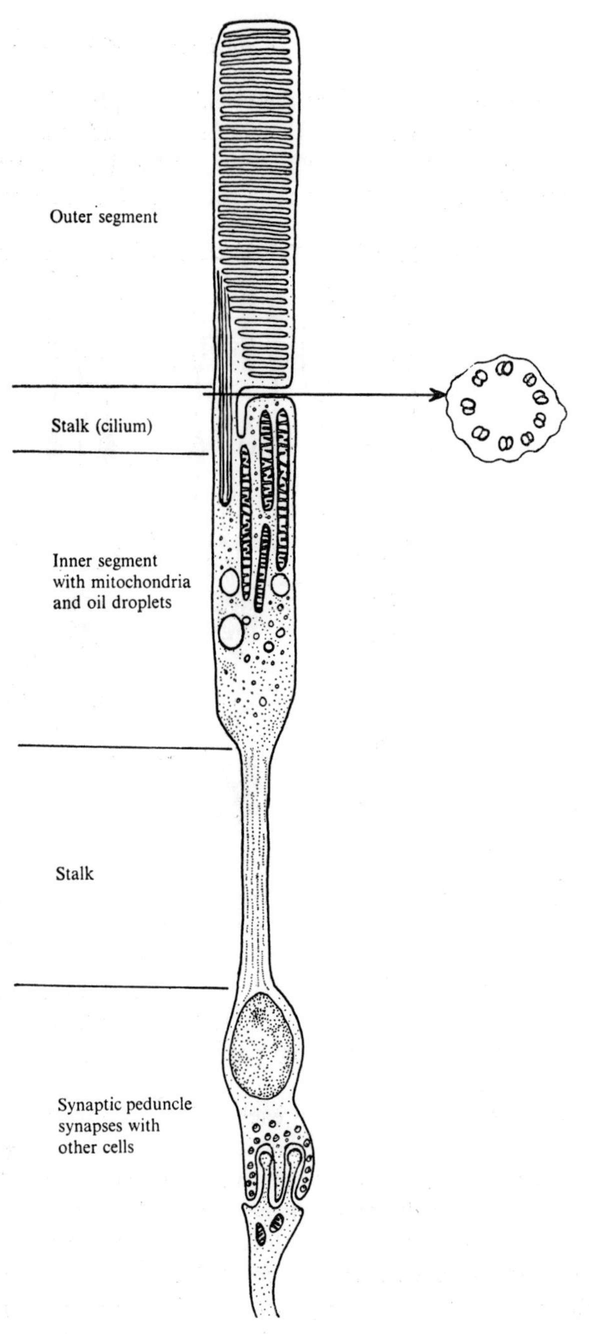

FIG. 6.13. A diagram of a vertebrate retinal element (rod or cone) in longitudinal section. Inset is a transverse section of the stereociliary region.

two Robertson membranes of the usual sort (lipid between two layers of protein) in near or partial contact in the middle and continuous round the edges (Fig. 6.14). Such a structure could be formed by invagination and pinching-off from the surface, and some electron micrographs, especially of amphibian and fish cones, show signs of

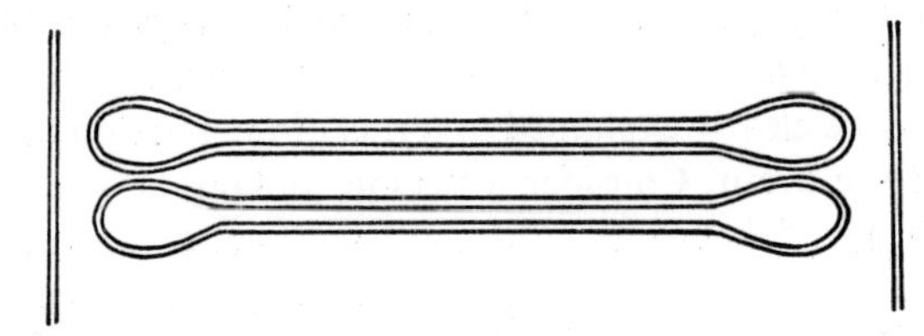

FIG. 6.14. Two discs from the outer segment of a vertebrate rod, in longitudinal section.

this. The cilium has a basal body in the inner segment. The synapses are peculiar in that the dendrons of the bipolar cells enter and branch in the peduncles of the rods. Otherwise the synapses are of the usual form, with a cleft about 10 nm across, and vesicles in the rod or cone. The rods are usually completely surrounded by a thin layer of cytoplasm of Müller's cells, which make up the packing glia of the retina.

In human eyes, and some others, there is a well-marked distinction in shape between rods and cones, but there is no essential difference between them in structure, and in many animals the distinction cannot be maintained.

The distribution of the rods and cones over the retina is not uniform. In man, the central part, the fovea centralis, where images are normally focused, contains only cones, and as the distance from this increases, the proportion of rods increases, until, at the periphery, there are scarcely any cones. The periphery of the retina is nearly colour-blind, so that it appears that cones are necessary for colour vision. Further, the periphery is much better at appreciating very faint light than is the fovea; a very faint star can often be seen only when one is looking not directly at it but slightly to one side. This suggests that the rods are concerned in appreciating low light intensities, that is, in twilight vision. The physiological distinction obviously cannot be made where there is no morphological distinction, and the distribution of rods and cones in mammals follows neither their diurnal or nocturnal habits nor their ability to discriminate colours. A fovea is known only in the primates.

Where the optic nerve enters the eye there are no sensory cells, so this gives rise to a small blind area, which, owing to the inversion by the lens, lies in the temporal half of the field. If the reader uses his left eye only, while reading *this* line, he will observe that the black dot on the left disappears at a certain point, only to reappear as the eye moves farther on. He may experiment with black, white, or coloured dots on a background of any colour, and always the blind area appears to be filled in with the appropriate colour. This apparent miracle may be clearly explained in terms of the relation between the retina and the brain. Consider a person looking at say a uniform green surface, and that while he does so a hypothetical operation is performed, whereby a slit is made in the retina by a single cut, and that the sides of the cut are pushed outward to leave an artificial blind gap. We will suppose no cells to be destroyed. All the cells of the retina continue to receive green light, the message conveyed by the optic nerve will be therefore unaffected, and consequently the impression received by the brain will also be unaffected, so that no blind spot will be noticed. We may suppose that the natural blind spot is not perceived for a similar reason, and this is equivalent to saying that there is no centre in the brain which represents the blind spot. If this explanation is correct it is not true to say that the brain 'fills in' the blind spot, but rather that it 'sews it up'.

The physics of the eye—the formation of the image—is relatively simple; the physiology is complicated and no complete account of vision can be given. Among the things that have to be explained are colour vision and the anomalies of colour blindness; simultaneous contrast—the fact that the colour or brightness perceived is not absolute but depends on neighbouring colours and brightnesses; and dark adaptation. There are many other phenomena only too often forgotten by theorists. For example, two lights (called metameric) may be matched even though their spectral compositions are different; if a black and white top is rotated it may appear coloured, even in monochromatic light.

Electrical investigations have been on the whole disappointing. Where recordings can be made from single cells (chiefly in the amphibian eye), some show in darkness a steady discharge that ceases when a light is switched on; others discharge at the beginning of illumination and continue to do so, though at a lower rate, until it ceases; a third set discharge only at the onset and end of illumination; and some respond only when the light is switched off.

The inner synapse layer is rich in acetylcholinesterase, and injection of acetylcholine produces spikes from the ganglion cells. These synapses therefore appear to be of the usual cholinergic pattern. The outer synapse layer is different, and the interaction between the light-sensitive cells and the bipolars may be purely chemical. Certainly the initial effect of the light is to induce a chemical reaction. The outer segments of rods contain a pigment called rhodopsin, or visual purple, which consists of a protein, scotopsin, and retinal (= retinene). Retinal is the aldehyde corresponding to vitamin A_1 (section M 2.142). If, as seems likely, rhodopsin is situated within the discs, its protein part would be in the protein portion of these, the retinal in the fatty layer. When light falls on it, the retinal, which is 11-*cis*, is isomerized to the all-*trans* form to make lumirhodopsin; one quantum can isomerize a single molecule, and this is enough to excite the rod. Since this is not enough energy to polarize a membrane, there must, as in the ear, be amplification. The subsequent changes—separation of the retinal from its protein, reduction of the former to its alcohol (retinol = vitamin A_1) and regeneration of rhodopsin—are enzymatic, and do not need light; they probably take no direct part in vision.

Rhodopsin is coloured, with maximum absorption of light at a wave length of 505 nm, which is about the maximum sensitivity of the retina. This has led physiologists to look for pigments that might have other absorption maxima, and so give a basis for colour vision. Young in 1802 suggested that the results of mixing coloured lights (whereby red plus green give yellow, red plus green plus blue give white, and so on) could be explained if there were three types of receptor, each responding to one of the colours red, green, and blue. There is now some evidence that human cones do contain three pigments, called cyanolabe, chlorolabe, and erythrolabe, with absorption maxima at 450 nm, 525 nm, and 555 nm respectively, of which only one is present in each cell. These would supply the physical basis for Young's theory. But there are still difficulties; brown, for instance, does not exist on its own, but is only seen as such if white or another colour is alongside it to provide the necessary contrast. The only mammals with colour vision, so far as is known, are the Anthropoidea and the Tupaiidae (tree shrews).

V 6.46. Vision in other Vertebrates

The eyes of other vertebrates are built on the same plan as those of mammals. Accommodation is brought about in several different ways. Birds and reptiles (except snakes) squeeze the lens by ciliary muscles, which are striped, and so thicken it; the ossicles in the sclera help to prevent distortion of the eyeball. Some birds such as hawks also change the shape of the cornea. In snakes, muscles in the iris put pressure on the vitreous which pushes the lens forward. Amphibia have little accommodation, but a protractor muscle pulls the lens slightly forward for near vision. Fish also move the lens, teleosts pulling it back for distant vision and elasmobranchs forward for near vision. Teleosts have no muscles in the iris and no pupillary reflex, but in sharks there is a slow change in the size of the pupil, the iris acting as an independent effector (section 7). The shape of a bird's eyeball makes it relatively free from the effects of spherical aberration. Binocular vision has been little studied; it occurs in some birds, such as hawks and owls, and in some fish.

The structure of the retina is much the same in all classes. A fovea is found only in some teleosts, in diurnal lizards, in a few snakes, and in birds. In the last there may be two foveae, one central as in man and one temporal. This arrangement, which is found chiefly in birds that catch their food on the wing, is probably used in binocular vision, an image of an object straight ahead being formed on the two temporal foveae simultaneously. The density of cones over much of a bird's retina is as high as in a man's fovea.

Full colour vision is found only in some diurnal teleosts, in urodeles, in tortoises, in diurnal lizards and in birds. The spectrum of sensitivity is generally similar to that for man, but some fishes and birds have a closed colour-circle, and fail to distinguish red from purple. Frogs have poor, and toads no, colour discrimination. No correlation can be made with the presence of rods and cones.

Birds and most marine fishes and adult amphibians, like mammals, have rhodopsin in their retinas and vitamin A_1 in their bodies. In freshwater teleosts and amphibian larvae vitamin A_1 is replaced by vitamin A_2, which is 3-dehydroretinol, and the corresponding retinal; this, in association with protein, forms porphyropsin, which has maximum absorption at a slightly longer wavelength than rhodopsin. It looks, therefore, as if rhodopsin were associated with marine and terrestrial life and porphyropsin with freshwater, and this is supported by the presence of both pigments in estuarine teleosts and in catadromous and anadromous fish and in lampreys. In the eel a changeover from porphyropsin to rhodopsin takes place before it leaves freshwater, and the reverse change occurs in the lamprey before it leaves the sea. But there are exceptions; some marine teleosts have porphyropsin, and both pigments are present in some freshwater fish and in *Xenopus*. *Rana temporaria* and *R. esculenta* have rhodopsin throughout their lives.

Whatever the pigments, they probably all act in the same way. Those from different species have different proteins and give different absorption spectra, but these tend to cluster round a few points, and a single species may have two or three pigments of closely similar spectra. A few other pigments have been claimed, especially iodopsin in the cones of the fowl.

J. 6.46. Vision in Invertebrates

Simple eyes, complete with lenses, have been described in many invertebrates, including dinoflagellates, medusae, and worms, but there seems to be no evidence that any of these can form an image, and in some animals, for example earthworms, the groups of cells classified as eyespots have been shown to have no sensitivity to light. The eyes of some lamellibranchs, for example *Pecten*, and of some gastropods, appear to be physically capable of forming images; that of the shore-snail *Littorina littorea* could do so only in air, but has muscular accommodation. The lateral ocelli of many insect larvae are sensitive to light and are largely used in orientation. The dorsal ocelli of adults, which have neurones rather similar to those of the compound eye (see below) are also sensitive to light, but their function is seldom clear. In cockroaches they are necessary for the establishment of the 24-hour rhythm (section G 9.34). Behavioural evidence that the eyes can form an image is available only for cephalopods and arthropods. The former have eyes superficially very similar to those of vertebrates, with a large lens which accommodates by moving forward when the eyeball is squeezed. *Octopus* can be trained to distinguish a rectangle with its long axis vertical from the same rectangle with its long axis horizontal, but it has difficulty in distinguishing a circle from a square of the same area. Such a deficiency is more likely to be due to the inadequacy of the brain than to inability to form the appropriate images. The only invertebrates for which colour vision has been proved are some insects, including bees, butterflies, many Diptera, a few beetles, and the bug *Notonecta*. These are, for the most part, species that visit flowers, and the advantage is obvious. Some differential sensitivity to wavelength has been claimed for a few spiders and crustaceans.

Behavioural responses suggest that cephalopods, the snail *Littorina littoralis*, and many arthropods can react to the plane of polarization, but some of the animals may have been reacting to differential light intensities caused by the varying reflections of light polarized in different planes. The ability is well established in *Carcinus*, in some insects (especially ants and bees), and in arachnids.

Wherever an examination has been made, eyes seem to depend on the presence of a pigment that is closely related to or indistinguishable from rhodopsin. In insects retinal is reduced to retinol, and there is some evidence that pigments with different absorption curves are present. Thanks to their large size, electrical recording from single cells of the retina is possible, and in the eyes of worker bees four types have been described, with maximum sensitivity to light of 340, 430, 460, and 530 nm, the first being in the ultra-violet. In the blowfly *Calliphora* cells with maxima at 470, 490, and 520 nm have been found, and all have a second maximum at 350 nm. These, in association with different pigments, would give a reasonable explanation of colour vision. In decapod crustaceans and cephalopods there are pigments, some at least of them related to vitamin A, which undergo changes in light. There are often layered arrangements, perhaps corresponding functionally to the discs of the vertebrate rods, in the receptor cells.

The adults of arthropods have faceted or compound eyes, each made up of a large number of ommatidia. Each of these has a refractive body or cone (which has no relation to the vertebrate cones), and below this a group of receptor cells, usually seven, making a retinula. The ommatidia generally radiate to give the whole eye the shape of a portion of a sphere, and are separated from one another by pigment. The compound eyes of crustaceans and insects are probably independent evolutions, in spite of their great similarity and the presence of similar eyes in Cambrion trilobites. Each retinula cell has a rhabdomere, which is an extension into the central axis of the ommatidium, and the rhabdomeres taken together constitute the rhabdome, in which is probably the photo-sensitive part, corresponding to the outer segment of the vertebrate rod. Each retinula cell has its own axon, but some of them synapse with those from neighbouring cones in the deeper layers of the eye.

The method of formation of the image by this complicated arrangement is by no means clear. According to the classical theory, in well-developed eyes such as those of bees and flies, when the pigment between the ommatidia is extended each receives light only from a narrow angle, and forms an erect image of a small part of an object placed in front of the eye. It can do this because of variations in the refractive index of the cone. The eye as a whole forms an erect apposition image, which is made up of the mosaic of the images of separate ommatidia. That such an image *can* be formed can be seen by looking at the inner surface of a compound eye through a microscope, and a photograph was taken in this way by Exner in 1891, but it is now known that the field of view of a single ommatidium in the bee and in three species of fly is about 20°, so that the mosaic must be blurred. A single receptor cell gives generator potentials when a point source of light is moved in front of its ommatidium for nearly this width, but they are at the maximum when the light is on the axis; presumably, therefore, the brain responds to this maximum stimulation.

The pigment is withdrawn, under the influence of hormones, in dim light, and the images formed by nearby ommatidia will then overlap, giving a blurred superposition image.

Electrical recording from the brain of a locust and of two blowflies shows that there are on and off discharges, and a continuous dark discharge which is inhibited by light. Experiments with a small moving light and with moving black and white stripes show that the minimum angular displacement that will produce a spike is about 0·3°, and occasionally as little as 0·1°. This is much less than can be accounted for on the mosaic theory. Direct observation under the microscope shows that there are not less than three images at different levels in the retina. Calculations show that all but the first are what are known as Fourier images, formed by diffraction, and a comparable set can be produced by an array, either of pin-holes or of lenses, of the same dimensions as the ommatidia. The third image in the locust gives a resolution of 0·35° and would be about adequate for the observed resolution of the eye. The details of this theory can explain how a bee can easily distinguish a hollow figure from a solid one of the same shape, but confuses a square with a triangle (Fig. 6.15). It is possible that the different receptor cells of the ommatidium have different functions. Six of the eight retinular cells of flies synapse with neighbouring cells and have a dark discharge inhibited by light; they respond predominantly by stepwise bursts of impulses to stepwise changes in light intensity, and could be used in detection of movement and recognition of pattern. The other two receptor cells do not synapse with other cells, have no dark discharge or off response, but produce impulses at a rate that varies with light intensity. They could

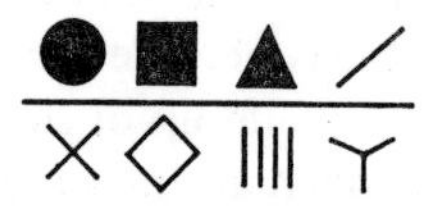

FIG. 6.15. A bee can distinguish any pattern in the top line from any in the bottom line, but confuses all those in each line. Redrawn from Hertz, *Z. vergl. Physiol.* **8** (1929).

produce the mosaic image. Differences between the fine structure of the rhabdomeres of these two cells might account for the perception of the plane of polarization.

V 6.47. The Electrical Sense

That man can feel the passage of an electric current, and that a current can stimulate nerves, have been known for centuries, but it is only within the last decade that sense organs and behaviour that depend on such currents have been discovered. They are, so far as is known, confined to fish, and are chiefly known in those that have electric organs (section V 7.5). Some of these are continually producing quite small electric discharges, which can only be detected with sensitive apparatus. They will surround the animal's body with an electrical field which is normally symmetrical; the near presence of any object of different electrical conductivity from water will distort this field, and it is this distortion to which the sense organs respond. Experiments have shown that the mormyrid eel, *Gymnarchus*, can distinguish between porous pots containing non-conducting material and those containing water; if its own current is fed back into the water it finds the electrodes and attacks them, while a bent copper wire placed in the water causes escape reactions.

It is probable that the receptors are various specializations of the lateral line system that are found in these fish. The ampullae of Lorenzini, part of the lateral line system of sharks and skates, are very sensitive to electric fields, and as the fish can detect those caused by the muscular contractions of other fish they may be used to detect prey.

7. Effectors

Effector systems are cells, collections of cells, or parts of cells (using the word in its old and broad sense) by which the organism acts in some way on the environment outside it, and it is by means of them only that an animal can respond to a stimulus or produce any change in its own position or in its own body. Put briefly, in the words of G. H. Parker: 'Effectors are the parts by which animals respond to changes in the world about them.' They are usually classified as follows:

1. cilia and flagella,
2. pseudopodia,
3. muscles,
4. glands,
5. electric organs (a specialization of 3),
6. luminescent organs (a specialization of 4),
7. nematocysts or urticators or cnidae,
8. chromatophores.

In addition, there are some minor types, such as the trichocysts and myonemes of some Protozoa. No animal has all of these, and most possess but three or four. Man, for instance, has only pseudopodia, cilia, glands, and muscles, and of these the last alone have any important connection with the outside world; all the rest serve only to maintain life inside the animal. Effectors are usually brought into action by stimulation either through the nervous system or by hormones, but there are cases of all classes, except electric organs, where the effector apparently reacts directly to a stimulus. Such an organ is called an independent effector, but its nature and reaction are not affected by the method by which it is activated.

Cilia, flagella, pseudopodia, and muscles are all concerned with movement, and their action is based on an alternation of contraction and extension. It seems that the basic chemistry of the change is always much the same, the contraction depending on the removal of phosphate ions from adenosine triphosphate or a related compound, with liberation of energy. The restorative processes, by which the energy is ultimately provided, vary widely.

MVJ 7.1. CILIA AND FLAGELLA

Cilia and flagella are contractile outgrowths from a cell, never withdrawn except at reproduction and encystment, and maintaining a shape and size that are nearly constant. The internal structure shows that there is no fundamental difference between cilia, which are generally many and short, and flagella, which are generally few and long. The distinction probably depends on the necessary mechanical differences between a short and a long structure, so that in the typical cilium the resultant force on the organelle when it moves is perpendicular to the long axis at the mid-point, while in the typical flagellum it is parallel to it. It seems likely that the rare examples of many long structures in a cell, as in the protozan *Multicilia*, are best considered as flagella. Flagella are found in the Protozoa, in sponges, in some of the endoderm cells of coelenterates, and in the sperms of most animals. Cilia are even more widespread: they form the locomotor organs of a whole class of the Protozoa, of ctenophores, of some planarians, and of rotifers, and of larvae of coelenterates, platyhelminths, annelids, molluscs, echinoderms, protochordates, and Amphibia. Internal ciliated cells are present in all Metazoa except Nematoda and Arthropoda; they serve to maintain currents of water for many purposes: for carrying food (Scyphozoa, lamellibranchs, protochordates), for carrying out excretory products (all groups with a true nephridium; section J 5.3), for carrying eggs (vertebrates), or for guiding sperms to the eggs (birds and reptiles). In the mammals cilia are present chiefly in the respiratory tract, where they help to keep out dust particles. They are found in all three germ-layers, and wherever they occur are very similar in structure and behaviour.

Both cilia and flagella are very thin structures, and they nearly always appear homogeneous under the light microscope. In electron micrographs a number of filaments can be seen running the length of the structure, and sections show that these are arranged with an axial pair and an outer ring of nine. This arrangement appears to be invariable in all the cilia and flagella that have been studied, whether they come from plants, Protozoa, or Metazoa. It is found also in sperm tails, and traces of it have been found in the sperms of insects, which are without functional cilia. The occurrence of a similar structure in sensory cells has been mentioned in the last chapter. A similar appearance in transverse section is shown also by many centrioles.

In most electron micrographs at high magnifications each filament can be seen to be hollow, and each outer one to be double, and this is probably the standard pattern. In some there is a third incomplete ring attached to each outer filament, giving an asymmetry to the

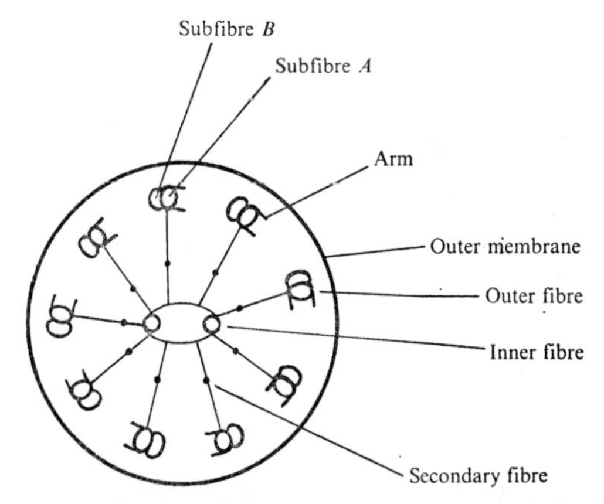

FIG. 7.1. Diagram of a transverse section of a typical cilium or flagellum, × *c.* 100 000. Based on an electron-micrograph in Yapp, *Borradaile's Manual of Zoology*, Oxford University Press (from Grimstone (1963) and others.

whole (Fig. 7.1). In some preparations each main filament appears to consist of 12–13 protofilaments, which may be hollow tubules, arranged in a ring. Various cross-fibres can be seen, and where the filaments run into the cell at their base they are modified and eventually join to form a continuous collar, which is the basal body (or basal granule, blepharoplast, or kinetosome). Especially in the Protozoa, deeper granules may be present, and these are connected to the flagella by fibres called rhizoplasts. There are a few minor variants on the general pattern; for example, in the lateral cilia of *Mytilus* and in mammalian sperm tails subfilament A appears to be solid. If the connection with the basal granules is broken, the cilia do not beat (Fig. 7.2). The electron microscope shows also that some flagella, such as that of *Euglena*, have a row of short spurs along one side.

Some large cilia, such as the laterofrontals of *Mytilus* (section G 2.21), have been shown to consist of several cilia beating together so that unless their coordination is upset they appear as one. In a similar way many undulating membranes consist of a line of cilia beating in order. When the rhythm is upset by placing a needle through the apparent membrane, the individual cilia can be seen.

The rate of movement of cilia appears high, but this is because, while velocity has dimensions of one in length and minus one in time,

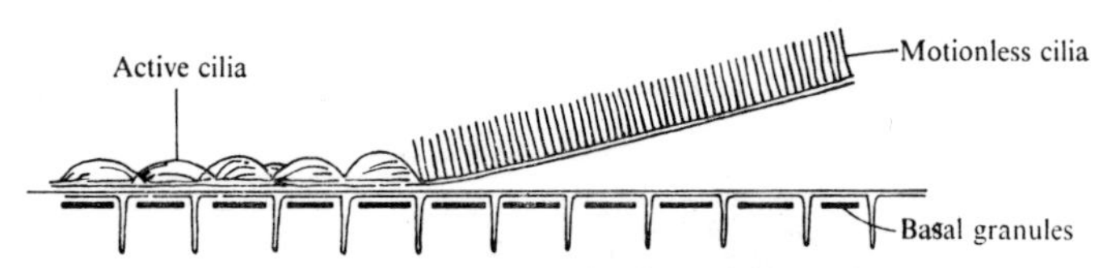

FIG. 7.2. Diagram of the lateral epithelium of *Mytilus* gill. The cilia are active as long as they are in organic communication with the cells. From Gray, *Texbook of experimental cytology*, Cambridge University Press (1931).

the microscope magnifies length but leaves time unaffected; it therefore magnifies velocity to the same extent as it does length. A cilium 10 μm long, beating at the normal rate of ten times a second and moving through an arc of π rad, has in its effective beat a velocity of only 1·5 mm s^{-1} at the tip. The speeds produced by cilia correspond to this; *Paramecium* moves at speeds of the order of 1 mm s^{-1}, and particles placed on ciliated epithelium never move at more than 3 or 4 mm s^{-1}. *Euglena* moves at about 0·2 mm s^{-1}. On account of their slow speed, and because they can only be present on surfaces, which bear a progressively smaller proportion to weight as size increases, cilia can only be used for efficient locomotion if the animal is small and of low density. Thus *Volvox*, with a radius of 0·5 mm and a specific gravity of 1·01, has a maximum velocity at 15 °C of 1 mm s^{-1}. At this speed practically all the external work is done against viscosity, and if all the cilia were to stop beating at once the kinetic energy of the animal would carry it only 0·05 mm—one-tenth of its own radius. In the same way full speed is attained almost at once.

The movement of cilia can be studied in three ways: they may be slowed down with drugs such as veratrin; they may be observed with a stroboscope; or they may be photographed with a cine-camera. It is obviously of doubtful legitimacy to argue from a drugged cilium to the normal, so that the first method, though useful, must be supported by others. The principle of the stroboscope is as follows. If a moving object is seen through a rapidly opening and closing diaphragm, it is observed in the positions which it occupies at times 0, δt, $2\delta t$, $3\delta t$, and so on, where δt is the interval between successive openings of the shutter. If δt is small enough the eye, owing to persistence of the retinal images, will appear to see continuous movement. If the object is not moving laterally, but is going through a cyclic movement like that of a rotating wheel, and if it takes a time t for one complete cycle, it is in the same position at $t+\delta t$ as it was at δt, the same at $t+2\delta t$ as at $2\delta t$, and more generally in the same position at $t+m\,\delta t$ as at $m\,\delta t$. For successive revolutions the same argument holds, and the wheel is in the same position at $2t+m\,\delta t$ as at $m\,\delta t$, and more generally in the same position at $nt+m\,\delta t$ as at $m\,\delta t$ (where n is a whole number). If, then, the diaphragm is arranged to open at intervals of $t+\delta t$, the object will be seen at times 0, $t+\delta t$, $2t+2\delta t$, and so on. It will therefore be observed in exactly the positions that it would have occupied at times 0, δt, $2\delta t$, etc., and so will appear to be in continuous movement at much less than its normal speed: how much less will depend on the relative values of t and δt. For investigating ciliary movement the cilium is viewed through a slit in a rotating disc, the rate of revolution of which is adjusted so that the cilium appears to stand still: at this speed the disc rotates once while the cilium goes through one or possibly more exact cycles. The speed of the disc is decreased slightly and the conditions discussed above become operative. The cilium is seen to move slowly. The speed of the disc must be such as to avoid flicker.

The third method consists simply in combining a cine-camera with a microscope. It is sometimes convenient to add a stroboscope to this arrangement.

All the methods of investigation agree in showing that the method of beat is that shown in Figs. 7.3 and 7.4. There is a rapid effective forward movement which takes (in frog epithelium) about 0·02 s, and a slower limp recovery which takes about 0·1 s. During the former the cilium is rigid and can be prevented from moving by a needle placed in its way, but during the recovery stroke it is flexible and simply passes underneath any obstruction placed in its path. Either it changes its state from one phase to the next, or it is capable of bending in one direction only, that is in the same direction as its effective stroke. A mechanical model of this sort can be made by gluing a row of small wooden cubes on to a piece of linen. If the cubes are in contact with one another bending is impossible away from the linen-backed side, but it is easy in the opposite direction.

Ciliated surfaces nearly always show metachronal rhythm, that is, each cilium is slightly out of phase with its neighbour, which is out of phase with the third cilium by the same amount, and so on. The result is that waves appear to pass over the surface; they may be with the direction of the beat (when each cilium is a little later than the

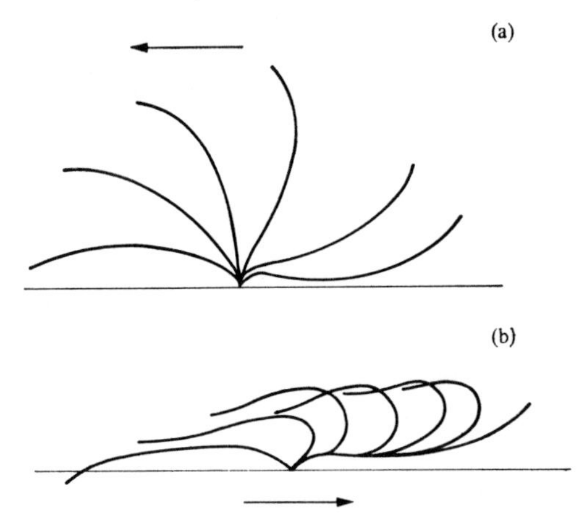

FIG. 7.3. Diagram illustrating thc form of terminal cilia of *Mytilus* during (*a*) the effective, and (*b*) the recovery, beat. From Gray, *Proc. R. Soc.* **B93** (1922).

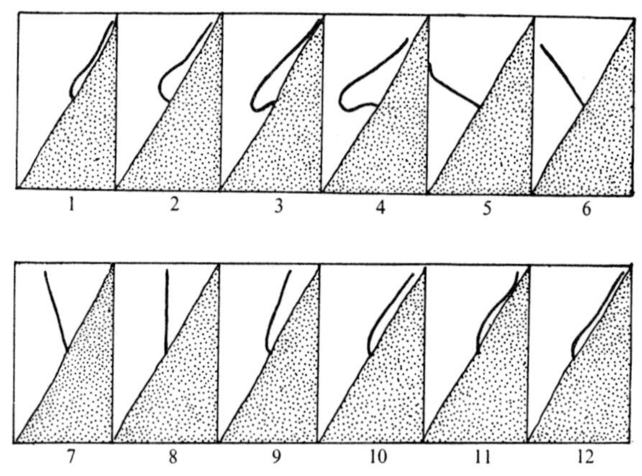

FIG. 7.4. The beat of a laterofrontal cilium of *Mytilus*. 1–5, recovery stroke, 6–12, effective stroke; interval 0·05 s. Drawn from a film by Sir James Gray.

one just behind it), against it, or across it, but the pattern is always the same for one cell or tissue. The mechanism of the rhythm is unknown; it could be by electrical control, or could be purely mechanical.

Flagella, on account of their length, are mechanically more complicated. They beat in a number of different ways. Some, such as the anterior flagellum of *Trichomonas*, beat more or less like a long cilium. Most have a complex beat, in which more than one curve is visible in the structure at one time, either in one plane or in a helix; the flagellum is then usually trailing, and much of the locomotion of the organism depends on the rotation of the whole cell, which is produced as a reaction to the contraction of the flagellum. The beating of a typical sperm tail is an almost perfect sine curve.

There can be little doubt that the filaments play an important part in the contraction, but whether by co-ordination, their own contraction or both is unknown. A mutant of *Chlamydomonas* which lacks the central pair of fibrils is non-motile.

Cilia and flagella are remarkably free from control by their owner. In the majority of cases they never reverse, and if a piece of epithelium from the roof of the mouth of the frog be cut out, turned through 180°, and regrafted in place, the cilia go on beating in their old way, although they now drive a current down the animal's throat. Reversal does, however, take place in the avoiding reaction of ciliates, in nemertines, and in the food currents of sea anemones such as *Metridium*.

Nervous excitation and inhibition of cilia have been described only rarely, chiefly in a few molluscs and protochordates, and even in these groups the majority are not under nervous control. The transmitter substances serotonin, acetylcholine, and adrenaline have been described as increasing the rate of various cilia, but there is no evidence that they act in this way in life.

The rate of beat of cilia is closely dependent on temperature, increasing up to a certain point (28 °C in *Mytilus*) and then falling off until death ensues. In absence of oxygen cilia will go on beating for 45 min and so can put up an oxygen debt. Chemical investigations have shown that various ciliated tissues, protozoan, invertebrate, and vertebrate, depend on adenosine triphosphate, and in molluscs, echinoderms, and vertebrates glycolysis, the tricarboxylic acid cycle and cytochrome have been demonstrated. One may therefore assume that immediate energy for the beat is supplied by active phosphate, and that this is quickly restored by the oxidation of carbohydrate. *Mytilus* gill tissue has a respiratory quotient of 0·8, suggesting that it uses largely protein. It is tempting to assume that there is some sort of reversible contraction of proteins, perhaps in the fibrils, comparable to that of muscle, and some confirmation of this comes from fractionation and chemical analysis of the cilia of the protozoan *Tetrahymena pyriformis*. Most of the adenosine triphosphate is present in the inner fibres and the arms of the outer nine, which appear to consist of a protein called 30S dynein, of molecular weight *c*. 5 400 000 and high phosphatase activity.

MVJ 7.2. PSEUDOPODIA

It is difficult to give an exact definition of pseudopodia, but the word is generally taken to mean those projections from a single cell which are of a temporary character and of not very definite shape. Some of the more extreme types exist for quite a long time and have something of the nature of specialized organelles, and all of them are limited in shape by the cell to which they belong. Pseudopodia are found in the Protozoa, in the cells lining the gut in some of the coelenterates, in the wandering cells (phagocytes) of the body fluids of most coelomate animals, and in similar cells in tissue culture. Those of the Metazoa are all of the same general type, short and blunt, but those of the Protozoa can be classified into four groups. Lobopodia are rounded blunt structures, the typical locomotor organs of *Amoeba*. Filopodia are fine and pointed, and are found in such amoeboid species as *Euglypha*. Rhizopodia are distinguished by the fact that they anastomose, and are characteristic of the Foraminifera. Axopodia have an axial filament, and are therefore at least semi-permanent in character: probably no satisfactory definition of pseudopodia could be made to include these; they are found in the Heliozoa. There are no sharp dividing lines between the classes; pseudopodia may be made more pointed by increasing the hydrogen-ion concentration or the osmotic pressure of the medium, and blunter by the reverse processes. It is possible that pseudopodia and flagella are fundamentally the same structures. The axial filaments of axopodia suggest this, and it is supported by the case of *Mastigamoeba* (= *Naegleria*) *gruberi*, whose organelles change from one to the other according to the external circumstances. In moribund specimens of the flagellate *Trichomonas* the undulating membrane appears to be replaced by pseudopodium-like projections which move rapidly over the surface. Pseudopodia are often assumed to be primitive, largely because they appear to be unspecialized and because it has been traditional to regard *Amoeba* as the lowliest of animals, but pseudopodial formation is certainly not simple, and most zoologists now consider some of the Mastigophora to be the most primitive protozoa.

The cell may bear one or many pseudopodia, and they are used for two chief purposes: for ingesting foreign particles and for locomotion. The particles ingested may be food, as in the Rhizopoda, coelenterates, and lamellibranchs; excretory matter in annelids; Bacteria in vertebrates; and tissues of the body that are digested and carried elsewhere during the metamorphosis of Amphibia. The different methods of ingestion of food particles in Rhizopoda are mentioned in section G 2.22. When pseudopodia are used for locomotion usually only a single cell is moved, and the expression 'amoeboid movement' may then be used to describe what happens; occasionally, however, as in *Hydra* and some other of the Hydrozoa, a larger animal is slowly carried by the pseudopodia of the cells of its base. Amoeboid movement has been studied chiefly in the Protozoa, and may be classified into four groups.

The first is that found in the Mycetozoa, which appear merely to flow over a surface as a liquid film of protoplasm, no well-defined pseudopodia being formed. The second is that exemplified by the normal movement of *Amoeba*, which has been much studied. The accounts differ so much that it is possible there is more than one type of movement. To study the movement of the surface of the animal small particles have been dropped on to the upper surface, and then watched. They have been seen to behave in three different ways (Fig. 7.5): sometimes they maintain their original position relative to the ground, which means that they gradually move to the posterior end of the amoeba; sometimes they maintain their position on the animal, so moving forward at the same speed; and lastly they sometimes move slowly to the front end of the pseudopodium. In this last case they may collect at the front end or they may return along the ventral surface, so that the amoeba has been described as 'rolling like a bag of oats'. The speed of *Amoeba* is of the order of $1\ \mu m\ s^{-1}$. When an amoeba divides, pseudopodia are formed at two opposite poles, and the two daughter cells are actively pulled apart, so that a waist is formed which narrows and finally separates. The point at which the separation takes place is always the posterior end of the animal, or uroid as it is sometimes called. Cells in tissue culture sometimes divide in the same way.

The third type is that in which the tips of pseudopodia stick to a surface and then the organelle contracts. If an *Amoeba* is suspended in water, pseudopodia are put out in all directions, and when one of them touches anything solid it attaches itself and the whole mass of protoplasm is drawn into it. Sometimes *Amoeba* moves by a looping

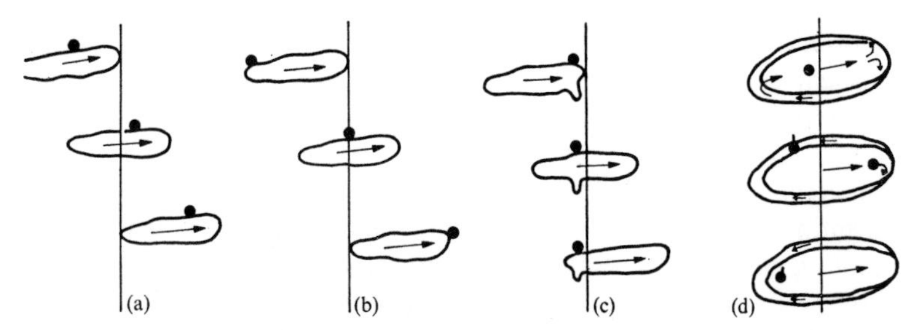

FIG. 7.5. Motion of *Amoeba* relative to the substratum and to an attached particle. In (*a*) the particle is moving at the same speed as the organism; in (*b*) the particle is moving forward over the surface of the amoeba, as it does in *A. discoides*; in (*c*) the particle is stationary relative to the substratum, and is moving backward relative to the surface of the organism; this occurs when the particles are heavy; (*d*), movement of ectoplasm in an amoeba suspended in a jelly. The vertical lines represent a fixed point in the environment. From Gray, *Textbook of experimental cytology*, Cambridge University Press (31). After Schaeffer, *Amoeboid movement*, Princeton University Press (1920).

movement; one pseudopodium is put out, and after it has stuck to a surface the whole animal moves after it.

The regular locomotion of some genera, such as *Difflugia* and *Polystomella*, is similar; when the animal has been drawn up somewhat, the same pseudopodia are extended again.

The fourth type is the Catherine-wheeling of the Heliozoa. Successive pseudopodia are put out, attached, and contracted, so that the animal rolls on a succession of spokes that it makes for itself. This is the quickest type of amoeboid movement, giving a speed about ten times as fast as that of *Amoeba*.

Any theory of amoeboid movement must explain all these observations and more, and it can be confidently said that no theory yet put forward will do so. The popular hypothesis of the sixties supposes that the movement depends on contraction of protein molecules, especially of the outer gelated part of the cytoplasm, or plasmagel (Fig. 7.6). According to some authors this pulls the inner plasmasol, which is a non-Newtonian liquid (section G 1.12) with perhaps a more solid core. According to others the contraction is at the uroid, and the plasmasol is simply squeezed forward. According to most observers there is change of liquid to gel at the anterior end, and the reverse in the uroid. Adenosine triphosphate injected at the uroid increases the speed of movement, at the anterior end it causes reversal. *Amoeba* certainly uses oxygen, and is cyanide-sensitive, so

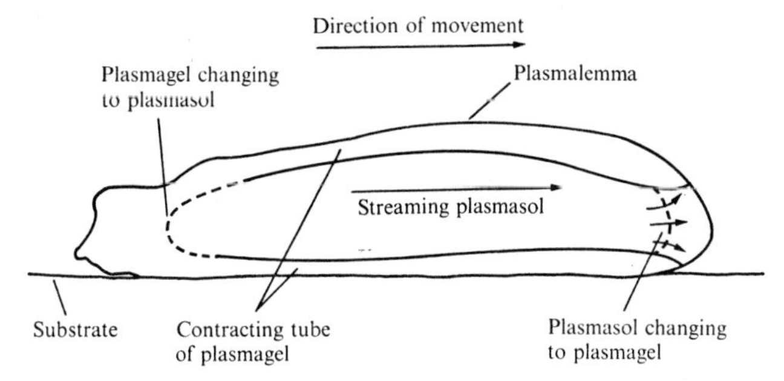

FIG. 7.6. Locomotion in a limax-type *Amoeba*. Slightly modified from Pantin, *J. mar. biol. Ass. U.K.* **13** (1923)

one may assume, as usual, the provision of energy by active phosphate, and an aerobic restorative process.

7.3. MUSCLE

MV 7.31. Vertebrate Skeletal Muscle

All types of muscle have great similarity, but it is easiest for the student to start with the type that is best known. This is vertebrate skeletal muscle, also called striped or voluntary muscle. It consists of spindle-shaped fibres, made up of a sheath, the sarcolemma, enclosing a liquid sarcoplasm in which are several nuclei and a number of myofibrils, which are the contractile elements.

The classical material for showing the contraction of muscle is the gastrocnemius (or calf) muscle of a frog, which is easily obtained (together with a piece of the femur on which it has its origin, its tendon, and a long length of sciatic nerve) from a pithed animal, and will remain active in Ringer's solution for an hour or so. The bone is pinned down, and if an electric shock is applied to the nerve (or even if it is pinched) the muscle can be seen to contract. It is usual to fix the tendon to a lever and make a trace on a smoked drum, apparatus that tells one more about the elasticity of string and the inertia of levers than about the properties of muscle. These can be demonstrated visually just as well and much more cheaply by running a string from the tendon over a pulley and attaching a weight that can be varied in size.

It is usual to distinguish between an isotonic contraction in which the muscle lifts only a light weight and contracts to its normal maximum, and an isometric contraction, in which the muscle is tied to

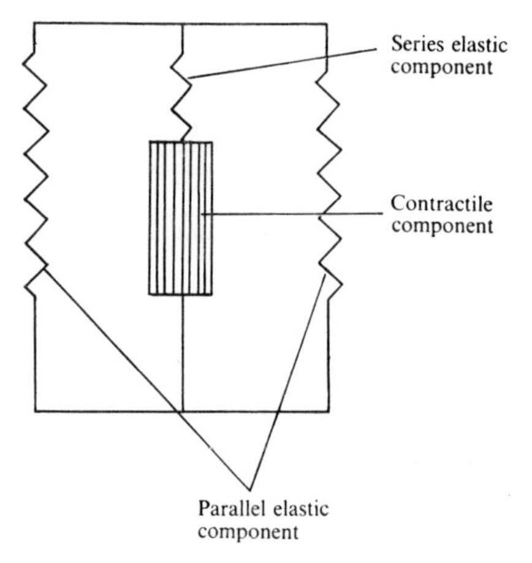

FIG. 7.7. Diagram to show the relationships of the elastic and contractile components of muscle. Contraction of the contractile component stretches the series elastic component but relaxes the parallel elastic component.

a heavy weight or a strong spring, and is supposed not to decrease in length. The distinction, though it represents two extremes, is a false one. The so-called isometrically contracting muscle visibly thickens; if the load is then suddenly reduced, there is a quick over-all shortening, followed by further isotonic contraction. The initial shortening is independent of stimulation and seems not to be metabolic; it is interpreted as being the adjustment of an inert part of the muscle which was under tension in the isometric contraction and is called the series-elastic component. (It is obvious that in life the slight elasticity of the tendon is acting in series in the same way.) The sarcolemma is also slightly elastic but acts in parallel with the fibres (Fig. 7.7).

A trace obtained with a simple lever-and-drum apparatus is shown in Fig. 7.8. The muscle has undergone a twitch—a contraction followed by relaxation and return to the resting length. More precise instruments show that at 0 °C the mechanical properties of the muscle begin to change about 3 ms after the stimulus is received, tension begins at about 12 ms, shortening is at full speed by 20 ms, is then constant for about 40 ms, after which there is a rapid return to the resting state. At higher temperatures these times are reduced.

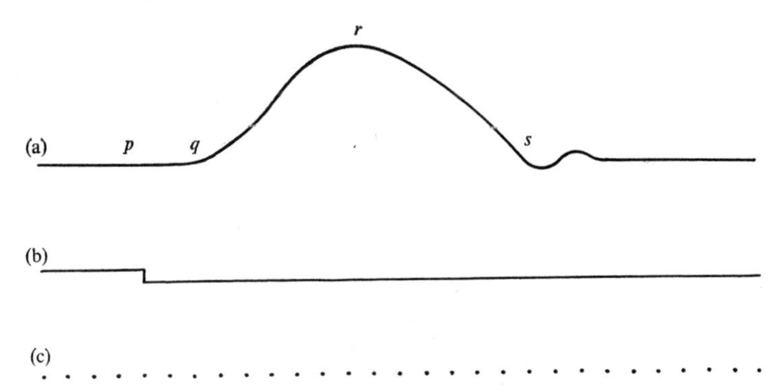

FIG. 7.8. Tracings of an isotonic contraction of gastrocnemius muscle of toad. (*a*) Curve of contraction of muscle; *pq*, latent period; *qr*, contraction; *rs*, relaxation. The deflections to the right of *s* are due to the imperfections of the recording apparatus. (*b*) Shows time of application of the stimulus. (*c*) Shows intervals of 0·02 s.

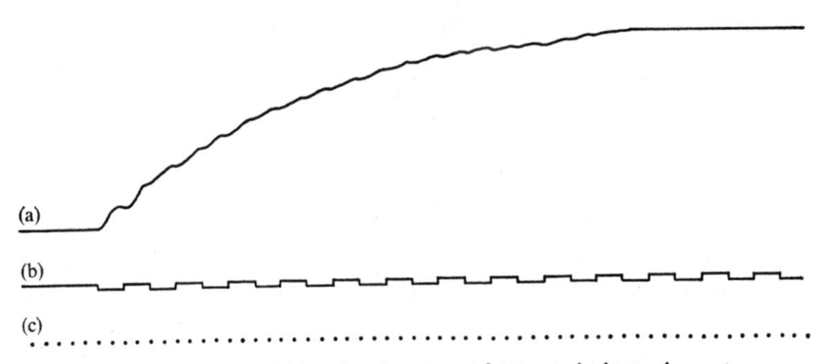

FIG. 7.9. Tracings showing development of tetanus in isotonic contraction of gastrocnemius muscle of toad; five shocks a second: (*a*) and (*b*) as in Fig. 7.7, (*c*) intervals of 0·1 s.

If two shocks are given very close together, the second has no effect, so that there is a refractory period, but at rather greater intervals one twitch is superimposed on another, so that with a series of quick shocks the muscle either goes into an oscillatory state called clonus, or remains permanently contracted in tetanus (Fig. 7.9). The strength of contraction of a muscle depends largely on the number of fibres contracting at any one moment.

The contraction of muscle is mediated through the nerves. A single motor nerve fibre branches and supplies a number of muscle fibres, which may be only a few or may be several hundred; perhaps 250 is

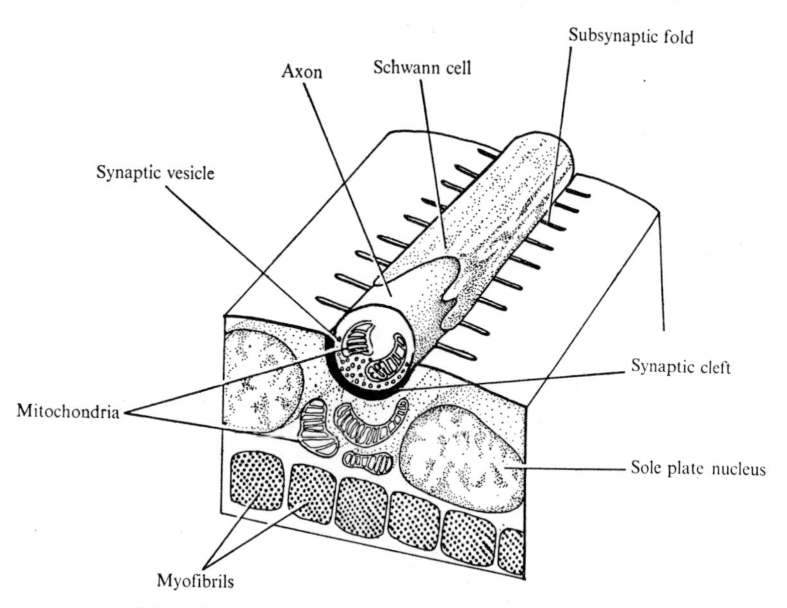

FIG. 7.10. Diagram of part of a nerve-muscle junction, ×20 000.

a rough average. A neurone and its muscle fibres are called a motor unit; the fibres of a unit are not contiguous but are scattered throughout the muscle, so that a light stimulus, which has started an impulse in only one neurone, will cause a slight but generalized contraction of the muscle. The nerve-muscle junction or motor end-plate has the same general type of structure as a synapse (Fig. 7.10). The myelin sheath of the nerve fibre ends, and the axon, still covered above by its Schwann cell, branches and runs in grooves on the surface of the muscle fibre. The relationship of nerve and muscle in these grooves is that of a synapse, the presynaptic membrane of the nerve being separated by a synaptic cleft from the subsynaptic membrane of the muscle. In the axon near the junction are many synaptic vesicles, and in both axon and muscle-fibre are many mitochondria. The subsynaptic membrane has subsynaptic folds in it, at right angles to the long axis of the groove. The outer surface of the myofibrils is about 0·3 μm below the junction.

If a microelectrode is placed in the muscle fibre, it shows small end-plate potentials near the junction, suggesting that there is some constant activity. When an impulse arrives these potentials increase to about 100 times their resting value, spread over the surface of the muscle, and cause a contraction. This can be imitated by applying a drop of acetylcholine to the outer surface of the sarcolemma. There is a potential, depending on the amount of acetylcholine, and when it reaches a high enough value a spike discharge is propagated over the surface of the fibre, which contracts. This is very similar to what happens in the stimulation of nerve, and may be explained in the same way in terms of permeability of the membrane. Acetylcholine is presumably slowly leaking out from the presynaptic membrane, so causing the resting potentials, and on arrival of an impulse enough is liberated to depolarize the subsynaptic membrane and cause a propagated discharge. If acetylcholine is injected inside the muscle fibre it gives only a graded response and there is no contraction.

While the energy relations of nerve are accidental, those of muscle are its most important characters. The function of a muscle is to do external work, and to do so it must convert chemical energy into mechanical; any heat that appears at the same time is, like that of a machine, a useless by-product, and lowers the efficiency of the system. All the energy comes ultimately from the oxidation of organic material, usually carbohydrate, but a muscle can contract without this, and, so far as is known, the only chemical reaction essential to muscular contraction is the hydrolysis of adenosine triphosphate.

Electron micrographs of muscle fibres and experiments with X-radiographs of isolated fibrils have enabled the chemistry and the contraction to be associated in a reasonably satisfactory way. A single striated fibre consists of alternate light and dark bands, which are further subdivided as shown in Fig. 7.11 (*a*). The dark A bands are anisotropic (doubly refracting) and the I bands isotropic, that is their refraction does not depend on the plane of polarization of the light transmitted through them. Each myofibril consists of two types of rod, one much thicker than the other, both embedded in sarcoplasm and arranged in hexagons in such a way that every thick rod is surrounded by six other thick ones and six thin ones, and every thin one by three other thin ones and three thick ones (Fig. 7.11 (*c*)). The thick rods consist of myosin, a protein of unit weight about 425 000, and the thin ones of actin, a protein of unit weight 74 000. Under certain conditions these two can combine to form actomyosin.

The rods are not continuous, but overlap as shown in Fig. 7.11 (*b*), so that the I bands have actin rods only, the H bands myosin rods only, and the rest of the A bands have both. In contraction, down to

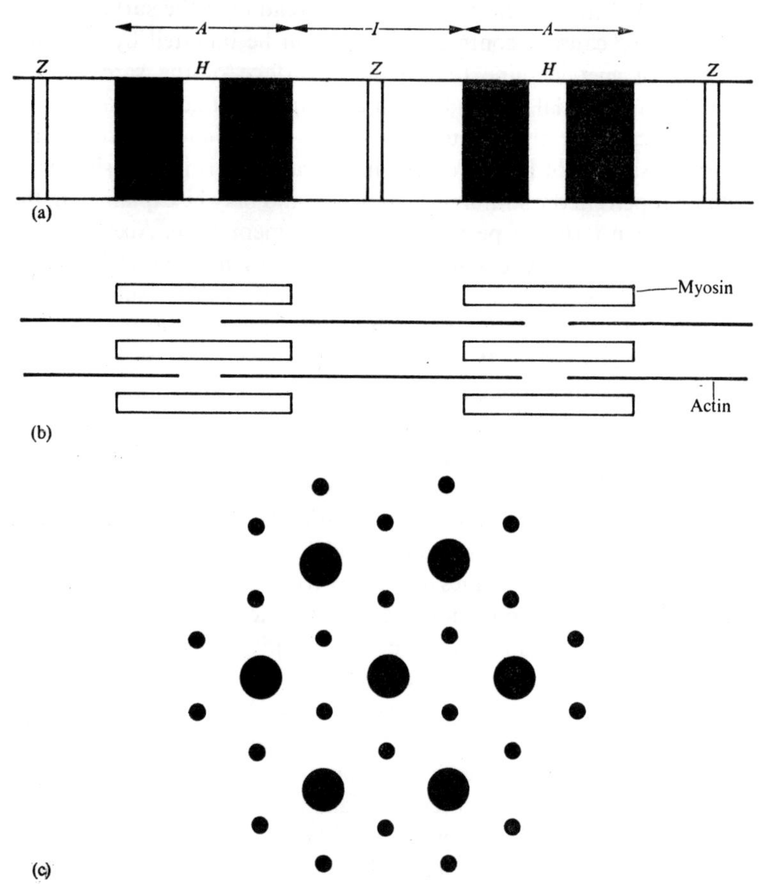

FIG. 7.11. Diagrams of the structure of voluntary muscle, based on A. F. Huxley: (*a*) longitudinal section, showing A and I bands, and H and Z lines; (*b*) longitudinal section (electron microscope), showing actin and myosin filaments corresponding to the bands and lines in A; (*c*) transverse section (electron microscope) showing the large filaments of myosin and small ones of actin.

85 per cent of resting length, both H and I bands shorten; at this length the H bands have disappeared. What has happened is that the actin rods have slid into the spaces between the myosin rods.

Further contraction is accompanied by further shortening of the I bands, during which the actin filaments fold or coil, until at 65 per cent of the resting length the I bands are obliterated. If there is further contraction still, the A bands shorten, and the myosin rods coil. The shortening probably takes place by a step-wise formation of actomyosin through cross-linkages, and the folding or crumpling is presumably caused by some sort of contraction and folding of the protein molecules, although the X-ray pictures have not given any certain evidence on this. The sarcoplasm has an anastomosing system of fine tubules arranged in a repeating pattern.

It seems that myosin is itself the esterase that removes the phosphate ion from adenosine triphosphate, and so liberates the energy that enables the actin and myosin to slide past one another and do work. Why this does not happen in the resting muscle, which has plenty of adenosine triphosphate in its sarcoplasm, is unknown. If this substance is added to isolated myofibrils they contract violently, so that an inhibitor must be present in the intact muscle, and an extract of sarcoplasm has this property. Its de-inhibition would seem to be the result of the changed permeability that occurs on stimulation, and since both calcium and magnesium ions are necessary for contraction it is possible that the entry of the former unbinds magnesium ions from association with the inhibitor so that they can mediate the hydrolysis of the ester.

Almost as soon as any adenosine triphosphate has broken down, it is restored by energy and phosphate derived from the breakdown of creatine phosphate, one of a class of compounds called phosphagens. This is itself restored by the energy of glycolysis (section M 4.12), which begins after only a few twitches. Even in the presence of oxygen, lactic acid is formed and an oxygen debt is put up; the rebuilding of glycogen from lactate almost certainly cannot go on in muscle. An outline of these complicated chemical changes is shown in Schema 9.

Except in a few details the above applies, so far as is known, to the simple muscle twitch in all vertebrates. In many of the fibres of the breast muscles of some birds, such as the pigeon, there is little glycogen, and fat appears to be the source of energy. Many fibres in the gastrocnemius of the fowl have several end-plates derived from different neurones, and the cholinesterases of birds (which destroy the acetylcholine and so prevent repeated contractions after one stimulus) are different from those of mammals; it may be that birds, with the peculiar requirements of flight, have developed features not possessed by other vertebrates.

In the myotomes of sharks and some other fish there are two types of muscle. Red fibres make up about 18 per cent of the total, and are

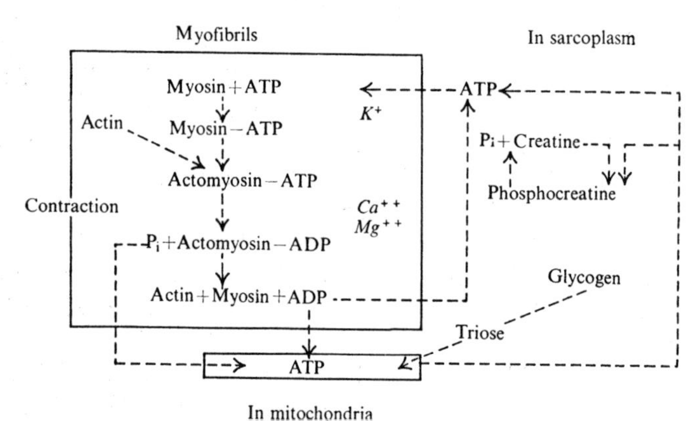

SCHEMA 9. The chemical changes in muscle. Start at the top left-hand corner. Some ATP is restored by the breakdown of phosphocreatine (top right), but a continued supply comes from glycolysis (bottom right).

alone active in slow swimming; they contain both glycogen and fat, but only the latter falls after prolonged activity. White fibres, with much less fat, are used in vigorous swimming, and lose glycogen by anaerobic glycolysis.

The simple scheme needs extension to cover the normal working of muscles. The end-plates described above are supplied by large axons conducting at 8–40 $m\,s^{-1}$ (in frogs) or 70–120 $m\,s^{-1}$ (in mammals), but there are also, in some muscles, simple nerve-muscle junctions, derived from smaller axons with a conducting speed of 2–8 $m\,s^{-1}$ which end in arborizations without the special type of synapse of the end-plate. Stimulation of these causes low potentials in the muscle, which are not propagated; several of them are needed for contraction, which is slow and prolonged instead of short and rapid like the normal twitch. Muscles of this sort have been demonstrated in the hag-fish, frog, reptiles, birds, and mammals, and are used where prolonged contraction is required, as in the maintenance of posture and in the amplexus of the frog.

Posture in general is maintained through the proprioceptor called the muscle spindle. This is a modified muscle fibre with a peculiar and complicated innervation, which may be represented diagrammatically by Fig. 7.12. It was shown long ago that the rate of production of electrical impulses in this depends on the force with which it is stretched (Fig. 7.13). In life such stretching will be brought about by the contraction of the antagonistic muscle. The result of the impulses is a reflex contraction of the ordinary fibres of the spindle's own muscle, which relieves the tension on the ends of the spindle and so stops the reflex. But the tension on adjacent fibres and their stretch receptors will be increased, and so they will in their turn contract. In this way some fibres of a muscle are always in contraction, and tone or tonus is maintained. The muscle spindle also contains contractile elements, the intrafusal muscle-fibres, in series with the sensory region. They are innervated by small motor nerve-fibres, separate from the large motor fibres that run to the extrafusal muscle-fibres that make up the rest of the working part of the muscle. An increase in the activity of the small motor fibres leads to a greater contraction of the intrafusal fibres, and so stretches the sensory region. The stretch reflex then results in contraction of the extrafusal fibres, and this reduces again the tension on the spindle. This reduction will not, however, be complete, and the tonus of the muscle will depend on the level of activity in the small nerve-fibres. Thus accurately graded and maintained contractions can result reflexly from activity of the small fibres without the co-operation of an antagonist, and direct stimulation of the large fibres by the central nervous system is found only in sudden and strong contractions.

Under normal circumstances stronger contractions of a muscle are brought about both by increased contraction of each fibre and by bringing more of them into use at one time. But it seems unlikely that true tetanus ever develops, since the rate of arrival of impulses is not fast enough. Fatigue, the phenomenon of a muscle failing to react to a normally adequate stimulus, could be due either to failure of the stimulating system, or to failure of the muscle itself to respond to the acetylcholine. It seems that the latter state can be reached in man by repeated voluntary movements of a muscle.

MV 7.32. Vertebrate Smooth Muscle

Plain, smooth, unstriated or involuntary muscle occurs chiefly in the internal organs such as the gut, arteries, bladder, and urinary and genital ducts. It has certain histological differences from skeletal muscle, but they are not clearly connected with the physiological

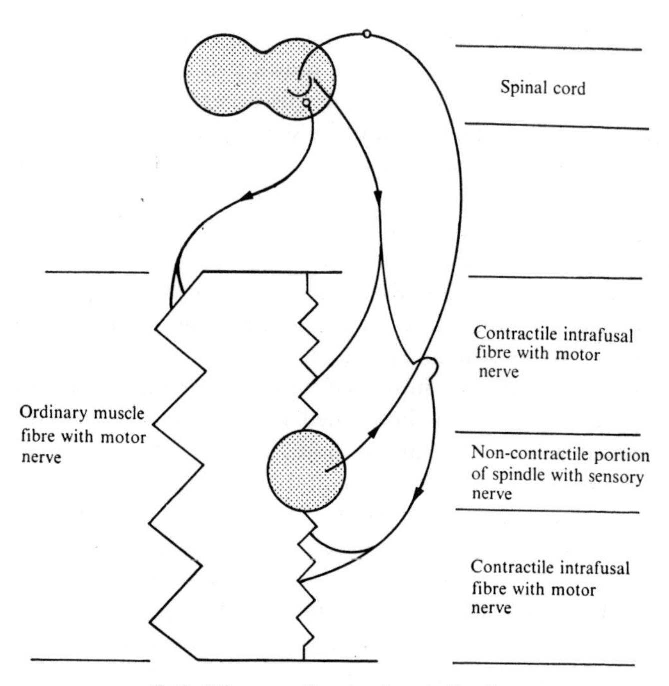

FIG. 7.12. Diagram of a muscle-spindle. See text.

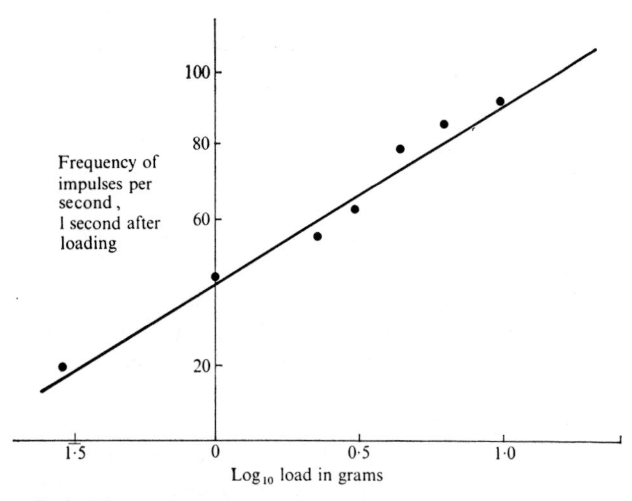

FIG. 7.13. The relationship between the frequency of impulses from a single muscle spindle of a frog and the applied load. Redrawn from Matthews, *J. Physiol.* **71** (1931).

differences. The chief of these is that smooth muscle possesses a much greater intrinsic irritability of its own, and with this is correlated its double innervation, one set of nerve fibres inciting the tissue to

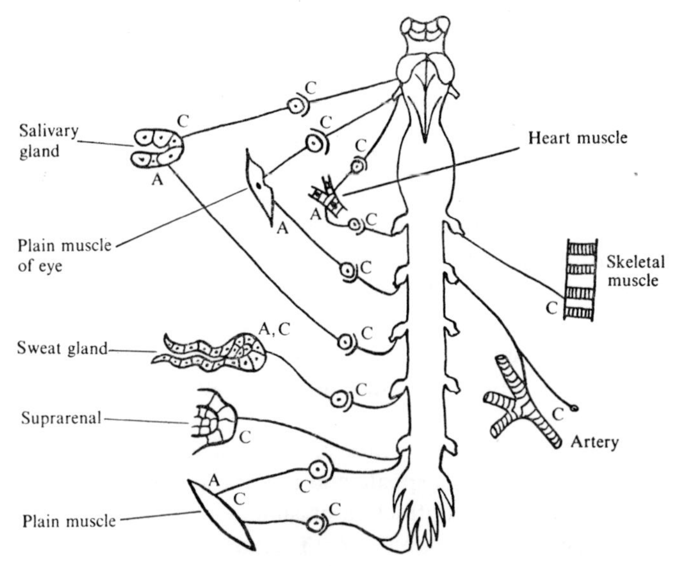

FIG. 7.14. Diagram showing points in the mammalian nervous system where there is evidence of cholinergic transmission (*C*) and adrenergic transmission (*A*). On the left, sympathetic and parasympathetic fibres and their ganglia; on the right a motor fibre with a peripheral axon branch. Modified from Dale, *Br. med. J.* (1934).

contract, and the other inhibiting it. Between these two types of stimulus smooth muscle carries out the details of its movements automatically, using this word in its strict sense. The motor nerve-fibres that supply smooth muscle are all autonomic (that is their cell bodies are all outside the central nervous system). The nerve-muscle junctions have the form of a fine plexus of non-myelinated fibres. In general, those that belong to the parasympathetic system secrete acetylcholine, which causes depolarization and contraction of the muscle fibre just as it does in striped muscle, while those of the sympathetic system produce noradrenaline, which causes hyper-polarization. There are, however, exceptions to this simple scheme, and noradrenaline causes depolarization and contraction of some muscles, and in others relaxation followed by contraction. The distribution of cholinergic and adrenergic nerves is shown in Fig. 7.14.

For experimental purposes nerve-muscle preparations may be made with smooth muscle just as with skeletal muscle. The contraction curves for the two are very similar, but the time relations of smooth muscle are very much slower. The apparent latent period, during which the ordinary arrangement of levers shows no change in the muscle, is from 0·2 to 2 s, and even the electrical changes do not start till 40 ms after the stimulation. Contraction itself may last for minutes, and the refractory period is long. The phenomena of summation of stimuli and summation of contractions are shown.

When a stimulus is applied directly to one point of smooth muscle, the contraction may spread over the whole tissue, a result that is very different from the highly individual contractions of fibres of skeletal muscle. Whether the conduction is carried out through the medium of protoplasmic continuity between the fibres, or whether it is merely the nerve plexus that is used, is unknown, but since the amnion of the chick contracts although it has no nerve supply, the former is possible.

The chemistry of smooth muscle, like that of striped muscle, is based on adenosine triphosphate and the breakdown of glycogen to lactic acid. The actual contraction presumably depends on a change of state in the actomyosin, but there is clearly not the neat arrangement of molecules that has been shown to exist in striped muscle. Electron microscopy generally shows some internal structure, so that a sliding mechanism similar to that in striated muscle is possible, but the details are unknown.

MV 7.33. Vertebrate Cardiac Muscle

Like the other types, cardiac muscle is made up histologically of fibres. Here they are cross-striped, and each consists of several cells placed end to end. Cardiac muscle differs, however, from the other two types in that its fibres are connected to one another by branches; this fact means that the fibre is not the effective unit, and accounts for many of the physiological peculiarities of heart muscle. The all-or-nothing rule applies to the heart as a whole, so that if any part of the intact organ contracts, the whole does.

For experimental purposes a complete heart, or a strip from some part of it such as the ventricle, may be used. The curve for contraction is similar in general form to that of skeletal muscle, but the reaction is slower. The response of the mammalian ventricle lasts 0·3 s, that of the frog 0·5 s. Rise of temperature quickens the response, a correct hydrogen-ion concentration must be maintained, and the heart is very sensitive to the other ions present in the medium. The refractory period is exceptionally long. The absolute refractory period lasts as long as the contraction, and this is followed by a period of subnormal sensitivity while the muscle relaxes. There is a short period of supernormal sensitivity before the normal state is reached. The effect of the long refractory period is that heart muscle cannot be tetanized, which is obviously of great value physiologically. The phenomenon of summation of stimulus is shown, repeated subminimal shocks finally causing a contraction.

The strength of the beat increases with distension of the heart because as in skeletal muscle the energy of contraction depends on the length of the fibres. The efficiency is about the same as in voluntary muscle, that is, about 25 per cent.

The chemistry of the contraction of cardiac muscle is not known so well as that of skeletal muscle, but there is evidence that in the frog it is somewhat similar. The respiratory quotient of the aerobic heart is 0·85, which suggests that fat or protein or both are used in addition to carbohydrate. The heart will go on beating in nitrogen for hours, although it is then much more sensitive to neutral or acid concentrations of hydrogen ion. Under these conditions the concentration of glycogen is reduced and that of lactic acid increases. In the aerobic heart, on the other hand, glycogen disappears only slowly. It seems therefore that when it is beating anaerobically the frog heart gets its energy as does skeletal muscle by glycolysis, but that under normal conditions there is some other source as well. This is confirmed by the action of iodoacetic acid, which would undoubtedly stop lactic acid formation. If oxygen is present when this reagent is added the heart goes on beating normally, but in its absence activity only lasts for a short time, about twenty beats. The details of the carbohydrate metabolism of the heart are unknown, and they may be different from those in skeletal muscle.

Creatine phosphate and adenosine triphosphate occur in cardiac muscle, though not in such high concentrations as in skeletal muscle, and it is at least possible that they act in the same way in the two situations. The small amount of phosphagen present would explain the short time for which the anaerobic heart can beat when iodoacetic acid is present.

Events in the mammalian heart may be similar, but judging from the effects of lack of oxygen it is unlikely that it ever beats anaerobically in the living animal.

J 7.3. INVERTEBRATE MUSCLE

All Metazoa have muscles, and in view of the remote phylogenetic connections between many of the phyla it is surprising that their muscles show as much similarity as they do. Nearly all that have been investigated give a twitch on stimulation which is of the same general form as that for vertebrates. The duration of the contraction varies greatly; in the sea anemone *Metridium senile* it may last for 6 s and in *Holothuria nigra* for 3 s, but in the flight muscles of insects it is complete in a few thousandths of a second, or even, in some flies, less than 1 ms. These compare with about 0·1 s for mammalian skeletal muscle, and 0·01 s for mammalian eyeball muscles, the fastest known in vertebrates. Many invertebrates can maintain many of their muscles in a contracted state for a very long time. Sea anemones and molluscs of the intertidal zone normally remain contracted for the whole time during which they are not covered by water, while the oyster and some other bivalves can remain closed for as long as 30 days. The muscles of molluscs may be taken as an example. When a lamellibranch with open valves is touched, the two halves of the shell are drawn together by the adductor muscles, and remain tightly closed until the animal spontaneously opens them again. If a solid object is placed between the valves before stimulation, they shut on this, and if it is gently removed they remain set until they open again or receive another closing stimulus. The tension necessary to open the valves forcibly when they are thus set is much greater than that which will just prevent closure; in other words the adductor muscles can support a weight that they cannot raise. Each adductor muscle contains two bands of fibres of different sorts. These can be separated, and it is found that their properties are very different. In *Pecten*, the one that is striated works at the same speed as frog muscle, and is used for the quick closing of the shell and for flapping the valves in swimming. The other band is much slower in its reaction, having in *Pecten* a relaxation period of about 30 s and in the oyster of from a quarter to one hour: it is this that maintains the tension. It may be regarded as acting in the same way as vertebrate smooth muscle that is maintaining tonus, but the tensions reached are very much greater.

Some of the slowing-down of contraction and relaxation is probably due to the deformation of the accompanying connective tissue.

In other muscles, such as the anterior retractor of the byssus of *Mytilus*, there is no separation of the fibres, but the two types of contraction are produced by different types of stimulation. A short shock produces an ordinary fast twitch, which is caused by synchronous contraction of many fibres. Sometimes, especially with stimulation by direct current, there is a slowly relaxing tonic response. During this the muscle is electrically active in short bursts of variable intensity, the larger ones corresponding to minute contractions of the muscle, and this goes on even though all the nerve supply is removed. Presumably the muscle continues to be automatically active, small groups of fibres contracting in turn. This hypothesis is strengthened by the observation that at some points on the surface of the muscle no electrical activity can be detected, and that if records are taken simultaneously from more than one site there is no correspondence between them. There are special inhibitory nerves from the pedal ganglion, stimulation of which hastens the relaxation of tonus, and abolishes the electrical activity associated with it. The tonic contraction of molluscan muscle thus appears to be very similar to that of smooth muscle in vertebrates.

Specialized nerve-muscle junctions comparable to those of vertebrates have been described only in some insect muscles and in polychaetes. Usually there are branched nerve-endings like those of vertebrate smooth muscle, and as in this there are often two different types of ending, or sometimes more. In nematodes there is a peculiar arrangement in which long tails from the muscle cells go to the neurones. Little is known of invertebrate transmitters. Acetylcholine is widespread, but it has been shown to be the effective neurohumour only in the motor nerves of leeches (and possibly some other annelids), and of holothurians, and probably in lamellibranchs. Adrenergic nerves are possibly present in leeches and molluscs, and other transmitter substances possibly exist, such as 5-hydroxytryptamine in molluscs and 5-6-dihydroxytryptamine in crabs.

In many invertebrate muscles the contraction, once started, does not, as it does in vertebrate muscle, spread throughout the fibre; in crustaceans, for instance, each fibre has many nerve-endings, and all of them must be stimulated for complete contraction. In the flight muscles of some insects, on the other hand, there must be not only spread of the contraction but myogenic control, because if nervous impulses are given at the rate at which the wings beat the muscles go into tetanus.

Where there is double innervation, one set of nerve fibres usually mediates a fast response, which is accompanied by a big electrical discharge and rapid fatigue, and the other a slower response with a smaller discharge. Control of this sort has been shown in annelids and molluscs.

The leg muscles of locusts have one set of axons that branch to give multiple innervation of every fibre. A single impulse produces depolarizations at every point of innervation and a twitch follows. Repeated impulses give tetanus. This system is used in hopping and jumping. Another nerve, which supplies only 30 per cent of the muscle-fibres, and gives ordinary end-plate potentials and a long-lasting depolarization of low value, is used in walking and tonic contraction. A third small nerve gives long-lasting hyperpolarizations that perhaps prepare the muscle for the quick contractions. The limb muscles of decapod Crustacea have three types of motor nerve. One appears to act in much the same way as vertebrate nerves, giving a large end-plate potential and propagated depolarization and contraction, while the second gives a small end-plate potential that causes only local contraction of the muscle-fibre, which may, however, be enough for slow movement and accounts for much of the normal functioning of the limbs. These small end-plate potentials may, on repeated stimulation of the nerve, build up to a threshold and give a propagated contraction. The third type of nerve is inhibitory, and acts to prevent response to the other two. Schema 10 shows these actions in diagrammatic form, and compares them with the vertebrate system. The action of the slow crustacean system is very similar to that of the vertebrate when drugged with curare. It is tempting to suppose that the three types of crustacean nerve act by liberating three different chemical substances, but there is no evidence for this.

The system of innervation often allows for a grading of response, which is increased by neuromuscular facilitation; one nervous impulse has no effect, but it enhances the influence of a subsequent impulse, so that if impulses follow rapidly enough there is response, and the higher the frequency the more complete the contraction (Fig. 7.15). Facilitation in the coelenterates produces two sorts of contraction; a volley of stimuli at intervals of less than 3 s produces a quick response, starting almost immediately after the first impulse is received but increasing in size with subsequent impulses, while

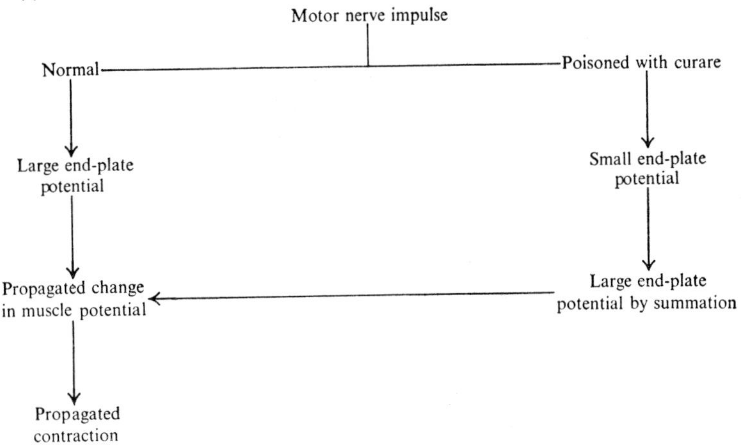

SCHEMA 10 (*a*). Amphibian nerve-muscle system. Modified from Katz, *Biol. Rev.* **24** (1949).

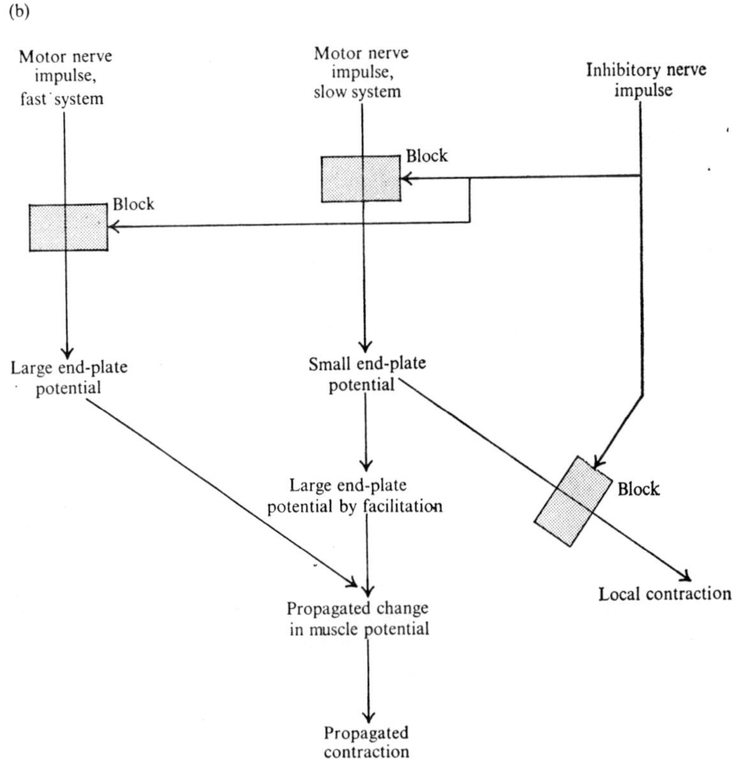

SCHEMA 10 (*b*). Crustacean nerve-muscle system. Modified from Katz, *Biol. Rev.* **24** (1949).

stimuli at longer intervals produce a slower smooth response that does not begin until several impulses have been received. The effect is thus similar to that of double innervation, but the place where the facilitation takes place is unknown.

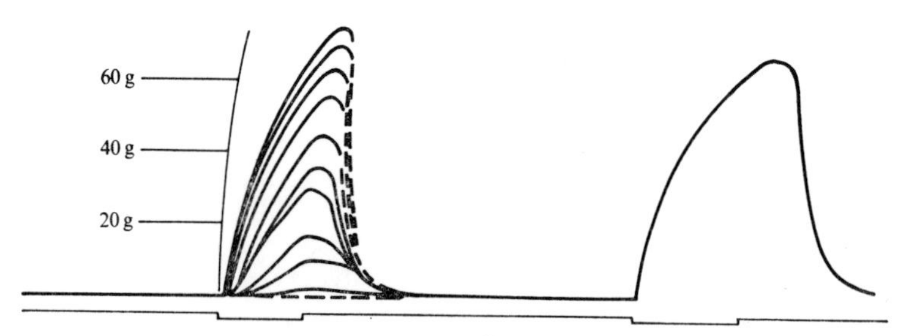

FIG. 7.15. Superimposed contractions of flexor of dactylopodite of leg of crab *Carcinus*. Frequencies of stimulation, left-hand curves, from below upward, 33, 44, 51, 67, 79, 94, 103, 137, 185, and 250 s^{-1} right-hand curve, 250 s^{-1}. Vertical scale, isometric contraction developed. From Pantin, *J. exp. Biol.* **13** (1956).

The tensions that can be produced in apparently weak invertebrate muscles are often high; the sea anemone *Metridium senile*, for instance, can exert 3·92 N mm^{-2}, which compares with only 0·35 N mm^{-2} in the frog sartorius, 0·6–1 N mm^{-2} for human muscles, and 0·46 N mm^{-2} for the leg muscles of a grasshopper. The flight muscles of a locust have a power per unit mass up to ten times that of human muscle; this is achieved mainly by the rapid contractions.

The hearts of molluscs have a myogenic pacemaker which, like that of cold-blooded vertebrates, is part of the contractile system. In arthropods the pacemaker is neurogenic.

Many invertebrate muscles, in medusae, annelids, molluscs, and arthropods, have a striped structure visible in the light microscope, and in general these are the faster-acting forms. In some annelids, cephalopods, and ophiuroids the myofibrils are wound helically round a sarcoplasmic core, and in some insect muscles the fibrils surround a central region containing the nuclei. In all the invertebrate muscles that have been adequately examined the electron microscope shows an arrangement that permits sliding, although the details differ. In *Balanus nubilis* and the larva of a blowfly (*Phormia*) the muscles shorten very greatly by overlap of the thick filaments, which pass through gaps in the Z-discs. In most, filaments are of actin and myosin, but in some muscles of molluscs there are also long filaments of paramyosin, of molecular weight about 134 000, which may be concerned with the polarized contraction.

At least where only actin and myosin are present the chemical change that provides the energy is probably the same as in vertebrates. The adenosine triphosphate is restored by a range of phosphagens. In some, including the ctenophore *Pleurobrachia*, many polychaetes, the Echinoidea and Ophiuroidea, and the protochordate groups Enteropneusta, Cephalochordata, and Tunicata, it is creatine phosphate, as in vertebrates. In most, including platyhelminths, annelids, arthropods, molluscs, echinoderms, and some tunicates and enteropneusts, it is arginine phosphate. In various polychaetes other compounds, phosphates of glycocyamine, taurocyamine, hypotaurocyamine, and guanidylethyl methylserine are present. The phosphagen in *Lumbricus* appears to be guanidylethylseryl phosphate. The distribution of these phosphagens can be related neither to phylogeny nor to habitat, and in the annelids different compounds may be found in species of a genus; some species have more than one. Each is accompanied by the appropriate enzyme for its hydrolysis.

The final source of energy is usually glycolysis of some sort. Flying *Drosophila* rapidly use glycogen, and flies flown to exhaustion can be made to fly again if they are given sugar solutions. Their fat reserves are used in starvation, but are mobilized too slowly to be used in continuous flight. Migratory locusts, which normally fly for longer periods without resting than does *Drosophila*, can use their fat reserves during flight.

The tube-feet of echinoderms depend on the contraction of muscles for their working, but their peculiarities justify special comment. Each is a nearly closed tube, having at its inner end a muscular ampulla. There is no circular muscle, and protrusion is brought about by contraction of the ampulla, which forces water into the tube and raises the pressure in it, which is initially near atmospheric, by about 20 mmHg. Bending of the tube-feet and their retraction are by longitudinal muscles. All these movements are controlled by the nervous system, and the contraction is roughly proportional to the size of shock given to the nerve. At each protrusion there is a slight leak of water, which is probably made up by osmotic absorption, the water in the tube having a slightly greater concentration of potassium than has sea water.

G 7.4. GLANDS

Glands, or organs that secrete, prepare chemical material and extrude it either by a duct or into the blood stream. The chemical portion of secretion goes on in cells, and the whole process can be divided into three phases. First, the raw materials must be taken into the cell, usually from the blood, and as this means the transfer of matter through a membrane energy may be needed. Second comes the chemical formation of the substance that is being secreted. In general this will need enzymes, and so will be dependent on the nucleus, without which little if any protein can be formed; adult mammalian erythrocytes cannot make haemoglobin, and Protozoa without nuclei can ingest food but cannot digest it. Energy also is usually needed for synthesis, and this is supplied by the oxidative enzymes in mitochondria acting on carbohydrates or other fuel. If the mitochondria of the pancreas are destroyed by X-rays, as they can be without any other apparent damage to the cell, no more secretory granules are formed until they reappear, which is a few hours later. Secreting cells have many mitochondria and use more oxygen when active than when resting; in salivary glands the rise is to five times the normal, and there is some evidence that they can put up an oxygen debt.

For the most part the actual site of synthesis is still disputed; it is not necessarily always the same. It may be the mitochondria, it may be the endoplasmic reticulum and its associated microsomes, and it may be the general cytoplasm, for some enzymes occur free in this. The secreted material often appears in the form of granules or droplets.

Lastly, there is the extrusion of the material, which is often accompanied by a measurable rise in oxygen consumption, heat production, and electrical changes.

The granules of secretion may be dissolved and go out in bursting vacuoles, as in the pancreas, or as much as half the cytoplasm may be lost with the granules. This happens in the mucus-secreting cells of the intestine of Amphibia, in the midgut and salivary gland of the snail, and in sweat and lacrimal glands. The nucleus and mitochondria remain. In some extreme cases, as in the midgut of the crayfish and in sebaceous glands, whole cells may be lost, so that each cell secretes but once, and the gland is only maintained by rapid mitosis. Extrusion depends on excitation either by a nerve or by a hormone. At the nerve endings in the adrenal medulla there are synaptic vesicles, and these increase in number when the nerve is repeatedly stimulated, so that presumably the activation of the gland is by acetylcholine, as is that of muscle. Much synthesis apparently goes on without any outside stimulus, but its rate is increased when the gland is exhausted and can vary with external conditions. Many glands, especially those connected with reproduction (section M 8.23) have a seasonal cycle, and may be completely inactive at some times of the year.

V 7.5. ELECTRIC ORGANS

Electric organs, which are sufficiently described by their name, are found only in fishes. They are known in some 250 species, some bony, some cartilaginous, and are particularly well developed in the three species *Gymnotus* (= *Electrophorus*) *electricus* (the electric eel), *Malopterurus electricus* (the electric catfish), and *Torpedo marmorata* (the torpedo or electric ray).

The organ is made up of a number of discs or compartments called electroplaques or electroplates which are derived from skin glands in the catfish and from striped muscle in the others. They are so arranged that the voltages due to the separate discs are largely summed; in short, most of the discs are in series and some in parallel. The voltage for each compartment varies, but is about 0·04 to 0·1, which gives a total potential difference for the whole organ of 200 V in *Malopterurus*, 300 in *Gymnotus*, and 30 in *Torpedo*. In the last two genera the shocks are used to stun prey before it is caught and eaten, and *Malopterurus* probably uses its electrical ability for defence.

The motor nerve divides and supplies each disc individually, the branches running on one surface of the disc only, which is posterior in the eel, ventral in the ray. At the junctions the axons make contact (except for the synaptic cleft) with short knobs on the disc. In these regions of the axon vesicles accumulate. The current in the organ flows from the innervated side to the other. (The direction of the current in *Malopterurus* is exceptional.)

Nerve-organ preparations, similar to the common nerve-muscle preparations, have been used for experimental study, and the results obtained are in a general way similar to those for muscle. There is a latent period of about 10 ms at 5 °C and 3 ms at 30 °C. The preparation can be fatigued, but is relatively indifferent to drugs such as curare and atropine. There is no spread of excitation from one disc to the next, other than by nerves, but each discharge is rhythmical, because the current produced by the first discharge excites the nerve and so stimulates a second discharge, and so on. There is thus, for one primary stimulation, a series of successive discharges at a frequency that varies with the temperature but is of the order of 100 per second. There is, however, a decrement in successive discharges, and the whole series seldom lasts more than half a second.

The ordinary reflex discharge in the intact fish is also a rhythmical one, but the rhythm here is due to the stimulus from the central nervous system, for its frequency depends not on the temperature of the organ, but on that of the ganglion from which the nerve starts. The reflex discharge, which in *Gymnotus* lasts 2 ms, is given on the receipt of various sensory stimuli by the fish, particularly by mechanical pressure on the skin. *Gymnotus* is stimulated to discharge when live fish are placed in the same tank; some of them are stunned, and the eel, which swims only slowly, is able to catch and eat them. Electric fish are themselves relatively insensitive to electric shocks, so that they are not adversely affected either by the voltage from their own organs, or from other fish in their vicinity.

Gymnarchus niloticus and some other species have an electric organ in the tail, from which they send out a continuous weak electric pulse. The use of this in obtaining information about the fish's surroundings is described in section V 6.47.

Of the details of the physiology of the organs little is known. The time relations are by no means clear, for while it is obvious that if the voltages of the separate plates of tissue are to be summed, as in fact they are, discharge in all of them must be simultaneous, it is impossible to see how the time of receipt of the stimulus can be the same for all of them, when the organ may be as much as 300 mm long and the nerve enters at one end. There is some evidence that the organ acts as a concentration cell, which agrees with its high efficiency: an average value in *Torpedo* is 60 per cent. The voltage of *Torpedo* is a development of that of the nerve-muscle junction, and can be excited by acetylcholine but not by direct electrical stimulation; that of *Electrophorus* is a development of the muscle action-potential, and can be excited either indirectly through acetylcholine or directly by an electric shock. The chemistry of the energy production in *Gymnotus* is fundamentally similar to that in muscle. Cretine phosphate is broken down, acetylcholine being formed at the same time, and there is subsequent formation of lactic acid.

VJ 7.6. LUMINESCENT ORGANS

Light that is within the range of wavelengths to which the human retina is sensitive is produced by representatives of about forty orders of animals, distributed through nearly all the phyla. It is found in the Protozoa (Radiolaria and Dinoflagellata); in all three classes of Coelenterata and in the Ctenophora; in Polychaeta and Oligochaeta; in the Ophiuroidea; in four classes of Crustacea, in the Chilopoda, and in many orders of Insecta; in the three main classes of Mollusca; in the Hemichordata and Urochordata, and in both Elasmobranchii and Teleostei. It is noticeable that all the forms which possess it, except for one glow-worm and a limpet, are either marine or terrestrial, but the significance of this is unknown.

Some squids, millipedes, and insects glow continuously, but in the majority there are intermittent flashes, and light emission begins or is increased on the receipt of some stimulus. Fire-flies (Coleoptera of the genera *Photinus* and *Photurus*) begin flashing when the general light intensity falls below a certain value. In many of the marine forms, including the Protozoa, general mechanical stimulation will cause the animals to light up; in the polychaete *Chaetopterus* the best stimulus is tension or pressure on segment twelve. In most of the Metazoa, and probably in all, there is nervous control; in *Chaetopterus* acetylcholine is weakly excitatory, so that the nervous control is probably of the usual pattern. Weak stimuli to this animal cause only local flashes, but stronger ones cause the light to spread, suggesting facilitation in the nervous system. The point at which the control acts varies; in the ostracod *Cypridina* and others granules of one of the chemicals concerned are extruded from the cells by muscles; in the boring bivalve *Pholas* and many worms active extrusion of material is started; and in many shrimps, fire-flies, squids, and fish, the luminescence that is begun is intracellular. Hormonal control has not been proved, but the slowness of the response in some fish suggests it, and in these species injected adrenaline causes flashing. Some cephalopods and fish have symbiotic luminescent Bacteria that glow continuously, but there may be nervous control of the emission of light by the opening and closing of shutters over the bright parts.

In glow-worms (*Lampyris noctiluca*) the light lasts for many minutes, sometimes varying in intensity, while in fire-flies the flash may be little more than a tenth of a second, and other animals mostly come in between.

The intensity of the light is low, the steady glow of the fire-flies being about 2×10^{-5} cd (candela) and their brightest flashes only 0·02 cd. As the area of luminescence is small the brightness of the surface is relatively high, ranging in various species of fire-fly from 1 to 143 cd m^{-2}. This compares with the brightness of the blue sky of 3000 cd m^{-2}, and of white paper suitably lit for reading of 12 cd m^{-2}, both these being considered as sources of light. The colours range from red to blue, and have usually short continuous spectra. Some fire-flies can produce more than one colour. The reaction is very efficient, from 20 to 90 per cent of the total radiation being luminous.

In some worms, in the fire-flies and glow-worms, in some fishes, and in some other animals, the light is produced chiefly or only in the breeding season, and from this and from the behaviour of the animals it is clear that it is a sex signal. The flightless female glow-worm attracts the male by her shining abdomen, and the two sexes of fire-fly flash in turn until they have found each other. Luminous animals living in the deep sea (shrimps, squids, fish) often have complex organs with reflectors and lenses, which suggests that the animal may be illuminating its surroundings. Presumably the light often conveys information, to neighbours, prey or the opposite sex, but how and what is mostly speculation. The light produced by planktonic forms, such as the dinoflagellate *Noctiluca*, when the sea is disturbed seems to have no function at all.

Robert Boyle showed in the seventeenth century that oxygen was necessary for the light of the glow-worm, and so it is for almost all the animals that have been examined. The exceptions, such as some ctenophores, are more apparent than real, since oxygen appears to be necessary for some preliminary reactions.

From many luminous animals (worms, crustaceans, insects, lamellibranchs, fish) it is possible to make two different extracts, which, when mixed in a test-tube, produce light. Since light is not produced if one of these extracts is boiled before it is added to the other, but boiling the second does not prevent the production of light, the first is presumed to be an enzyme and is called luciferase; the other is called luciferin. As a general scheme one may say that light is produced when luciferin is oxidized by molecular oxygen with the help of luciferase. The luciferases are non-specific within a narrow range, working usually only with luciferins from animals of the same family.

Beyond this, there is little or no uniformity in the chemistry. The luciferase of fire-flies is a protein of molecular weight about 100 000 and containing about 1000 aminoacids, while their luciferin is a relatively simple 3-ring thiazole compound of molecular weight 279. The light-producing reaction needs magnesium ions and adenosine triphosphate as well as oxygen; if LH_2 stands for luciferin and A for luciferase, the first stage may be summarized:

$$LH_2 + A + ATP \rightarrow ALH_2AMP' + 2 \text{ phosphate.}$$

The complex of enzyme, substrate, and adenosine monophosphate is in an excited state, with energy which it gives out as light ($h\nu$) on oxidation:

$$ALH_2AMP' + \tfrac{1}{2}O_2 \rightarrow ALAMP + H_2O + h\nu.$$

The luciferase of *Cypridina* also is a protein, but its luciferin is quite different from that of fire-flies, containing no sulphur. The reaction that produces light does not need either magnesium or adenosine triphosphate. About 100 molecules of oxygen must react to give one quantum of light, which means that only one collision in 100 produces a molecule that is in the necessary excited state to give out light. The luciferase of *Pholas* will flash with a mixture of flavin mononucleotide and nicotinamide-adenine dinucleotide (section M 2.141), so that its luciferin may be, or contain, these vitamins. Some coelenterates produce light by the reaction of a protein with calcium ions. Its is obvious that no single scheme can represent the chemistry of animal luminescence.

J 7.7. NEMATOCYSTS

More fairy-stories have been written about nematocysts, or cnidae, 'the stinging cells' of invertebrates, than about any other biological structure, and there is still no agreed account of how they work. According to the usual story, they are produced in special cells called cnidoblasts, but this has been denied. They are not themselves living, for they can be stored and made to discharge long after they have been separated from their owner, so that their claim to be considered as effectors is doubtful.

When fully formed, a nematocyst consists of an ovoid sac or capsule which has a tough wall of a material somewhat similar to cartilage in chemical composition, which at one pole is invaginated to form a tube; the latter has a wide basal portion, and then a longer coiled narrow thread. The inside of both parts of the tube usually bears spines, arranged in whorls of three at different levels. The base of the tube, where it is continuous with the walls of the capsule, is closed off from the outside by an operculum. The capsule is filled with liquid which is perhaps a mucoprotein. After the nematocyst has been discharged its structure is not essentially altered, but the operculum has burst and the tube has been evaginated, so that it projects into the surrounding medium and its spines are now on the outside. The thread is usually said to evaginate also, but this has been denied on the basis of electron micrographs, as has the existence of nematocysts with no spines. In an average nematocyst the tube is 700 μm long, but there is much variation. Seventeen different varieties based on this general plan have been recognized, and of these three or four are present in *Hydra*.

At the top of the operculum or by its side there is sometimes a short stiff process or cnidocil, which according to tradition is a sense organ. It is claimed that some electron micrographs show that it has the typical basal body and nine fibrils of a stereocilium, but these are visible only to the eye of faith. Neither the best histological preparations nor the electron micrographs show any connection between the nematocyst or cnidocil and the nerve-net, and since cnidocils are present only in the Hydrozoa they cannot be necessary for discharge.

The process of discharge is always an evagination, and takes from 1 to 3 s. It can be slowed down and observed under the action of chemicals, and a cinematographic record of it has been made. The appearance of the wall of the capsule does not change and its volume, although it usually alters slightly, does not change in any regular way. The tube, on the other hand, increases in size, sometimes to double its resting diameter and length. The spines may remain attached or they may fall off, and several may stick together as they fall, forming a sort of harpoon. Sometimes the contents of the capsule can be stained with neutral red, and the discharge, which is then slowed down, can be watched under the microscope. No liquid comes out of the tube until evagination is complete, but before this the contents, both of capsule and tube, break up into coloured and colourless drops. Liquid then comes slowly out of the end of the tube; the colourless parts immediately disappear in the water, but the coloured drops run together and dissolve only slowly. The volume of liquid which goes out is greater than that of the original capsule, and some coloured liquid remains inside. In those nematocysts that have a blind tube no liquid leaves it, but the capsule swells on discharge.

In the living coelenterate the normal stimulus for discharge is nearly always some chemical derived from an animal, followed by contact, but occasionally a high intensity of one of these may cause discharge by itself. Chloride is necessary. The nature of the chemical substance is unknown; it is not a protein or a carbohydrate but may be a lipid. Isolated nematocysts can be discharged by hypotonic solutions, by many anions, and by extremes of acidity or alkalinity, but these do not affect the structure *in situ.* Most surface-active agents cause discharge of both isolated cnidae and those *in situ,* which, taken with the natural discharge, suggests that the effect of a stimulus is to cause a change in the surface of some part of the structure, as a result of which water enters and causes a swelling that bursts open the operculum, so that the wall contracts and pushes out the tube and its liquid contents.

The precise function of the nematocyst also remains uncertain. Cnidaria are surely on the whole able to sting, but the stinging properties of the jelly-fish as judged subjectively by man seem to be quite independent of the discharge of nematocysts, and it is at least possible that the poison is not produced by these effectors at all. Some animals are immune to the attacks of coelenterates and even feed on them, but it is not known whether this is true chemical immunity or whether these animals just do not stimulate the cells to discharge. The spines of the large nematocysts of *Hydra* and some others pierce chitin mechanically, and the tube enters the perforation, which is enlarged by solution (Fig. 7.16). The corkscrew type and the small form with spines possibly assist in holding the prey and the small spineless type is used for attachment of the tentacles when the animal walks. Release in all types is produced by the discharge of the nematocyst from its cell. Some types of nematocyst, with a closed tube and no spines, have been said to have no conceivable function whatever.

G7.8. CHROMATOPHORES

It is convenient to include under the name of chromatophore all cells near the surface of an animal by which it can make temporary changes in the colour or shade of its skin. The word is not applied where a colour change is caused, as in blushing, by changes in blood-content of the superficial vessels, nor where a prolonged change such as the tanning of the skin on exposure to the sun is brought about by the development of fresh pigment. A chromatophore is a cell that contains a quantity of pigment that varies in its disposition. Where the pigment is black the name melanophore

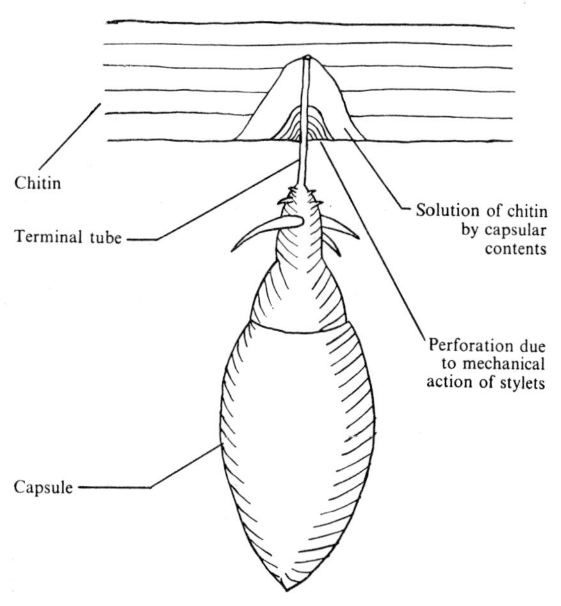

FIG. 7.16. Nematocyst of *Hydra* with the tube perforating the chitin of an arthropod. Redrawn from Toppe, *Zool. Anz.* (1908).

is often used, and corresponding words derived from the Greek have been used for other colours.

V 7.8. CHROMATOPHORES IN VERTEBRATES

Many cold-blooded vertebrates change their colour by the movement of pigment within the cytoplasm of branched cells, or sometimes syncytia, which are the chromatophores (Fig. 7.17). These are derived from the neural crest, and so are presumably homologous with the melanocytes of mammalian skin that have the same origin and shape, contain the common pigment melanin, but do not move it about within the cell. Mammalian skin changes colour, if at all, by the manufacture or destruction of pigment, and although this can happen in chromatophores also as a long-term change, it is not that which gives them their special character.

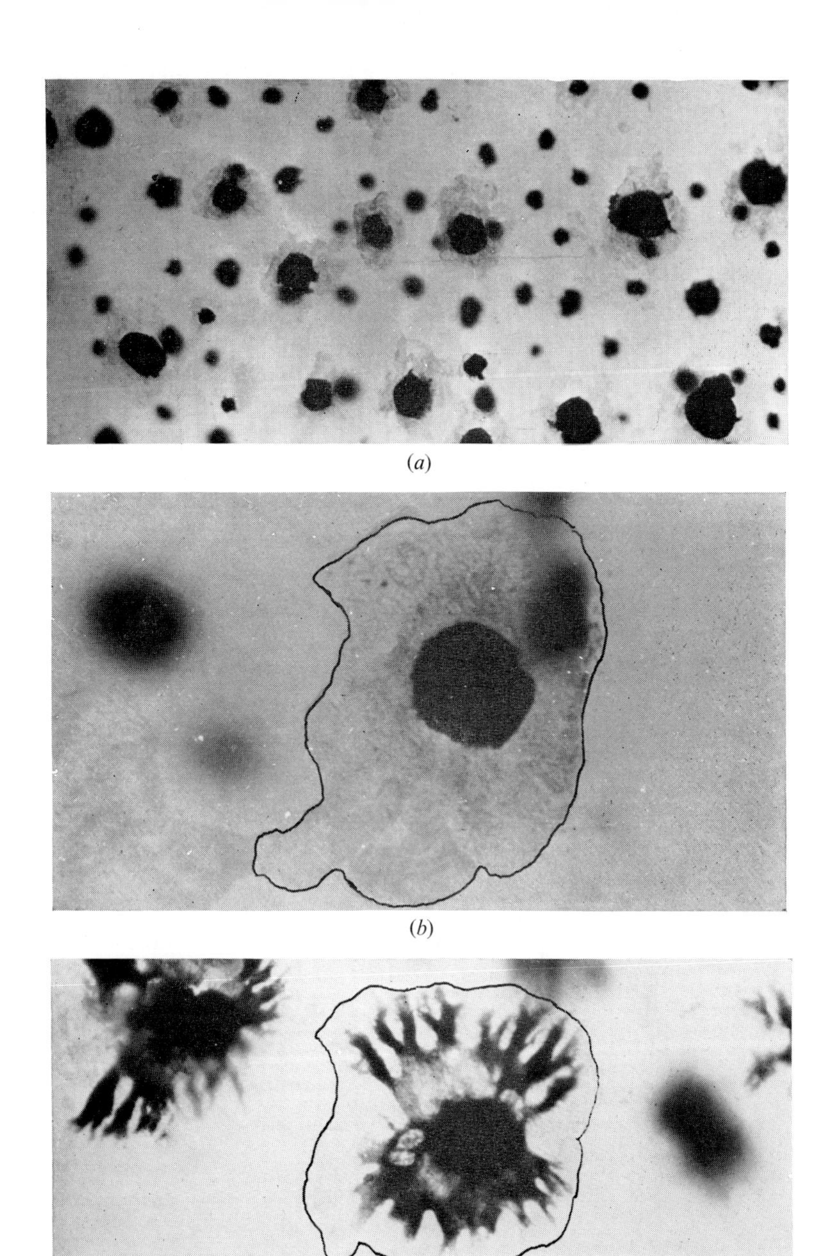

FIG. 7.17. Chromatophores of minnow, *Phoxinus phoxinus*. (*a*) General view of the skin, with pigment concentrated. (*b*) A single chromatophore with pigment concentrated, at a higher magnification; a second chromatophore, out of focus, overlies it. (*c*) A single chromatophore with pigment dispersed, at the same magnification as (*b*). The diameter of the cell is about 150 mm. In (*b*) and (*c*) the outline of the cell has been emphasized. Photographs by O. W. Harry.

The effect of concentration of the pigment in the centres of all the cells is to make the animal look light; dispersion makes it look dark. If the only pigment present is melanin, it can only vary from black through grey to white, but when, as in many teleosts, several pigments are present in different cells, the independent movement of these may cause brilliant colour changes. Some amphibians and reptiles also have chromatophores of more than one colour, but in the latter change of colour is brought about also by melanin moving to uncover fixed masses of pigment in a deeper layer of cells. The pigment granules of isolated chromatophores can be made to move about by various reagents; it seems that dispersion is generally associated with a lowered viscosity of the cytoplasm, and there are accompanying changes in the cell wall. Electron micrographs of the chromatophores of the teleost *Lebistes reticulatus* show that the pigment granules are enclosed in a sac consisting of a membrane 8 μm thick more or less concentric with the cell wall. Fibres outside the sac, but not attached to it, apparently contract tangentially to compress and concentrate the pigment.

In life, the response is to the environment, chiefly to the colour or luminous intensity of the background or to the general illumination, but also to temperature and humidity. Fish generally resemble their background, and the assimilation may be very close, some flat-fish being able to make themselves into a fair imitation of a chess-board. On a dark background, a fish is dark, but in complete darkness it is pale. Most amphibians resemble their background, but a dark skin is produced also by low or zero general illumination, moisture, and low temperature, and pallor by the reverse (Fig. 7.18, frontispiece). When the different conditions are in opposition to one another one is preponderant or a balance is struck. Most of the reptiles that can change their colour are lizards, and in addition to assimilating to the background they may, like the chameleon, change colour when they are handled. In general, pallor is produced by a light background, high illumination, low temperature, and contact or other noxious stimulus.

In many larval amphibians there is a response, sometimes called primary, which differs from that of the adult; they are pale in darkness and dark in bright light, so resembling fish. This response is not abolished if the larvae are blinded. The normal response of fish is completely, that of amphibians and reptiles nearly, abolished by blinding. In the adults therefore the light must act through the eyes;

in the larvae and in fish in the dark the chromatophores may be acting as independent effectors, but there is no proof of this. There is good evidence that they do so in *Xenopus* and the lizard *Phrynosoma.*

The control of chromatophores may be nervous or hormonal or both. In the cyclostome *Lampetra planeri* there is a single hormone from the posterior lobe of the pituitary, which causes darkening; light stimuli, received by the pineal and parapineal in the larva and by these and the lateral eyes in the adult, inhibit the pituitary and cause pallor. Intraperitoneal injection of extract of mammalian 'posterior lobe' causes darkening, and by analogy with the amphibians one may assume that the hormone concerned is similar to intermedin (see below).

The majority of elasmobranchs that have been investigated have hormonal control. A hormone from the intermediate lobe, which may be called intermedin, causes darkening; comparable extracts from the mammalian pituitary have the same effect, and the removal of the pituitary causes pallor. The response is slow, most species taking a few hours to adjust to their background and *Scyliorhinus caniculus* as many as 100 h. Some workers think that in *Mustelus canis* pallor is caused by stimulation of nerve fibres, but there is no direct evidence of this, and on anatomical grounds, since the skin has no sympathetic innervation, it seems unlikely.

The teleosts are more various. Pituitary hormones seem to play some part in most of them, but no consistent story can be made. Extracts of telost pituitaries cause dispersion of pigment in amphibians, suggesting that intermedin is present, but when the reverse experiment of injecting extracts from amphibians into teleosts is carried out, there may be disperson, as in *Ameiurus*, or no effect as in most species. Removal of the pituitary may produce slight pallor as in *Ameiurus* or none at all as in *Fundulus*. The colour changes of teleosts are much quicker and more exact than in elasmobranchs, which suggests control by nerves; this has been conclusively demonstrated, though the details remain obscure. The chromatophores are supplied by branches of the sympathetic system, and stimulation of this causes pallor in appropriate parts of the skin. Cutting the nerves causes darkening, so that dispersion of the pigment may be the resting state, the opposite to the case in elasmobranchs. According to some workers at least in some fish such as *Fundulus* darkening also can be produced by nervous action, which suggests double innervation similar to the sympathetic/parasympathetic control of glands and smooth muscle in mammals; this hypothesis is supported by the facts that adrenaline causes concentration of pigment in most chromatophores and acetylcholine dispersion. The difficulty, which remains unresolved, is that parasympathetic fibres have not been demonstrated in fish. As in amphibians, the ratio of reflected to incident light is important.

The Amphibia behave broadly in the same way as elasmobranchs; concentration of pigment is probably the resting state of the cells, and a hormone intermedin, or melanophore stimulating hormone (MSH; formerly B-substance) from the intermediate lobe causes dispersion and darkening. The release of this depends on the albedo, or ratio of light from below (light reflected from the background), which will fall on the dorsal part of the retina, to that from above (the general illumination), which will fall on the ventral portion of the retina. Impulses to the brain presumably initiate changes in the hypothalamus, but beyond this the nervous system plays no part in colour change. Melanophore-stimulating hormones extracted from the intermediate lobe of mammals have been purified and shown to consist of about twenty aminoacids (the exact number depending on the species) of which seventeen or eighteen are constant. It therefore fits into the general pattern of adenohypophyseal hormones, which are, as we have seen in section MV 6.131, polypeptides. There are some indirect arguments for the presence of a second, melanophore concentrating, hormone (W-substance) in amphibians, perhaps secreted by the tuber of the hypophysis, but its existence has not been proved and seems doubtful. In a few species there is evidence for slight control by nerves.

In the lizard *Anolis* control is by the single dispersing hormone, there is no nervous control, and adrenaline does not produce pallor. The only chelonian known to change its colour (*Chelodina longicolis*) appears to be similar. In *Phrynosoma* (miscalled the horned toad) there is double control, pallor being produced either by sympathetic stimulation or by the liberation of adrenaline into the blood, and darkening by intermedin. Pallor can be induced in limited areas of the skin of the chameleon by stimulation of the appropriate nerves, but the very complicated changes of colour of this animal have not yet been satisfactorily explained.

It is possible to see the outlines of a reasonable phylogeny in this complicated story. The earliest vertebrates had chromatophores that reacted directly to light, and this state remains in larval cyclostomes and amphibians. Very early, before the Gnathostomata arose, control by a pituitary hormone that disperses melanin was superimposed on this, and such control persists, possibly aided by a second hormone, in elasmobranchs and amphibians and to a lesser extent in teleosts and a few reptiles. At least twice, in the teleosts and reptiles, the hormonal system has been supplemented or nearly replaced by nervous control. It has been suggested that this is most marked in the families of teleost that are unknown before the Pleistocene. The mammals and birds have lost the power of moving the pigment in their cells, but the former at least have retained the melanophore-stimulating hormone that is now apparently functionless. Adrenaline often causes pallor, in amphibians as well as fish, but whether it has any natural function, except in reptiles, is unknown. There is also some evidence that the pineal (the receptor for colour change in larval lampreys) produces hormones that play a part in colour change. Removal of the pineal from sockeye salmon (*Oncorhynchus nerka*) causes dispersion, and melatonin extracted from mammalian pineals concentrates the pigment of melanophores of the frog.

J 7.8. CHROMATOPHORES IN INVERTEBRATES

Chromatophores of the same general type as in the vertebrates are found in polychaete larvae, leeches, echinoderms, crustaceans (almost exclusively the Malacostraca), and a few insects. In view of this scattered distribution it seems unlikely that they are all homologous. In leeches, polychaetes, and sea urchins only the primary reaction by which the cell responds directly to illumination has been described, but in a starfish neurohumors appear to be used.

Control is best known in the crustaceans. In the shrimp *Crangon* (= *Crago*) there are four different pigments, brown, white, yellow, and red, which are distributed amongst the chromatophores in all possible ways except that brown is always present. By appropriate movements of the pigment the animal matches itself to any background that comes towards the red end of the spectrum, and *Palaemonetes* does even better. Other Crustacea, such as some crabs and isopods, are similar, but not so complicated. If both eyes of one of these animals are excised there is no response to a changed background, the chromatophores remaining permanently in dispersion, so that the eye is clearly the organ that receives the stimulus. In these and other Crustacea the movement of the pigments is controlled by hormones that are formed in various parts of the nervous system and stored in the sinus gland and elsewhere. It is impossible to give any simple account of the matter, because no two genera seem to be quite alike. In general, antagonistic dispersing and concentrating hormones are present, but any one of these may have different effects on different pigments and different effects on the same pigment in different types of chromatophore. Most investigators think that there are probably four hormones. Investigation is difficult because there is often a daily rhythm that persists throughout the experiments and because some chromatophores seem to be able to respond directly to illumination, concentrating their pigment in bright light. In this they may be acting as independent effectors, but more probably they are merely made more sensitive to the hormones. The control of the chromatophores of the isopod *Ligia* is by hormones, probably two.

Carausius (*Dixippus*) *morosus*, one of the few insects with reversible colour change, does not have typical chromatophores; instead, pigment moves in ordinary epidermal cells, coming to the surface to make the animal darker and retreating to make it pale. The animals are generally pale by day and dark by night, and this response is abolished if the eyes are removed, but dampness also causes darkening. Most of the regulation is carried out by a hormone, arising in the brain, which causes darkening; it is said to concentrate the pigments of *Crangon*, but not of *Palaemonetes*. A second hormone may be present.

These invertebrate hormones have not been purified, but that of *Carausius* is said to disperse the pigment of amphibians, and an extract of the eye-stalks of prawns is said to concentrate that of plaice. It seem likely that any such cross-effects are accidental, and might be expected if the invertebrate hormones are polypeptides.

A chromatophore of a cephalopod is a quite different sort of structure. At rest the cell is a small sphere, but attached to its circumference in the plane of the surface are radiating muscle fibres, which on contracting pull the cell out into a disc and expose more pigment. Differently coloured cells may be present, so that varied colour patterns can be introduced. The muscles are controlled by nerves in the ordinary way, but there is possibly hormonal control as well.

8. Reproduction

Reproduction means in biology the production of an animal or plant that is in some way a different individual from the parent or parents that gave rise to it. An exact definition is difficult, but the general meaning of the word is simple. It is convenient and fairly logical to divide the methods of reproduction used by animals into sexual and asexual, but it must be noticed that the meaning of the second of these words does not correspond exactly with the use a botanist would make of the term; most of the asexual means of reproduction found in animals correspond more to the vegetative methods of propagation of plants than to the asexual method by means of spores.

G 8.1. ASEXUAL REPRODUCTION

Asexual reproduction cannot be defined in any positive way. It is reproduction in which there is neither any sexual process, nor anything that can be interpreted as a reduced sexual process. It can take place in a number of ways, mostly not primitive, and the placing of these together under one head is more convenient than logical.

G 8.11. Fission

In its more restricted meaning fission implies the division of a cell, and consequently it can only be a method of reproduction in the Protozoa. In them it is the normal and characteristic method by which increase in number is brought about. It consists in a division of the cytoplasm of the animal, preceded by an orderly division of the nucleus. In a few genera, such as *Actinophrys* and *Amoeba*, it has been possible to make out that there is a mitosis similar to that of metazoan cells, and differing from this only in that the nuclear membrane remains intact for most or all of the division. In other species either the individual chromosomes are too small for it to be certain that longitudinal division takes place, or the constancy of their number has not been established, but since there is general resemblance to mitosis the process is regarded as a reduced form of this and is called cryptomitosis. True amitosis has been generally admitted for the meganucleus of the Ciliophora, but in some, such as *Euplotes*, a wave of change can be seen starting from each end of the nucleus, so that the two meet in the middle. During this time the quantity of deoxyribonucleic acid in the nucleus is doubled, so that it may be that the waves represent mitoses in a large number of subnuclei.

When fission results in the existence of two daughter individuals each about half the size of the parent it is said to be simple and binary. Sometimes there is more than one division before the products separate and start to lead independent existences, so that several small individuals result. This end may be brought about in two ways. In repeated fission there is more than one ordinary binary division in quick succession: an example is *Vorticella*, which produces eight small motile forms by three divisions. In multiple fission there are several divisions of the nucleus so that a syncytial condition is temporarily produced, and then the cytoplasm falls apart round the nuclei. Multiple fission is associated with a peculiar condition of the nucleus known as polyenergid in which several sets of chromosomes appear to be present; when the nucleus divides, these sets separate. Multiple fission always occurs in the formation of spores, and is then called sporulation. Spores, in the strict sense, are the products of reproduction of a zygote (section G 8.21), so that this type of reproduction is closely comparable to the asexual reproduction of plants. It occurs in many Protozoa, but is best seen in some of the parasitic forms such as *Monocystis* and *Plasmodium*.

Budding is the name given to fission in which one of the products of the division is much smaller than the other. It has no relation to budding in the Metazoa. It occurs, for example, in *Arcella*.

G 8.12. Fragmentation

Fission is sometimes taken to include not only the division of the Protozoa, but also all cases where a metazoan divides into two or more parts without there being any very special structures concerned in the process. For this it is better to use the word 'fragmentation'. *Hydra* is said occasionally to divide into two, and division of this sort is normal in many other coelenterates. The scyphistoma of jellyfish such as *Aurelia* produces a number of medusae by successive transverse divisions (Fig. 8.1). Actinozoa such as *Metridium* divide in two longitudinally, and the majority of the corals do the same, but the polyps remain attached, so that a colony is produced. The process nevertheless belongs here, as it is quite distinct from budding.

Fragmentation has been observed in a species of *Convoluta* and is a normal method of reproduction in some oligochaetes, particularly the freshwater families Aeolosomatidae and Naididae; in some genera of the latter sexual individuals have never been found. Division takes place at a definite point called a fission zone, which is usually formed at a fairly characteristic level (*n* segments from the front, where *n* is approximately constant). This is a region where several new segments are interpolated, and where reconstruction of tissues takes place, the activity being primarily epidermal. Usually the breaking of the animal into two parts does not take place till some reconstruction has occurred, but in some species it precedes even the formation of new segments. Sometimes new fission zones appear before fragmentation occurs, so that a chain of as many as eight parts is formed. The whole is coordinated and swims as one animal until just before separation (Fig. 8.2). In some polychaetes there is regular and spontaneous fragmentation into short lengths, usually of two, three, or four segments, although in one species single segments which regenerate fore and aft are separated. The pattern of breaking is constant for each genus. In the syllids there is also another type of fragmentation, in which new heads are formed before fission, and this produces special reproductive individuals which have no function but to swim to the surface of the sea and liberate gametes, so that it alternates with sexual reproduction. An example is the Palolo worm mentioned below in connection with seasonal spawning (section J 8.23). In some genera, such as *Nereis*, the reproductive individual is rather different in appearance from the normal, and has been given its own generic name.

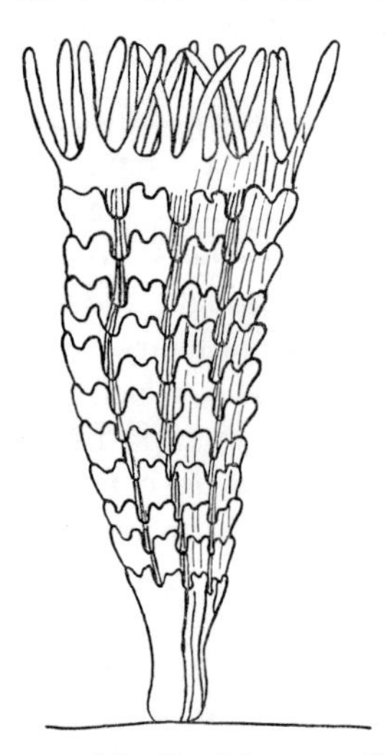

FIG. 8.1. Scyphistoma of *Aurelia aurita*, dividing transversely to form medusae. The process of division is called strobilization, and the successive medusae which split off from the top, turn over, and swim away are ephyrae. $\times 10$.

The type of protozoan fission called plasmotomy is probably best considered here. It is a division of a multi-nuclear organism independently of its nuclei, and may be subdivided according to the number and type of the products. It occurs in *Opalina*, where it is binary, and in the Mycetozoa, where it is multiple. The reverse process, plasmogamy, in which two or more cells flow together without fusion of their nuclei, is also known, and in a few sponges comparable larvae are formed.

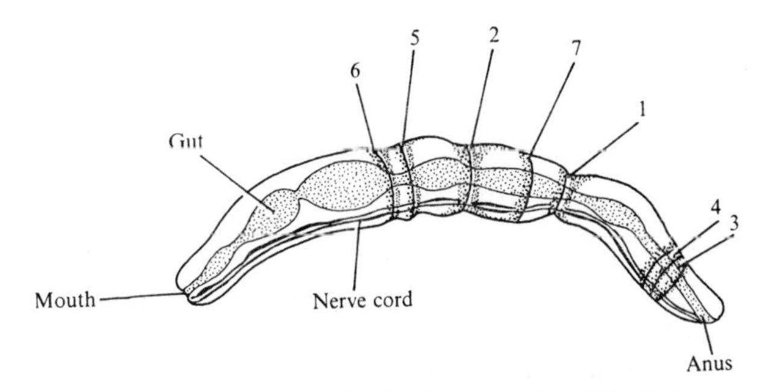

FIG. 8.2. *Chaetogaster* reproducing by fragmentation. The fission zones are numbered in the order of their appearance. ×20. Redrawn from Stephenson, *The Oligochaeta*. Clarendon Press, Oxford (1930), after Wetzel, *Z. wiss. Zool.* **72** (1902).

G 8.13. Gemmation or Budding

Gemmation or budding may lead primarily either to an increase in the number of individuals or to the production of a colony. It is difficult to separate it formally from fragmentation, but in practice there is seldom likely to be confusion. A bud is much smaller than the parent, is lateral, and is formed as a whole from a small group of embryonic cells. Budding is characteristic of the coelenterates and the ascidians. Lateral buds grow on *Hydra* when the food-supply is good, particularly (in the laboratory) in April and May. Secondary and even tertiary buds may be formed before separation begins. This last process is carried out by the bud seizing on to some solid object with its tentacles, and pulling itself apart from its parent. Except that it may grow one or two more tentacles, and will at the proper season form gonads, the bud at the time of detachment is a perfectly formed *Hydra*. In most of the other coelenterates, whether Hydrozoa or Actinozoa, the buds do not become detached, so that a colony is formed. Budding occurs only occasionally in the Scyphomedusae, where the scyphistoma sometimes multiplies in this way.

There are several species of compound ascidians, which bud in different ways. In the larger number of genera, of which *Clavellina* is an example, a hollow median stolon grows out from the ventral side of the abdomen, and on this buds grow. Blood vessels and other structures grow into these and are common to the whole colony. In a few genera, such as *Botryllus*, the buds are formed from paired outgrowths of the atrium, and in *Doliolum* the buds are formed at one point, become detached, and then migrate to another place where they become attached again.

8.2. SEXUAL REPRODUCTION

G 8.21. Gametes and Fertilization

The essence of sexual reproduction is that two cells come together and fuse. The fusing cells are called gametes, the resulting structure is a zygote, and the orderly process by which the cytoplasm of the two mixes and the nuclei become combined to form one is called syngamy. 'Fertilization' and 'syngamy' may be taken as synonymous, but the former is generally used to cover the process whereby the sperm approaches and enters the egg and initiates further development, while syngamy is used for the mixing of the two cells, and never unless the two sets of chromosomes combine to form a single nucleus. The fact that sometimes the nucleus of one of the fusing cells divides before nuclear fusion takes place but after the two cytoplasms have mixed, makes a formal definition of gametes and of fertilization practically impossible, for in these cases the cytoplasm that fuses is that of a different pair of cells from those to which the fusing nuclei belong.

It is obvious that syngamy by itself leads to a reduction in the number of individuals. By itself, then, it does not lead to a perpetuation of the race unless the gametes exceed in number the parents that produce them. This is normally the case.

It follows from the definition of syngamy as the fusion of two cells that the Protozoa are the only animals in which it can take place between adults. In this phylum there are a few species where ordinary full-grown individuals may fuse, as in *Polytoma*, and the partner that bears all the cytoplasm in the syngamy of ciliates may perhaps be regarded as full-grown: it is certainly full size. Such gametes are called hologametes, and the small specially prepared ones merogametes. Outside protozoology the terms are never used since they are unnecessary.

The Protozoa also differ from the Metazoa in that the two fusing gametes may be both alike so far as the microscope can distinguish. They are then called isogametes. More often in the Protozoa, and everywhere outside this phylum, the gametes differ in size or form or both, and are called anisogametes. The smaller may be called simply a microgamete and the larger a megagamete, but more often the one which is small is also more active, and is called a male gamete or a spermatozoon, and the other, large and sluggish, is called a female gamete or an ovum. The first term is often shortened to sperm, and the second translated as egg. At this point, where dimorphism of gametes begins to be developed, sex, as distinct from sexual reproduction, begins. The differentiation of the sexes is another thing still—the existence of two types of adult to produce the two types of gamete. Although primarily the difference need concern only the reproductive organs (such is the case in the coelenterates) it may affect accessory structures, as in the frog, the whole external appearance, as in many birds, or the general mental outlook, as in man. An animal that produces both male and female gametes is a hermaphrodite. Much has been written about the possible reasons for the existence of sex and sexual reproduction, but all the alleged explanations are teleological rather than physiological.

In the Metazoa the preparation of the gametes involves a meiosis, that is a process of cell division in which the chromosomes are reduced to the haploid number. In the Protozoa this is difficult to make out, but it has been observed in widely separated species—in *Actinophrys*, *Paramecium*, and some amoebas—so that it is probably the normal thing. In most Sporozoa and in the Volvocina meiosis occurs immediately after syngamy, and the animal lives nearly all its life in the haploid state. (This is, of course, also the case in the gametophyte generation of plants.)

Except for being haploid, and having more stored food material than most, the egg does not differ greatly from the standard animal cells of the textbooks. The sperm, on the other hand, differs from most metazoan cells in being motile, and much attention has been given to studying its activities. In all phyla except the Nematoda and the Arthropoda it has a single long tail, which is a flagellum (section MVJ 7.1). Traces of fibrillar structure, suggesting a flagellar origin, have been found in the processes of the sperms of insects, but not in the peculiar stiff processes of those of crustaceans or in the amoeboid sperms of nematodes. In the typical sperm there is also a head, consisting of little more than a nucleus, with an anterior cap, the acrosome, and a middle piece, with the basal body of the flagellum.

In general, sperms are inactive in the testis, and only start to swim when they come into the fertilization medium. The semen of the salmon and some other fish contains something that inhibits the movement of sperms, for they start to swim when the semen is merely diluted; it is possible that the inhibition is simply too high a concentration of potassium ions. In other fish, and in starfishes, sea urchins, and many worms, some organic substance derived from the egg is said to be necessary, for the sperms begin to swim only after they make contact with water that has passed over unripe eggs. This egg-water activates them, raises their respiration rate, increases their life span, and may make them clump together. Names such as fertilizin and gynogamone have been given to a substance, supposed to be a mucopolysaccharide, produced by the egg, which is said to react with a protein called antifertilizin or androgamone produced by the sperm, but it is unlikely that all the effects of egg-water are produced by a single substance, and the possible effects of changes in ionic balance or concentration have not always been eliminated in the experiments. There are great specific variations, the sperm of *Echinus esculentus*, for example, having its activity increased four times by egg-water, while that of *E. miliaris* is scarcely affected.

The metabolism of mammalian sperms is broadly similar to that of other tissues, but the cells contain very small reserves and most of their energy is obtained from external fructose which is present in the secretion of the seminal vesicles. Experimentally they can use glucose and mannose as well when these are added to the medium surrounding them. Glycolysis appears to be of the normal type, the citric acid cycle operates, and cytochromes are present. Fructose is also present in the semen of the rough-hound and the locust. Sea-urchin sperms use mainly phospholipids, and these are used also in mammals, particularly when the sperms are still in the epididymis.

In nature, sperms probably seldom live for more than 24 h after leaving the testis, and fertilizing power is lost sooner than this, but in a few species, such as bats and some snakes, they are stored in the female for months, or even, as in the honey bee, for years.

The mere prodigality in the number of sperms probably ensures that many of them are brought close to an egg, although in the mammals that have been examined (rat, rabbit, ferret, ewe) it seems that of the 100 or 1000 million sperms in one ejaculate, usually less than 100, and never more than a few thousand, reach the Fallopian tube, where fertilization occurs. In man the distance which the sperms have to travel to meet the egg in the tube is about 250 mm; their speed of movement is 2–3 mm min^{-1}, so that their journey would take them about 2 h, which is well within their life span of up to 40 h. It has been shown that in a number of mammals, the time from copulation to their appearance at the site of fertilization is very short, so that they must be sucked up by the female tract. In the cow, even though they are placed in the vagina, they arrive in the upper oviduct within 4 min.

A chemotaxis, by which a sperm in the neighbourhood of an egg turns and swims towards it, has been claimed for a few species, but if it occurs at all, it is not common. If by chance a sperm touches the jelly that surrounds an egg it nearly always burrows in normally to the surface. The penetration is assisted by the enzyme hyaluronidase and others, and the egg may put out a process that helps to draw the sperm in. Since fertilization is specific, or nearly so, there must be a chemical compatibility of egg and sperm that is probably similar to the normal immune reactions of cells. In some mammals the sperms do not acquire the capacity to enter the egg until they have been in the female's body for some hours.

In elasmobranchs, urodeles, reptiles, birds, some molluscs, and many insects several sperms enter the cytoplasm, but when one of the male pronuclei has begun to fuse with that of the egg, the others degenerate; some evidence suggests that this is caused by a substance diffusing from the successful pronucleus. In other species, the first contact of egg and sperm causes a change in the cortex of the egg which passes over the surface in less than 2 s, and reduces the chance of entry of another sperm to one-twentieth of normal. Further changes, taking about 1 min, render the egg quite impermeable, by the formation of a fertilization membrane. This, in some animals, appears to be merely the pre-existing vitelline membrane which has been lifted up and made visible, and in all cases the membrane is only a sign of much more fundamental changes. There is, however, no regularity in the changes in oxygen consumption, respiratory quotient, and so on, since these depend also on the stage of development of the egg at which fertilization takes place.

G 8.22. Parthenogenesis and Gynogenesis

Teleologically, fertilization has two aspects: the initiation of changes in the egg so that a new adult is produced, and the combination of two haploid nuclei to form a single diploid nucleus, with all the genetic consequences that follow from this. Under certain circumstances each of these may be upset.

Experimentally, the eggs of several animals, such as sea-urchins and frogs, have been made to develop by the action of various chemicals or by pricking them with a needle. When this is successful, the cells of the resulting embryo are usually haploid, and the adult is dwarfed, but sometimes there is a failure of cytoplasmic division after the first induced nuclear division, and so the diploid state is restored. This artificial parthenogenesis demonstrates the independence of the developmental and genetic aspects of fertilization.

Sperms have little food material, and no one has ever induced them to develop on their own; nor has male parthenogenesis been recorded in nature except in a few Protozoa, such as *Actinophrys*, where the difference between the two gametes is slight. Female parthenogenesis, the development of a cell which by origin is clearly a female gamete, is fairly common.

In some flatworms, rotifers, and wasps and saw-flies (Hymenopetra), parthenogenesis is the only known form of reproduction, and males have never been seen. Some earthworms are polyploid species, and have solved the problem that this presents by becoming parthenogenetic. More often, parthenogenesis and ordinary sexual reproduction go on together. In *Paramecium* and some other ciliates a type of parthenogenesis called endomixis, in which the meganucleus is destroyed as in conjugation, has been described, but its existence has been denied by later workers. In bugs of the family Aphidae the wingless females which hatch in spring from the eggs that have survived the winter reproduce parthenogenetically, and several similar wingless parthenogenetic generations, all females, follow. Finally in autumn winged females and males are produced, and together they give rise to fertilized eggs that survive the winter and begin the cycle again. There is no constancy in the number of parthenogenetic generations, and the production of males and winged females appears to be connected with falling temperature or shortage of food or both. The Cladocera such as *Daphnia* and many rotifers are similar.

In some insects (Cynipidae or gall-wasps) there is a regular alternation. For example, the individuals of *Neuropterus lenticularis* that appear in March are all females. They lay parthenogenetic eggs in buds of the oak, and in the galls so formed the larvae of the next generation develop. The images that emerge from the galls in June are very different from their parents, and were formerly designated by a different generic name. Some are male, some female, and between them they produce fertilized eggs which are laid in oak leaves. The pupae remain in galls on the dead leaves until the following spring, when they come out as the next parthenogenetic generation.

Lastly, in the bees parthenogenesis is facultative, that is, it appears to be under the control of the female laying the eggs. The queen (the female) copulates only once in her lifetime, and stores sperms in a receptaculum seminis for months or sometimes years. Only if the eggs are to be fertilized are sperms allowed to escape on to them as they are laid.

Parthenogenesis is rare in the vertebrates; it occurs in a few lizards, and eggs of the turkey are said occasionally to develop without fertilization. In the teleost fish *Mollienesia* (= *Poecilia*) *formosa* it seems likely that although a sperm must enter the egg before development can take place, there is no fusion of nuclei, a phenomenon called gynogenesis. Males of this fish are very rare indeed, and it seems probable that the penetration of the egg is by sperm of other species of the genus. Gynogenesis is known also in a beetle, *Ptinus latro*, where, as in the fish, foreign sperm is used, and in a few planarians.

It is obvious that in the absence of any special arrangement parthenogenesis would mean a halving of the chromosome number in each generation. Almost all the conceivable ways by which this can be avoided are used by some animal or other. In Crustacea and Aphidae only one polar body is given off at maturation, and there is no reduction division; in bees the eggs are formed normally, but those that develop without fertilization all form males and there is no reduction in spermatogenesis. The cells derived from parthenogenetic turkey eggs are diploid, and there has perhaps been reduction followed by fusion of the egg with one of its polar bodies. This has been observed in some invertebrates, including starfish. A comparable fusion of nuclei derived from the same parent occurs in some ciliates and other protozoans, where it is called autogamy.

Paedogenesis is the name applied to parthenogenesis when it occurs in a larval form. It is found in Cecidomyidae (Diptera) such as *Miastor*. It should strictly be distinguished from neoteny, which is ordinary sexual reproduction by a larva.

M 8.23. Sexual Reproduction in Mammals

Before fertilization can take place the eggs and sperms must be produced by the parents, and must be brought into close proximity. In most mammals the production of gametes is seasonal, and the sperms are always, as in most land animals, placed within the female body by the act of copulation or coition. Both gametogenesis and copulation are largely under the control of the endocrine system.

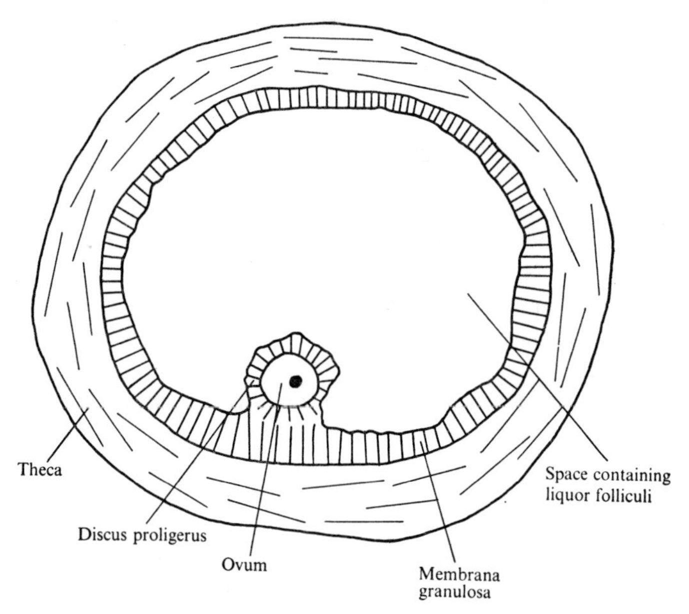

FIG. 8.3. Diagrammatic section of a nearly ripe Graafian follicle of a rabbit, $\times 125$.

The essential part of the ovary is the Graafian follicle. When this is ripe (Fig. 8.3) it consists of a vesicle with an outer wall or theca and an inner one called the membrana granulosa. Attached to the latter at one side is a mass of cells, the discus proligerus, which contains the ovum. The cavity of the vesicle contains a fluid, the liquor folliculi. The theca is formed from the general stroma of the ovary, while the membrana granulosa, discus proligerus, and the ovum itself are developed from the original cubical epithelium on the surface of the fetal ovary. It is improbable that this epithelium sinks in to form any new follicles after birth and in almost all the mammals which have been examined oogenesis is confined to the period shortly before and shortly after birth. Exceptions in which it has been observed in the adult are the cat, the nine-banded armadillo, and a few lower primates. Ovulation consists in the rupture of the follicle and the liberation of the ovum into the body cavity. Only a few follicles burst together; in the rabbit the normal number is five or six on each side, and in the human being but one, the ovaries apparently working randomly.

After ovulation the cells of the membrana granulosa and of the theca increase in size so that in the place of the follicle there is a dense mass of cells known as the corpus luteum.

The process of growth and regeneration of the follicle, with its accompanying changes in other parts of the body, is called the oestrous cycle.

The simplest sexual rhythm in mammals is annual, for most wild mammals breed at one season only of the year. Traditionally this is the spring, and most mammals breed about April in the northern hemisphere, and September in the southern. In most Artiodactyla, however, the rutting season is the fall, which, since they have a long period of gestation, leads to the birth of young in the spring. Some mammals with a short true gestation period also copulate in autumn, but the development of the embryo is delayed. Either, as in the weasels and bears, there is delayed implantation, the blastocyst remaining quiescent in the uterus, or, as in bats, the sperms are stored until spring. When an animal is moved from one hemisphere to another its breeding season may change so that the animal is still sexually active at a time of increasing or decreasing light intensity, as the case may be. Red deer that have been imported into New Zealand continue to rut in the fall, but this is 6 months different from their time in Europe. Sheep taken from Britain to South Africa have changed similarly, sometimes during their first year in the new home.

The oestrous cycle fits into this annual breeding cycle, but is not necessarily synchronous with it, for although some animals, such as the fox, are monoestrous, that is have one ovarian cycle in a breeding season, others are polyoestrous, and have several cycles fitting into the year. There is no essential difference between the two cases, but polyoestrus is best regarded as a secondary rhythm imposed on the main one. There are four phases of the cycle.

1. Dioestrus, a quiescent period when there is no sexual activity. The long period of quiescence between breeding seasons is called the anoestrus.

2. Pro-oestrus, a preparatory phase during which the follicle is developing.

3. Oestrus, the period of heat. At this time there are usually

changes of hypertrophy in the external genitalia, and it is only at this time that the female shows any sexual desire: at other times males are repelled. There are also changes in the walls of the vagina, which become cornified.

4. After oestrus there is a degenerative metoestrus (or postoestrus), but this is generally merged into the state of pregnancy (when fertile union has occurred) or pseudo-pregnancy (when it has not).

In many species, including man, ovulation occurs spontaneously during oestrus, but in others it occurs only after copulation; which is the commoner cannot be said, but it would seem that precise association of insemination and the liberation of eggs would increase the chances of fertilization. In the rabbit, ovulation takes place 10 h after copulation. The half-dozen ova are packed, with some other cells from the follicle, in albumen which clots round them when the liquor folliculi escapes. The mass is carried to the ostium abdominale by the cilia of the body cavity, and it then enters the Fallopian tube, forming a plug at the top of this. By this time the thickest density of the up-swimming sperms has nearly reached the top of the tube, and the chance of fertilization is high. The sperms liberate a protease that dissolves the albumen, so that when the eggs are fertilized they can be carried down to the uterus by the cilia of the tube. Fertilization has occurred within 16 h from copulation. Division of the egg starts in the Fallopian tube and continues in the uterus. The beginning of the attachment of the fetus to the latter starts on the eighth day.

It is likely that the course of events in other mammals is similar, but the time relations will naturally be different.

Changes in the endometrium (the lining of the uterus) have begun even before the embryo has arrived, but they become more marked after this. The superficial epithelium degenerates, and the embryo sinks in against the connective tissue. By a growing together of tissues from both embryo and mother the placenta is formed, and to make the maternal part of this the uterus grows out into villi, and its blood vessels, particularly the veins, increase greatly in size. As pregnancy continues the uterus increases in size, and its cavity becomes greatly enlarged to hold the developing embryos. The vagina also increases in size. Other changes occur in pregnancy, particularly the development of the mammary glands so that milk can be supplied to the young at, or (in man) a day or two after, birth. In the later stages of pregnancy the animal often behaves differently from the normal by retiring from the herd, making a nest, and so forth. Seals migrate long distances to land for parturition to take place.

Pseudopregnancy is a state that occurs at the end of the oestrous cycle in many mammals when fertilization has not taken place. It consists in the same phenomena as pregnancy, but they are developed to a lesser degree. The changes in the uterus begin, but cannot go to the stage of formation of the placenta, and there is some enlargement of the mammary glands. The bitch collects material for a bed, and the doe rabbit plucks her fur as she does to shelter her young in the ordinary way. In most mammals, but not in the rat or mouse, dioestrus is simply a short and not very marked pseudopregnancy.

The growth of the Graafian follicles is not itself the factor that determines the other changes, since oestrus occurs when all the follicles have been destroyed by X-radiation. It does not occur, however, if both ovaries are removed in their entirety. It is therefore likely that an oestrus-producing hormone is manufactured by some part of the ovary other than the follicles. Such a hormone was first discovered in extracts of ovaries in 1912, and has since been found in the placenta and in urine of both males and females, but particularly in that of females at about the time of ovulation and when pregnant. Extracts of all these induce oestrus within 2 days after being injected into spayed mice, but neither the urine nor any organ normally has oestrogenic properties unless either an ovary or a placenta be present. The primary product liberated by the ovary is almost certainly oestradiol, and the other less active substances with oestrogenic properties, such as oestrone, oestriol, equiline, and equilenin, are probably derived from it. Collectively all are called oestrins or oestrogens; they are steroids (section M 6.142), and are not specific. A number of related synthetic substances have been shown to produce oestrus.

Oestrus is usually followed by pregnancy or pseudopregnancy, and the change-over to this phase is marked by the development of the corpus luteum. In those animals, such as mice, in which there is scarcely any pseudopregnancy, the corpora lutea develop very little and soon degenerate unless copulation occurs. If a female mouse is mated with a sterile male, fertilization cannot occur, but the corpora lutea persist much longer than usual, and there is a marked pseudopregnancy which postpones the onset of the next sexual cycle for 7 days. (Normally the whole cycle only takes 5 days.) Pregnancy or pseudopregnancy is associated with the production by the corpus luteum of a steroid called progesterone (or progestin, or luteosterone) which in addition to sensitizing the uterus for the attachment of the embryo inhibits oestrus and ovulation. Oestrus can therefore occur again only when the effects of the corpus luteum have worn off.

The discovery of oestrogen and progesterone gives an immediate cause for oestrus and pregnancy, but it does not fully explain either the rhythm or the first onset of oestrus at puberty and in each annual season. Two hormones from the adenohypophysis (section MV 6.131) are at work here. Injection of suspensions of the anterior pituitary into adult females causes ovulation, followed, as would be expected, by formation of corpora lutea. Similar treatment of immature females leads to a precocious oestrus and ovulation; conversely, removal of the pituitary leads to complete cessation of sexual activity. Follicle-stimulating hormone (FSH), which is a glycoprotein, induces ripening of the follicle, while luteinizing hormone (LH), a mucoprotein, transforms it into the corpus luteum. Both are necessary for ovulation.

The interaction of the anterior pituitary and the ovary seems to be mutual, for in castrates the gonadotropic activity of the pituitary increases. Moreover, it can be lowered again by the administration of oestrogen.

While the mother is pregnant her mammary glands develop. The first phase, the growth of the tubules, is produced by the combined action of oestrogen, progesterone, and the growth hormone of the pituitary, and is also helped by the corticosteroids from the adrenal. The chemical secretion of the milk follows on the production of another hormone from the pars distalis, prolactin (or lactogenic hormone, or mammotropin), which is a polypeptide. The maintenance of pregnancy depends on the continued production of progesterone. In the sheep and man the corpora lutea have a natural life-span, in the rat and mouse their life is maintained by prolactin, which is therefore called also luteotropin (LTH). In the pig the corpora lutea are finally destroyed by their own progesterone. In some animals, such as man, horse, cat, and rat, they degenerate, or may be removed before pregnancy is ended, without causing abortion; the production of progesterone has been taken over by the chorion of the placenta, which is also able to make some of the other sex hormones.

The cycle may then be pictured as follows (Schema 11): first, the pituitary produces follicle-stimulating hormone, which induces the ovary to form oestrogen. The latter brings on the characteristic phenomena of oestrus, and reduces the activity of the pituitary. Oestrus is followed by ovulation and luteinization as a result of the presence of luteinizing hormone. (It is obvious that corpora

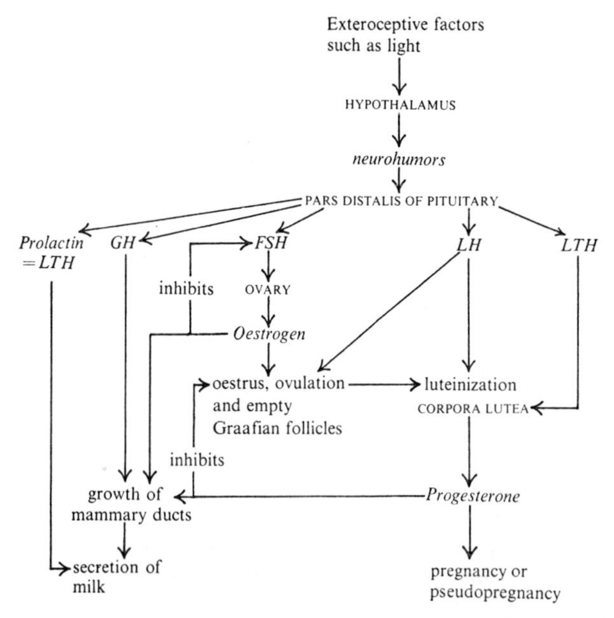

SCHEMA 11. The sexual cycle in female mammals (corticosteroids omitted). Small capitals indicate endocrine glands, italics indicate hormones.

lutea cannot be formed until ovulation has occurred.) Meanwhile the inhibition of the pituitary by oestrogen has removed the stimulus for any further ovulation or production of oestrogen, and then with the depressant factor removed the pituitary becomes active again. The ovary is again stimulated to produce oestrogen, and a second cycle has begun. Where the FSH/LH ratio is high, as in the mare, oestrus is long and ovulation occurs before it is over. Where the ratio is low, as in the cow, oestrus is short and ovulation does not take place until it is finished. Cold and other factors may upset the balance, so that oestrus and ovulation do not go together, with sterility as the inevitable result.

To make the story reasonably complete we must know the immediate cause of the secretion of gonadotropic hormones by the anterior pituitary, and here the oestrous cycle can be linked up with the longer seasonal cycle. If doe ferrets are exposed to electric light in winter, they come to full heat in 38–64 days, although they are not normally in season until April. It is the visible and particularly the ultra-violet rays which are effective, and similar results are obtained for a spread or a concentrated dose of the same total magnitude. Animals from which the anterior pituitary has been extirpated do not respond. Raccoons also have been induced to breed 6 weeks before their normal time by exposing them to an increased light intensity.

In sheep, the effective influence is the change from an increasing to a decreasing light ration, and this produces oestrus just as easily when it occurs at an experimental day-length of 13·5 h as when it occurs, under natural conditions, with one which is much longer. Clearly, if this were not so the system could only work at a restricted latitude. While all the mammals that have been investigated are dependent in some way on light, the exact form of the stimulus differs from one species to another so as to be ecologically suitable. There is even some slight evidence that the sexual success of nocturnal Malayan forest rats is increased by the full moon.

In these experiments the effect of temperature was eliminated, but it has been shown that in some rodents a variation of temperature is also necessary to maintain the cycle. Temperature may possibly be an important accessory factor in nature, and food may also have an influence. The oestrous cycle of mice is shortened by the proximity of a male, and may be suppressed, in the absence of a male, by the presence of other females. In domestic animals, such as the cow, that may breed at any time of the year, it seems likely that excess of food or some other factor has eliminated the necessity for light or any other external stimulus to start the cycle in operation.

Oestrus has been produced experimentally by electrical stimulation of the hypothalamus, and it can be prevented by cutting the blood vessels from this to the adenohypophysis. It is therefore probable that light causes the hypothalamus to send hormones to the hypophysis, which then secretes the gonadotropins. There is some evidence that there are separate factors for luteinizing hormone and follicle-stimulating hormone, and that the former is a polypeptide. In an experiment a blind ferret did not respond, so that the stimulus is probably received by the eyes, but in rats light can penetrate to the hypothalamus and direct stimulation is not impossible. In those animals in which ovulation occurs only after copulation, this causes a rapid activity of the hypothalamus, for if adrenergic blocking agents (which are known to inhibit nervous pathways) are applied more than 3 min after copulation they do not inhibit ovulation. Hypophysectomy, done within 1 h, does prevent ovulation, but after this time does not, so that the pars distalis must by then have produced and released enough follicle-stimulating hormone. How the hypothalamus is stimulated is unknown, for ovulation takes place even if copulation is carried out with the vagina and vulva anaesthetized, the ovaries transplanted, and the sympathetics cut.

In the Old World monkeys and in man the oestrous cycle appears to be replaced by the menstrual cycle, which is marked by a flow of blood and mucus from the uterus. By investigations on monkeys, together with some evidence from man, it has been shown that the two cycles are closely interlocked. First, there is the interval dominated by oestrogen, when the follicle is maturing, leading up to ovulation. During this time the endometrium is unspecialized and thin. After ovulation, a corpus luteum is formed, and progesterone becomes the dominating hormone; under its influence the endometrium thickens and becomes vascular and glandular, the premenstrual period. This represents a preparation for the ovum, and is obviously a condition of pseudopregnancy. When fertilization does not take place the corpus luteum degenerates, and so both hormones are reduced in quantity. As a result of this the endometrium breaks down and the blood and cells resulting are discharged in menstruation. There is a brief post-menstrual period before the next follicle begins to mature. Thus the periods of pro-oestrus and oestrus are represented by the interval, pseudopregnancy by the premenstrual period, and the dioestrus by the menstrual and post-menstrual periods. In the human menstrual cycle ovulation probably occurs between the thirteenth and seventeenth days, counting from the beginning of the menstrual flow (Schema 12), with modifications when the cycle is of unusual length. Ovulation can be suppressed by progesterone, which is the basis of contraceptive pills.

Birth is under hormonal control, partly by the cessation of supply of progesterone, and partly by the contraction of the uterus under the influence of oxytocin (section M 6.132). In many mammals it is made easier by a softening of the pubic symphysis and the iliosacral suture, which is perhaps induced by a hormone called relaxin made by the corpus luteum. It appears to be a polypeptide.

In most mammals the males also show a seasonal rhythm; in some the testes become small in the off season, and are withdrawn into the coelom.

The interstitial cells of the testis produce a hormone called testosterone which is closely allied to the ovarian hormones (androsterone is less active and is found only in urine). Injected into castrated rats testosterone prevents degeneration of the accessory sexual organs—vasa deferentia, prostate gland, etc.—which they would otherwise

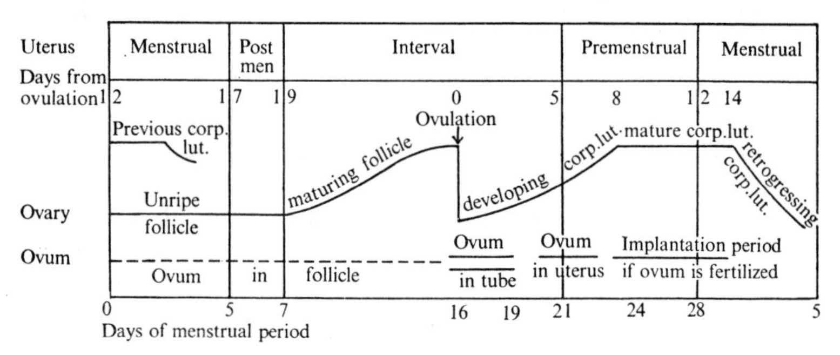

SCHEMA 12. The relationship of the human menstrual cycle to the normal mammalian cycle. (Slightly modified from Corner.)

suffer. Rats castrated before puberty show no sexual behaviour, but if they are injected with a derivative of testosterone their accessory organs develop and they copulate when presented with females. Testosterone is not formed if the anterior pituitary is removed, but hypophysectomized animals do make it if they receive injections of luteinizing hormone. Buck ferrets that receive artificial light in winter become sexually interested and their accessory organs develop, but the unions are sterile, because spermatogenesis is incomplete. Cotton-tail rabbits (*Sylvilagus transitionalis*) treated in the same way produce sperms. These observations suggest that the male is influenced similarly to the female, but the details of the process are even less well known. There is no evidence that males undergo rhythmical changes similar to the oestrous cycle of females, and there is no reason why they should. In most mammals (the rabbit is an exception) the female is only willing to receive the male while she is on heat—that is just about the time of ovulation—and provided that the male is then able to discharge active sperms fertilization is reasonably certain.

This account of the control of reproduction is simplified; there are many specific differences and a given hormone often has different effects when it is injected into different animals. The steroids from the adrenal cortex have sexual functions, and others with oestrogenic or androgenic properties are known both in invertebrates and in plants; this is not surprising in view of the chemical relationship between the sex hormones and cholesterol, which is a universal constituent of animal cells, and ergosterol and other steroids which are common compounds in plants. Young grass is rich in such substances, and may affect the lactation and the composition of the milk in cows that feed on it.

Finally, both androgens and oestrogens play some part, though exactly what is not clear, in the general psychic and emotional life. Androsterone probably raises and oestrogen lowers the position of chimpanzees in the social hierarchy. The evaluation of the function of these hormones is made more difficult by the fact that both occur to some extent in both sexes.

V 8.23. Sexual Reproduction in other Vertebrates

The seasonal breeding of birds is even more obvious than that of mammals, and in northern latitudes is often associated with a greater or lesser degree of migration. Equally, that birds are sensitive to light is shown by the 'dawn-chorus'—the outburst of song that occurs just before sunrise. Some thirty species of several families (but mostly passerines) have now been investigated more or less scientifically, and the general picture is closely similar to that in mammals. The annual cycle is most marked in the male, where the testes show a great variation in size and are at their minimum about midwinter. As the days lengthen they begin to enlarge, and the cells become active and produce spermatozoa. They continue like this for a few weeks or months, and then begin to shrink, and by autumn are small and inactive again. The secondary characteristics of breeding condition—plumage changes, song, and courtship—follow this cycle. The ovaries also follow a seasonal change in size, which in some species at least is a little later than that of the testes.

Various forms of sexual activity, and sometimes, especially in the male, growth and gamete-production in the gonads, can be induced at any time from late autumn onward by increasing the length of day. There are differences between species, but in general it seems that there is no need for the increase to be gradual; it is the number of hours in which the light is above a certain minimum of intensity that matters, although in some a split ration of light—two short periods with in 24 h—has more effect than one continuous period of the same total length. At the end of the breeding season there is in most species a refractory period during which no manipulation of the light has any effect. The causes and the significance of this are disputed, but in general it seems to be produced by the relatively short days of autumn. Many birds show a little sexual behaviour in autumn, marked in the males by a renewal of singing, and in a few species this has been shown to be connected with the end of the refractory period and a temporary increase in the size of the testes.

There are still many difficulties in this explanation. Gametogenesis may begin in England at the end of the first week in January, when the days are only 14 min longer than at their shortest a fortnight before, and the increase in the gonads of birds that migrate from the southern to the northern hemisphere begins before they start their journey. Many tropical birds have breeding seasons (sometimes of 9 months), even though the maximum change in day-length is very small. There is some ecological evidence that breeding seasons may be connected with food supply or rainfall, and in a careful experiment breeding was induced in the tropical red-billed weaver-finch, *Quelea quelea*, by feeding it with green grass. Rain plus dry grass had no effect, nor did rain added to green grass increase the induction of breeding. It has also been suggested that changes in the composition of the light, including the ultra-violet, may be important in the tropics (where many plants also are seasonal). Temperature is undoubtedly often an accessory factor, for sexual activity may cease altogether in cold weather, as may often be seen in an English spring, when May frosts cause the cessation of pairing and the abandonment of the nests of many birds.

The light acts on the hypothalamus, either through the eyes or directly. In the domestic duck direct stimulation is adequate, and in normal circumstances red light, but not blue or green, probably reaches the brain in adequate quantity; this would explain why red light is more effective than green in inducing sexual activity. The adenohypophysis is then stimulated by neurohumors, as in mammals, but the hormones that it produces have not been adequately characterized. Probably two gonadotropins are produced, but the relationship of these to mammalian follicle-stimulating hormone and luteinizing hormone is unknown, since the mammalian hormones injected into birds have similar, but not identical, effects to those obtained with avian extracts. As in mammals, one component causes the formation of sperms, the other induces the interstitial cells of the testis to secrete an androgen, which is responsible for many of the male characters of plumage and behaviour. The influences of the two hormones in the female are not so clear as in mammals, but the ovary is stimulated to produce oestrogen, which determines many of the female characters. There is evidence that ovaries can produce androgens and testes oestrogens, and both also produce a progestin.

Ovulation does not usually take place unless a male bird is present, and in most species there is a series of behaviour patterns, best known as foreplay, in which both sexes participate, which must take place before coition or ovulation will occur. It seems likely that foreplay induces secretion of follicle-stimulating hormone, or its equivalent, and that luteinizing hormone is induced by other external factors.

Domestic pigeons will not ovulate if they are alone, but will do so if two hens are confined together, or even if one is provided with a mirror. A few birds, such as domestic ducks and hens and grey parrots, need no such stimulation.

Generally one nest is built, and a characteristic number of eggs are laid; if the nest is destroyed another will be built, and if eggs are removed as they are laid, more than the normal number may be produced. The house sparrow, for instance, normally lays a clutch of four or five, but has been induced to lay as many as fifty. Stimuli from the eggs when the bird sits on them cause the adenohypophysis to secrete prolactin, which inhibits the secretion of gonadotropins and so stops the laying of further eggs. It causes (in cooperation with oestrogen) broodiness, and the development of brood patches on the abdomen—bare, vascular, areas of high temperature, through which the eggs are kept warm. In some species these occur in the male as well as the female, and in both sexes of pigeon prolactin causes the secretion of pigeon's milk from the crop.

The effect of the sex hormones on psychic life is comparable to that in mammals. Androsterone raises and oestrogen lowers the position in the social hierarchy just as it does in apes.

Little work has been done on reptiles, but in many species reproduction is seasonal. The testes of some snakes have an annual cycle of size and spermatogenesis similar to that of birds, and, as the increase begins in early spring, light in some form is probably the stimulating factor. Hypophysectomy or ovariectomy in pregnant viviparous snakes causes resorption or abortion of the embryos. Progesterone has been extracted from the ovaries and blood of viviparous snakes.

If specimens of the South African clawed toad, *Xenopus laevis*, are brought into the laboratory, they show no sexual behaviour, even during the mating season. Amplexus can, however, be induced at any time by injections of anterior pituitary extract. Ovulation follows, and the larvae from the unions have been reared through to a late tadpole stage. In the female, anterior pituitary extract can be replaced by progesterone, and sometimes this will induce mating and ovulation even in hypophysectomized or immature individuals. Several other frogs and toads are broadly similar, and further experiments suggest that luteinizing hormone, or its equivalent, is required for sperm formation, and both this and follicle-stimulating hormone for the production of eggs. In *Rana temporaria* spermatogenesis is seasonal, but artificial light in December makes the frogs try to spawn. *R. esculenta* in Mediterranean countries, on the other hand, produces sperms throughout the year, and although those living in colder climates do not, they can be made to do so at any time simply by raising the temperature. American spade-foot toads (*Scaphiopus*), which breed in temporary pools, produce eggs or sperms at any time within a few hours of heavy rainfall, provided that the temperature is more than about 10 °C.

The sexual seasons and migrations of some fish are as striking as those of birds. Little experimental work has been done, but in lampreys, elasmobranchs, and teleosts hypophysectomy prevents sexual maturation. Both androgens and oestrogens have been found in the blood of teleosts, so the general pattern is probably the same as in other vertebrates. In the minnow a day of 17 h will induce gametogenesis provided the temperature is above a certain minimum, while in the stickleback temperature alone can be effective.

J 8.23. Sexual Reproduction in Invertebrates

The sexual patterns of invertebrates are as various as their structures. In the majority the gonads are ripe only at one season of the year, the advantage of this presumably being that if the same number of gametes are produced in a shorter time the chance of fertilization is higher. In many species, if young were produced in unfavourable seasons they might not be able, through low temperature or lack of food, to survive or grow. If there is a season, it is essential that eggs and sperms should be produced at the same time, and this can be achieved if both sexes respond to the same environmental stimuli. In *Planaria alpina* the gonads develop when the temperature falls below 10 °C, in the oyster high temperatures are necessary, and a sudden rise causes liberation of gametes. In the pond snail *Limnaea palustris* eggs are produced when the length of day is 13·5 h or more. In some parasites the stimulus comes from the host; nematode parasites of the sockeye salmon probably depend on the host's sexual hormones, and full maturation of the eggs of the rabbit flea (*Spilopsyllus cuniculi*) only takes place in pregnant does.

In the sea, gametes may simply be shed into the open and depend on large numbers for the chance of fertilization, but in many polychaetes the two sexes undergo a metamorphosis and swarm together, so that the eggs and sperms are shed in close proximity. In the Palolo worms, *Leodice* (*Eunice*) *fucata* in the Atlantic and *L. viridis* in the Pacific, the swarming time is very limited—the last quarter of the October–November moon in *viridis*, and the last quarter of the June–July moon in *fucata.* The worms live in tubes, from which the heads normally protrude. The night before the swarming the animals change their position so that the tails protrude, and as soon as the sunlight touches the water on the morrow the tails break off and swim by themselves to the surface. They emit eggs or sperms from the anterior wound where they broke off from the main body of the worm, and then die. The metamorphosis of *Platynereis dumerillii* into the sexual heteronereis is brought about by increasing day-length, but is for a time inhibited by a hormone from the cerebral hemispheres. The value of lunar periodicity, which occurs also in some sea urchins and molluscs, is doubtful, since closely related species may reproduce well enough without it, and a species that has a lunar rhythm in one place may be without it in another. Even Palolo worms from some areas do not swarm, but if those from the swarming areas are kept in the dark in the laboratory, they swarm at the same time as their relatives that have been exposed to the moonlight. No satisfactory analysis of the components of lunar rhythm has ever been made.

There is some evidence that reproduction in molluscs and crustaceans is controlled by hormones, but very little experimental work has been done. In the slug *Arion* (which is hermaphrodite) there may be two hormones, one from the brain causing development of eggs, and one from the eye-stalks causing development of sperms. In the unisexual *Octopus* a single hormone from the eye-stalks initiates development in both sexes. In many decapod crustaceans there is an ovary-inhibiting hormone in the sinus gland (which is dependent on the light ration) and a small gland on the vas deferens without which the testis cannot develop. In the insects by contrast, though metamorphosis is under endocrine control, once this is achieved the further development of the gonads follows inevitably.

The actual release of the gametes may depend on the presence or activities of the opposite sex. *Odontosyllis enopla*, the Bermudan fire-worm, swarms on the second, third, and fourth days after full moon throughout the year, at 55 min after sunset whatever the state of the sky. When the females swim to the surface they begin to glow, and then, as each one starts to release her eggs, she swims in circles of 2 or 3 inches in diameter and becomes highly phosphorescent. A male, which has been waiting 10 or 15 ft below, swims straight for the centre of the circle, flashing as he does so, and the two swim together in slightly wider circles emitting eggs and sperms. If the female ceases to be phosphorescent he does also, and he is incapable of finding her. The worms are not luminous at other times.

In some species the shedding of gametes is determined by some substance released by the opposite sex. Under normal circumstances in *Nereis limbata* (an American Atlantic form) several males swim round a female and shed sperms, and she then liberates eggs. If the sexes are confined in the laboratory, the gametes are retained, but the addition of water that has contained females causes the release of sperms, and sperms cause liberation of eggs. A similar type of mechanism operates in the oyster, and it is probable that something of the kind is widespread.

The chance of fertilization becomes even greater if the sperms are placed inside the female; such copulation is essential in terrestrial species, since sperms can move only in water. It is found in aquatic species such as turbellarians, cephalopods, gastropods, and crustaceans, chiefly where strong currents would be liable to sweep freely shed gametes away.

In some groups, such as turbellarians and gastropods, there is a special intromittent organ or penis. In *Octopus*, a specially modified tentacle, the hectocotylus, places the semen in the mantle cavity of the female. The male spider sheds a drop of semen on to a leaf, or sometimes on to a small web that he spins specially for the purpose. To this he applies the end of his palp, which has the terminal joint modified to contain a tubular seminal vesicle. The palp containing the sperms is then inserted into the genital opening of the female. This might almost be called artificial insemination. Male and female scorpions perform a dance, a 'promenade à deux' in which the male holds the chelicerae of the female. He extrudes a packet of sperms, or spermatophore, and guides his partner over this. She presses down until its valves are within her genital aperture, and stands immobile until the semen is absorbed. She then leaves it and the pair separate. In the insects special parts of the exoskeleton hold the male and female together in coition.

G 8.24. Hermaphroditism

It is not necessary to sexual reproduction that there should be separate sexes, and in many species both eggs and sperms are produced by the same individual; that is, the animal is hermaphrodite. In many of the Protozoa, such as *Paramecium*, hermaphroditism may perhaps be primitive, but in the majority of animals in which it occurs it seems to have some connection with the environment, for it is common in parasites and in the fresh-water and terrestrial forms of some groups, but is rare in the sea. It is obvious that in parasites and in animals living in small isolated ponds hermaphroditism accompanied by self-fertilization would greatly increase the chance of the production of offspring, and might therefore have been evolved by natural selection, but in fact self-fertilization is rare, and ova and sperms are not usually produced at the same time. An animal that forms first sperms and then ova is protandrous, and if the order is reversed it is protogynous. In many animals, such as the snail and the earthworm, this alternation of sex is seasonal, but in some others there is no retrogression to the original sex once the change has taken place. The term hermaphrodite is not very appropriate for these, as there is just one very real, though perhaps slow, change of sex during the life-history. The best-known example is the mollusc *Crepidula fornicata*, the slipper limpet. The larvae of this are male when newly hatched, but if they settle down on the floor of the sea they become females. Others settle on top of them, produce sperms, and fertilize the females below. After a time these males begin to change their sex, and other males settle on their backs. This goes on till a pile of half a dozen or more molluscs is formed, those at the bottom being females, those in the middle hermaphrodite, and those on top males, of which the later ones are waiting for the animal below to change its sex. Each individual is at first functionally male, then hermaphrodite, and finally female, with the exception of the first in a chain, which is female only, and the last, which is male only.

The determination of sex is, in general, genetic, one sex having an additional chromosome, but in vertebrates the embryo is potentially hermaphrodite. The gonad is at first indeterminate; if the cells of the cortex develop first they become eggs, and suppress the cells of the medulla, but if these are ahead they become sperms. Experimental manipulation can produce an animal whose sex is opposite to that which its genes would normally produce. Once the sex of the gonads is decided, hormones take over. As the gonad develops into one sex or the other it produces something that may induce like development in other rudiments of the opposite sex grafted near it. This effect has been shown in amphibians, birds, and mammals, and in some species the male is dominant, in others the female. In general the Müllerian duct develops without hormonal control, but if a testis is present it is suppressed; conversely, the Wolffian duct does not develop unless a testis is present. Whether these effects are due to the adult hormones, or whether they are produced by special ones, is not known.

All the secondary sexual characters that have been tested can be maintained in male castrates by injection of testosterone, which may therefore be the only hormone concerned in their production. For example, castrated deer do not grow antlers, and the fructose normally added to the semen of the rabbit and bull by the accessory glands is not formed after castration, but both antlers and fructose are restored by testosterone.

The secondary effects of oestrogen are not so clear. Injected into castrate adolescent kangaroos it causes the scrotum (which is in front of the penis in marsupials) to change into the typical female pouch. With this sort of control it is not surprising that errors sometimes occur, and intersexes, having both male and female characters, are produced. The cyclostomes and a few teleosts are functionally hermaphrodite, and sex-reversal in old age is known in birds.

Something similar probably occurs in crustaceans. In the males of Malacostraca (except for the isopods) there is a small androgenic gland attached to the vas deferens. Its presence is necessary for the development both of the secondary male characters and of the testis itself. If it is transplanted into a female she becomes masculinized, and may even produce fertile sperm. It therefore presumably makes a male-determining hormone. Although in general the female seems to be an indifferent form (i.e. one that has not been determined to become male), the ovary itself produces a hormone that is necessary for the development of some secondary female characters, such as the brood-pouch. There is probably also an ovary-inhibiting hormone in the sinus gland.

The sex of insects is much more stable. In some species, however, intersexes or reversal of sex can be produced, for example in some Hymenoptera by parasitism. This may act by reducing the food supply, and is perhaps comparable to the situation in some parasitic nematodes, where a high rate of infestation causes more of the eggs to become male.

G 8.25. Viviparity; birth rate

Where there is internal fertilization it is possible for the eggs to develop within the body of the mother. Such a condition, called ovoviviparity, has been achieved by a number of invertebrates, and by scattered fishes and reptiles. Complete viviparity, in which the embryo is nourished as well as kept within the body, occurs in mammals (except monotremes), probably a few Squamata, and in many sharks.

It is obvious that if the population is to remain stable, the birth rate must equal the death rate, and one of the most difficult problems of physiological ecology is to decide how the equivalence is maintained. The simple theory of natural selection suggests that unless a high birth rate has special adverse effects, for example by reducing the food available for each sibling in a family, as happens in some birds, it ought always to confer an advantage on its possessor, and so be selected. Obviously this is not always so; many large sea birds, which have plenty of food, few enemies, and long life, lay few eggs and do not breed in every year, so that their birth rate appears to be adjusted to their low death rate. Experimentally the fecundity of many animals can be altered by manipulating external conditions, but these have little or no relevance to the main problem.

9. Regulation

It is characteristic of living things that they maintain within themselves conditions different from those of the outside, physical, world. They are nevertheless affected by external conditions to a greater or lesser degree. Animals, much more than plants, are able to maintain their internal state under widely varying external circumstances, and this chapter is about the ways in which they do so. We have already seen in earlier pages several examples of control of rates of bodily processes, and these and others contribute to the total picture.

The subject has drawn round itself a large vocabulary of technical terms, jargon, and ordinary words used in restricted or special meanings. Most of these terms will not be used here, but a few, especially of the first class, are necessary. If an animal maintains its internal condition constant, or nearly so, in the face of changing magnitude of some factor of the environment, it is said to regulate. It is described by an adjective compounded of 'homoio-' (= 'homoeo' or 'homeo') or 'homo-', and a suffix indicating the factor. Since the regulation is seldom or never perfect, 'homoio-', meaning 'similar', is more accurate than 'homo-', meaning 'same', but the shorter form, as in homothermic, maintaining a constant temperature, is commoner. If an animal's internal state follows that of the environment it is described by a corresponding word beginning poikilo-. Just as regulating mechanisms are not perfect, so there are few animals that do not at least delay the change in their bodies, but the distinction is a useful and practical one.

Animals that cannot regulate are generally able to live satisfactorily over a range of magnitudes for each physical factor. If this range is narrow, they are described by an adjective beginning 'steno-', with an appropriate suffix; if the range is wide, the prefix is 'eury-'. If an animal is exposed, experimentally or otherwise, to different environments it may change the range within which it can live, either by increasing its tolerance of extreme conditions, or by increasing its power of regulation; it is said to acclimate, or acclimatize. An attempt is sometimes made to distinguish between these two words, but as they have existed side by side with the same meaning since the early nineteenth century this is unwise.

In addition to, or instead of, having regulatory mechanisms, many animals behave in such a way as to remove themselves from the influence of changing conditions; instead of facing adversity they flee from it.

G 9.1. CHEMICAL REGULATION: THE COMPOSITION OF BODY-FLUIDS

G 9.11. Water

Most animals in the sea have little difficulty with water, and those in streams and rivers have too much of it. Since the problems are associated with those of osmosis and ionic concentration they will be considered in the next section.

Land animals, or at least those living in dry situations, constantly face the danger of drying up. Since in these circumstances they cannot get more water, they must reduce its loss, often very markedly. All truly terrestrial animals have a cuticle that is relatively impermeable, for example a layer of keratin in amniotes and a superficial layer of wax only 0·25 μm thick in insects. Terrestrial animals with a less waterproof outer layer, such as earthworms and frogs, are confined to damp places.

One of the difficulties of an animal in a dry environment is that it still needs oxygen, and since the diffusion of all gases follows the same physical laws, anything that restricts the loss of water vapour will also restrict the entry of oxygen. The best that can be done is to close off the respiratory surface when it is not much needed; insects, for example, keep the spiracles nearly closed except in flight. Mammals and birds have a further difficulty in that they use evaporative loss of water through the skin as a means of regulating their temperature (sections M, V, 9.2). There has been selection for an ability to do this with the minimum quantity, for the evaporative water loss from desert-living rodents is less than that from laboratory species of rats and mice.

All this adds up to the fact that the chief way of reducing water loss when less is available must be by producing less urine. The vertebrate kidney (section MV 5.1) has the ability to reabsorb water, and as less water becomes available it does so more and more, producing urine with a greater concentration of solids. It is a common experience that in hot weather, when more water is lost in the sweat, unless excessive quantities are drunk less urine is produced.

Birds and the Squamata convert their nitrogenous waste into uric acid, which can be precipitated, so that even more water can be absorbed. Since the urine is always semi-solid, they have little scope for any further saving of water in exceptionally dry conditions.

Onychophora, myriapods, most insects, and many gastropods (especially terrestrial species), also produce uric acid, but the extent to which an individual can adjust its loss of water is little known. The desert locust, *Schistocerca gregaria,* absorbs much water from the rectum.

G 9.12. Salts

G 9.121. Osmotic Regulation. The chief functions of biological membranes are to keep one set of things in and another set out, or in other words their most important property is to have a considerable degree of impermeability. The extent to which they succeed depends on many things, but to a first approximation they may be considered as semipermeable, allowing the passage of water but restricting that of solutes. Hence if a cell is in water, it is only in equilibrium if the osmotic pressure, minus any hydrostatic pressure, is the same on each side of the wall. Solutions of the same osmotic pressure are called by physiologists isosmotic; in classical physical chemistry they were called isotonic, but this word is now used by physiologists of two natural fluids, or a natural fluid and an environmental one, which are in equilibrium across a membrane. In view of this contradiction the word isotonic is not used in this book.

Most marine invertebrates have blood and body fluids isosmotic or nearly so with sea water, with a depression of the freezing point of about 2·0 °C. They are poikilosmotic, and if the salinity of their environment alters, for example if it is lowered near the mouth of a river by a flood of fresh water, they take up or lose water accordingly. They may be stenohaline or euryhaline. Others, and especially those living in estuaries, have some power of regulation, and so are homoiosmotic (Fig. 9.1). If an animal such as the flat-worm *Gunda ulvae* or the polychaete *Nereis diversicolor,* is placed in dilute sea water, it at first swells, but later may return more or less to its original size. It must be pumping water out against the osmotic forces, and so is doing work. This is shown by the accompanying increase in its respiratory rate (Table 6). If cyanide is added to the water it cannot obtain energy and swells passively. The chief way in which water is lost is as hypotonic fluid through the nephridia. The crustaceans continue to excrete urine isosmotic with their blood, but increase its quantity. This means that they lose more salts, and these are taken up from the medium by active transport across the gills.

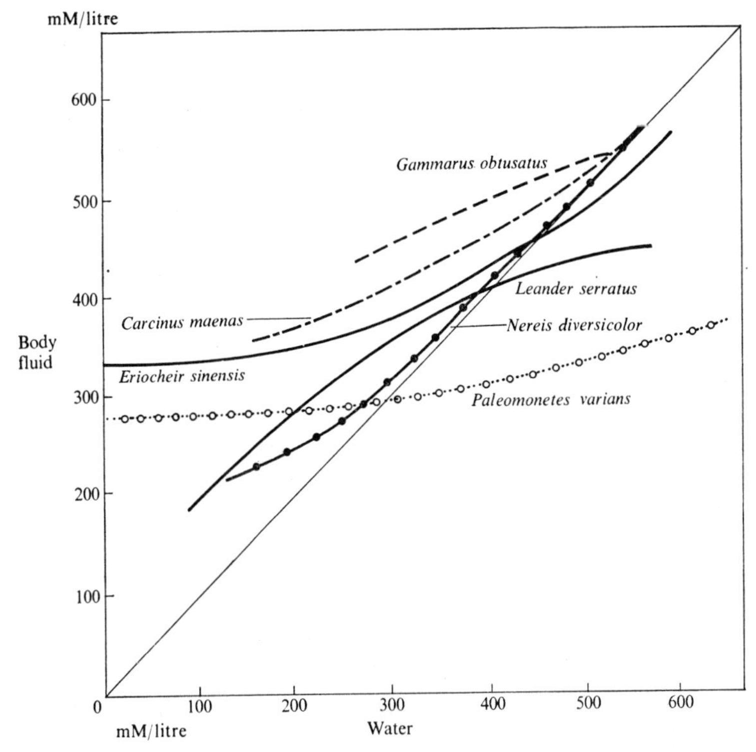

FIG. 9.1. Relationship of the internal to the external medium in various brackish-water invertebrates. Re-drawn from Beadle, *Biol. Rev.* **18** (1943).

Marine teleosts are peculiar in having body fluids with an osmotic pressure less than that of the medium; the depression of the freezing point of their blood is about 0·8 °C. To make up for the water that goes out by osmotic leakage they drink sea water and absorb it, and get rid of the excess salts that are thus passed into the body by excreting chloride, sodium, and potassium ions chiefly through special cells on the gills, and other ions by the kidneys. The production of urine, which is isosmotic with the blood, is low, so that very little water is lost in this way. A few invertebrates, especially the decapods *Leander serratus* and *Palaemonetes varians*, also have hyposmotic body fluids (Fig. 9.1).

TABLE 6. *Oxygen consumption and water absorption of* Nereis diversicolor *in dilute sea water. Two specimens were used for each length of time (from Beadle)*

Time in hours in 25 per cent sea water	Percentage increase in weight	Percentage increase in respiratory rate	Percentage of sea-water equivalent to body fluid
4·25	47	32	55
4·25	54	26	54
22·25	87	42	37
22·25	135	0	38
50·75	74	136	38
50·75	35	26	42

The blood of the hagfish is nearly isosmotic with sea water. So is that of elasmobranchs, but in these a large part of the internal osmotic pressure comes from trimethylamine and urea, nitrogenous excretory products that are retained in the blood. There is therefore a problem of imbalance of ions, which is met by the excretion of salts through the rectal gland; in *Squalus acanthias* it eliminates sodium chloride at twice the concentration present in the plasma. There is also some secretion through the kidneys. *Latimeria* also has much urea in the blood, and an osmotic pressure near that of sea water.

In fresh water, all animals have body fluids of a higher osmotic pressure than the medium, and so tend to take up water and must regulate by excreting it. The depression of the freezing-point of natural fresh waters varies, but is in general about 0·003 °C. Characteristic values for the blood of the animals inhabiting it are:

	Δ °C
Anodonta	0·09
Astacus fluviatilis	0·8
Salmo fario	0·5

In general it seems that the organs that morphologists have, rightly or wrongly, called excretory organs, are used for regulation. The contractile vacuole of Protozoa, the terminal vesicle of the nephridia of flat-worms, and the excretory ampulla of nematodes, have all been shown in various species to alter their rate of pulsation inversely as the external osmotic pressure. Contractile vacuoles are absent from nearly all marine and parasitic Protozoa, and the excretory systems of marine and parasitic flatworms and roundworms are often simpler than those of freshwater forms.

Earthworms secrete a hyposmotic urine from the nephridia, and crayfish (*Astacus*) from the antennary gland. This has three parts, a coelomic portion concerned with filtering, a secretory labyrinth, and an absorbing tubule. In the crayfish the first and third are large, while in the closely related marine lobster *Homarus* the tubule is almost missing but the labyrinth is large. A similar relationship holds in other arthropods and in some other phyla.

Little is known of the control of osmoregulation in invertebrates. In *Limnaea stagnalis* there is evidence of control by a neurohumour from the pleural ganglia.

In the vertebrates the excess water is excreted through the kidneys, and in freshwater teleosts and Amphibia the glomerulus is well developed. Other mechanisms are at work too, particularly a reduced permeability of the skin. The gills of the eel *Anguilla* are freely permeable and excrete chlorides, etc., when the fish is in the sea, so that the osmotic pressure of the blood is kept lower than that of the medium. In fresh water they become impermeable to both chloride and water, so probably making it easy for the fish to pass from the sea into rivers. Cormorants, penguins, other sea birds, and falcons eliminate much sodium chloride through the nasal glands, and turtles and probably other marine reptiles do so through orbital glands.

Some species of toad (*Bufo*) are euryhaline, adjusting the concentration of both ions and urea in the plasma, while others are stenohaline, and rapidly die in sea water. Turtles appear to be unable to regulate, but because of their thick shell can withstand changes for months or years.

The water and salt balance of vertebrates is controlled chiefly by hormones from the neurohypophysis and adrenal cortex, as described in sections M, V, 6.132 and M, V, 6.142. The nervous system is also involved, for the permeability of the skin of a pithed frog is increased to five times its normal value.

The ability to regulate may be a property that the animal carries with it from the egg up, or it may be acquired during ontogeny. The fish are examples of the first. There is no appreciable change in the depression of the freezing-point of tissues of *Salmo* from the eggs in the oviduct through all the embryonic stages up to the blood of an adult. The Amphibia (frog, toad, newt) are different. The egg in the oviduct has a depression of the freezing-point of 0·48 °C, about that of the blood of the adult, but on fertilization the depression drops to 0·045 °C. It rises during development at first, and more slowly after gastrulation. The pronephros is active well before hatching, so that perhaps regulation develops as an organ for carrying it out is formed. Amphibian eggs will not develop in water isosmotic with the serum of the adult. Cladoceran eggs are similar to those of the frog, showing a fall of osmotic pressure on fertilization and a subsequent gradual rise. The ability of *Nereis diversicolor* to regulate improves as it grows older, young worms being less sensitive to changes of osmotic pressure than are larvae, and adults less sensitive than the young forms.

G 9.122. Ionic Regulation. Osmotic equilibrium is not the same as ionic equilibrium, and it seems that no animal, marine or otherwise, has body fluids that agree exactly in composition with the external medium. A small part of the difference is due to a Donnan effect introduced because there are present in the plasma proteins to which the membranes are impermeable, but if body fluids are dialysed against sea water, so that this effect is eliminated, differences are still found. Some representative values are shown in Table 7. Sulphate is almost universally low, but the other ions are more variable; the differences are large only in the decapod Crustacea and in the cephalopods. A peculiar and perhaps important fact is that magnesium is very low in the active crustaceans and much higher in the sluggish forms.

Analysis of the urine of arthropods and molluscs shows that much of the difference is produced by differential excretion. Thus an ion such as magnesium, which in decapod plasma is hypotonic to sea water, is hypertonic to plasma in the urine. There must in addition be uptake of some ions against a concentration gradient, probably by the gills. Isolated gills of the crab *Eriocheir sinensis* absorb sodium chloride from a solution only one-fortieth of the concentration of that of the animal's blood, and this absorption stops in the absence of oxygen or in the presence of cyanide, so that it is, as usual, an active metabolic process.

Within the animal there is not uniform distribution of ions, potassium being in general in excess within the cells and sodium outside. The special case of this in nerve cells has been described in section MV 6.31. The external regulation of ions may be regarded as an extension of this internal regulation, which is a fundamental and necessary property of living membranes.

TABLE 7. *Ionic regulation in some marine invertebrates (from Robertson)*

	Concentrations in plasma or coelomic fluid as percentage of concentration in the same fluid dialysed against sea water					
	Na^+	K^+	Ca^{++}	Mg^{++}	Cl^-	SO_4^{--}
Coelenterata						
Aurelia aurita	99	106	96	97	104	47
Echinodermata						
Marthasterias glacialis	100	111	101	98	101	100
Tunicata						
Salpa maxima	100	113	96	95	102	65
Annelida						
Arenicola marina	100	104	100	100	100	92
Sipunculoidea						
Phascolosoma vulgare	104	110	104	69	99	91
Arthropoda						
Maia squinado	100	125	122	81	102	66
Dromia vulgaris	97	120	84	99	103	53
Carcinus maenas	110	118	108	34	104	61
Pachygrapsus marmoratus	94	95	92	24	87	46
Nephrops norvegicus	113	77	124	17	99	69
Mollusca						
Pecten maximus	100	130	103	97	100	97
Neptunea antiqua	101	114	102	101	101	98
Sepia officinalis	93	205	91	98	105	22

G 9.13. Hydrogen-ion Concentration

The more primitive animals are very sensitive to changes in acidity, but all animals which possess a body fluid, that is, all the coelomates, maintain an internal environment of approximately constant hydrogen-ion concentration, and they do this by the same method as is used by the biochemist in the laboratory. Almost any organic acid will act as a buffer; that is, if its degree of dissociation (y) is plotted against the pH of the corresponding solution (x) a curve of the type shown in Fig. 9.2 is given. Along the straight part of the curve a big change in degree of dissociation corresponds to a small change in pH. If hydrogen ions (in the form of another acid) are added to the solution, they will, by the law of mass action, drive back the dissociation of the acid originally present, and along the straight part of the curve they must do this to a considerable extent before the final pH is very much altered. The solution thus resists change of pH, and is said to be buffered. The position of the straight part of the curve depends on the particular acid used, the hydrogen-ion concentration corresponding to its mid-point being numerically equal

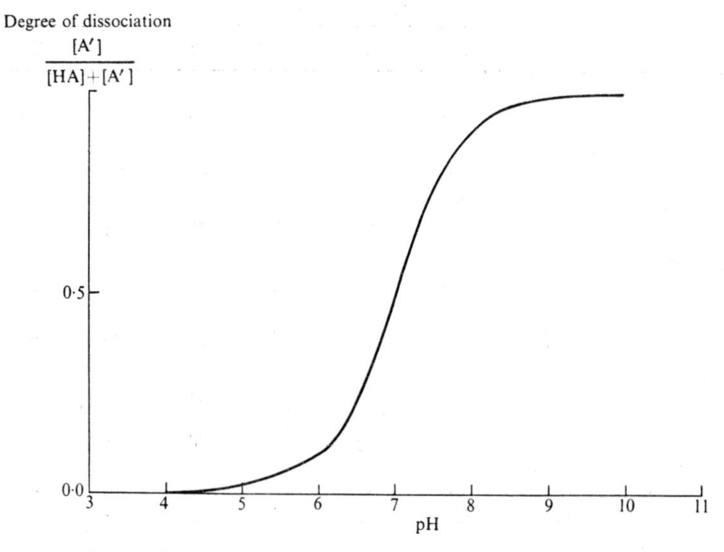

FIG. 9.2. Graph relating the dissociation of an acid of dissociation constant 10^{-7} to the pH of the resulting solution. Such an acid will act as a buffer between pH 6·5 and 7·5. Other acids will give similar curves shifted either to the right or to the left according to the value of the dissociation constant.

to the dissociation constant. Thus acetic acid, with a dissociation constant of 10^{-7}, buffers best at a hydrogen-ion concentration of 10^{-7}, i.e. at pH 7, and is effective from about pH 6·5 to pH 7·5. The degree of dissociation corresponding to the middle of the straight part of the curve may be obtained by adding a salt of the acid, which depresses the ionization in the usual way. It is customary, but incorrect, to name the buffer by the salt. Strong acids are not considered as buffers, because their dissociation constant is very high, and so they only work in very high concentration of hydrogen ion where buffering is seldom important.

Since the product of the concentrations of hydrion and hydroxyl is constant, a similar argument to the above can be applied to bases.

All animals buffer their blood and other fluids by proteins, bicarbonate, and phosphate. The normal pH varies from one animal to another but is usually between 7 and 8.

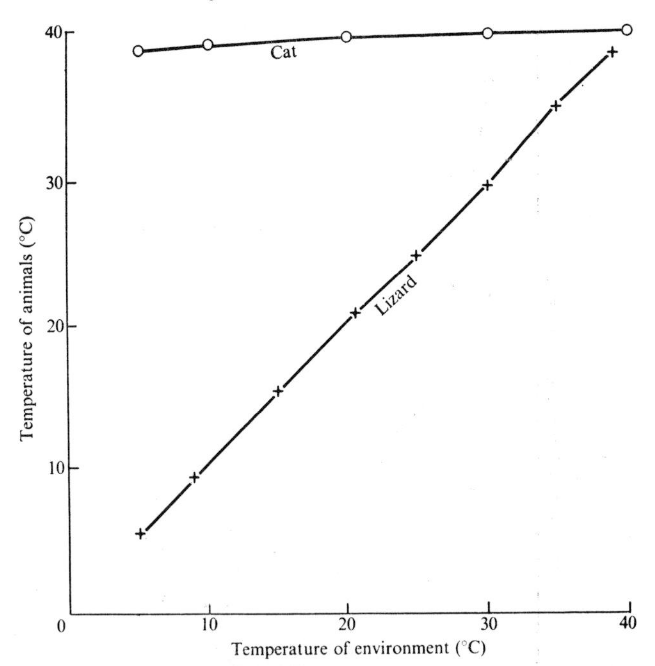

FIG. 9.3. Body temperatures of cat and lizard (*Cyclodus gigas*) in different environmental temperatures. Redrawn from Martin, *Phil. Trans. R. Soc.* **B195** (1903).

G 9.2. TEMPERATURE

In the majority of animals the body temperature is not greatly different from that of their surroundings, and so the rate of respiration (apart from specially stimulated activity) depends on the latter. Figs. 9.3 and 9.4 show this. The temperature of the lizards is almost the same as that of their surroundings, and their carbon dioxide production corresponds. Such animals are called cold-blooded or poikilothermic. They are handicapped by both low and high temperatures. In the former their activity is greatly slowed down, and in the latter they have no means of preventing themselves from becoming overheated and so dying.

The actual temperature of cold-blooded animals depends on the balance between the receipt and loss of heat. The chief ways in which heat is received are by the metabolism of the body and by radiation from the sun, and the chief forms of loss are by conduction and con-

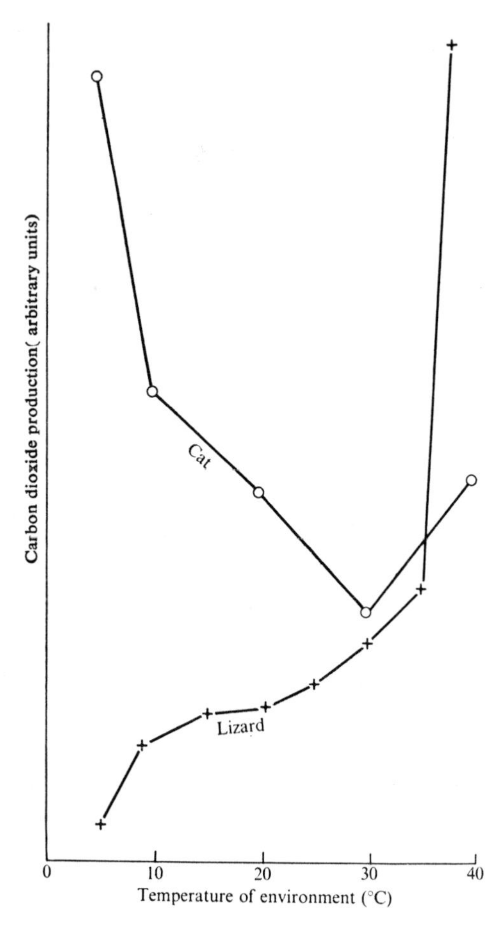

FIG. 9.4. Carbon dioxide production of cat and lizard at various environmental temperatures. Cf. Fig. 9.3. Redrawn from Martin, *Phil. Trans. R. Soc.* **B195** (1903).

vection and by evaporation. In water, radiation is of little importance, except perhaps for large animals near the surface, and evaporation of none. The animals rapidly lose heat to the water, which is a good conductor, and maintain a temperature a little above that of their surroundings, like any other body which is producing heat.

On land two types of situation are important. Animals that have a large damp surface, either internal or external, may lose so much heat by evaporation that they are cooler than their surroundings, and since the rate of evaporation depends not on the relative humidity but on the vapour pressure deficit, the difference will be larger the drier the air and the higher the ambient temperature. Earthworms, molluscs, insects, and Amphibia all show this effect, and some of them may have temperatures as much as 10 °C below that of the atmosphere. This is the same phenomenon as that of the wet-bulb temperature in a hygrometer, but the final temperature reached in the animal is affected also by the metabolism, and may be slightly (in molluscs) or considerably (in Amphibia) above that of the wet-bulb.

The other case is that of relatively dry-surfaced animals in sunlight. They may absorb enough radiant heat to raise their temperatures well above that of the atmosphere. Insects and reptiles show this effect very well, but it is purely physical, and applies also to inanimate objects and warm-blooded animals; it allows the possibility of sunbathing in Switzerland when the air temperature is well below freezing-point.

Both cooling by evaporation and heating by insolation may be of some slight importance in protecting the animal from extremes, but neither can be called regulation.

M 9.2. TEMPERATURE CONTROL IN MAMMALS

Mammals are warm-blooded or homoiothermic. Their resting metabolism, that is their energy production irrespective of exercise, goes up as the temperature goes down, in such a way as to keep their body temperature constant. They are in fact living thermostats. The actual temperature maintained depends on the species, and is subject to slight variation with the individual and with activity. Fig. 9.3 shows how the body temperature of the cat remains approximately constant in varying environmental temperatures. The graphs in Fig. 9.4 show that the different behaviour of the cat and of the lizard is correlated with an increased carbon dioxide production at the lower temperatures in the former; at 35 °C the regulating mechanism is beginning to break down. The chief seat of the increased respiration is the muscles, and in the extreme case this is obvious when shivering takes place. Some authors have denied the existence of non-shivering thermogenesis, but it has been demonstrated in rats, mice, and dogs.

In the non-acclimatized animal it makes up about half the extra heat production, but on acclimatization the proportion rises, and may reach 100 per cent. The same thing is obvious in man every autumn. Muscle, brown fat, liver, and brain all contribute to the non-shivering production of heat. This type of regulation is called chemical, and can deal only with a lowered external temperature, but there is another type, rather badly called physical, which adapts its possessor to higher temperatures than normal by control of heat loss. This takes different forms. In man, of the total heat produced 5 per cent goes in heating food, drink, and respired air, 15 per cent in evaporating water from the lungs, and the rest in cooling and evaporation from the skin. When the external temperature is increased, or when as a result of exercise more heat is produced, the rate of heat loss is raised in two ways. The cutaneous blood-vessels are dilated so that heat is taken to the surface more rapidly, and the sweat glands are activated so that evaporation is increased. Another device, not used by man, is to increase the rate of breathing, and so of evaporation from the buccal cavity and lungs; a development of this is tachypnoea, in which the animal breathes in rapid short pants but takes little extra air into the lungs.

In some species the basal metabolic rate remains constant over a certain band of environmental temperatures, called the range of thermal neutrality, and under these conditions all regulation must be physical. The point at which metabolism begins to increase as the environment is cooled is called the lower critical temperature, that at which metabolism begins to rise as the environment is warmed (and so chemical regulation begins to break down) is the upper critical temperature.

The degree to which temperature control exists, and the forms that it takes, vary widely. The Prototheria, Edentata, and lemurs, and some species in other orders, have a variable temperature and little control, maintaining their body a few degrees above ambient. The camel when short of water allows its temperature to rise so that it does not evaporate so much. The temperature of many bats and of some insectivores and rodents drops when they are sleeping; such animals have been called heterothermic. A similar range of abilities is found amongst the marsupials.

There is difficulty in defining a sweat gland, but glands which, by varying their range of activity, contribute to temperature control, are found in primates, dogs, horses, and artiodactyls, and probably more widely. Increased respiration occurs in rabbits, pigs, cattle, and sheep, and tachypnoea is marked in the dog. Control is helped by the presence of a thick layer of fur or fat. The long wool of the merino sheep, which is adapted to hot climates, actually seems to prevent the animal from overheating, for when it is shaved off, at an air temperature of 38 °C the skin temperature rises by 3 °C and the rate of panting is doubled. The man who wore an overcoat in summer because 'what keeps out the cold will keep out the heat' is thus vindicated.

At low external temperatures heat is often saved by lowering the surface temperature of the skin, which reduces loss by radiation and by conduction to the air. This reduction is most marked in the limbs and extremities. A calf, for example, in an ambient temperature of 5 °C and with a rectal temperature of 39·5 °C, had the skin of its chest at 31·2 °C, that of a foot at 10·1 °C, and that of the pinna at 7·0 °C.

The scrotum of mammals acts as a very efficient local thermostat, and this is perhaps its primary function. Sperms develop well over a rather narrow range of temperature, which is slightly below that of the body. The position of the testes outside the body cavity provides this, and the contraction of the muscles of the scrotum, aided by some vascular reaction and sometimes by sweating, maintains the temperature constant within a very narrow range. The temperature of the testicles of a bull varied only between 35·0 and 37·0 °C when the temperature of the air ranged from 15·0 to 37·8 °C.

From the few species that have been examined it seems that mammals which are born open-eyed and active can generally regulate at birth, but those that are born blind and helpless do not acquire the ability until some days later, the physical and chemical methods coming independently. Some examples are given in Table 8.

Thermoregulation depends on the nervous system, for mammals paralysed with curare behave like lizards. Sweat glands can be stimulated to operate through the sympathetic system, and shivering depends on nervous stimulation. Non-shivering heat production is probably increased by a greater production of catechol hormones by the adrenal medulla. It is not so easy to say how the necessity for a change in heat production or loss is first recognized by the body, and it may be that the systems are different in different mammals. Sweating can be started reflexly in man by heating the skin, and panting in sheep by heating the scrotum or the inside of the mouth, so that peripheral receptors appear important. There is also response to changes in deep body temperature, for changes of 0·2–0·5 centigrade degrees in the blood flowing through the hypothalamus will start the appropriate regulatory responses in a number of species

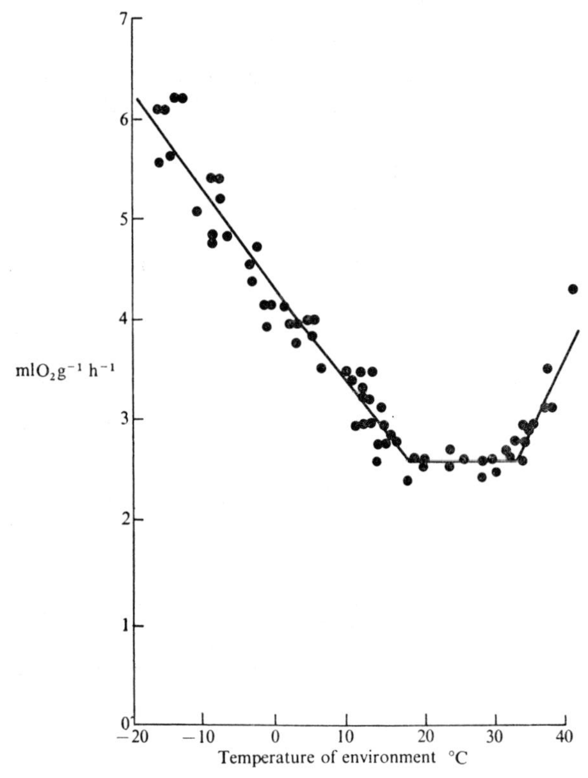

FIG. 9.5. The oxygen consumption of the cardinal, *Richmondena cardinalis*, at various environmental temperatures. Twenty-two birds were used. Redrawn from Dawson, *Physiol. Zoöl.* **31** (1958).

(dog, rabbit, ox, cat, rat, monkey). Non-shivering thermogenesis is the chief control so induced, while shivering, and its substitute non-shivering that follows acclimatization, are more responsive to the temperature of the skin. There are still difficulties in the explanation, and no evidence that significant variations of internal temperature occur naturally on cooling. A full explanation must also account for the way in which the setting of the thermostat varies cyclically, as in man, through the day.

V 9.2. TEMPERATURE CONTROL IN BIRDS

The only other fully homoiothermic animals are the birds, and, as in so many things, they closely parallel the mammals. They have good chemical regulation, but little or no non-shivering thermogenesis,

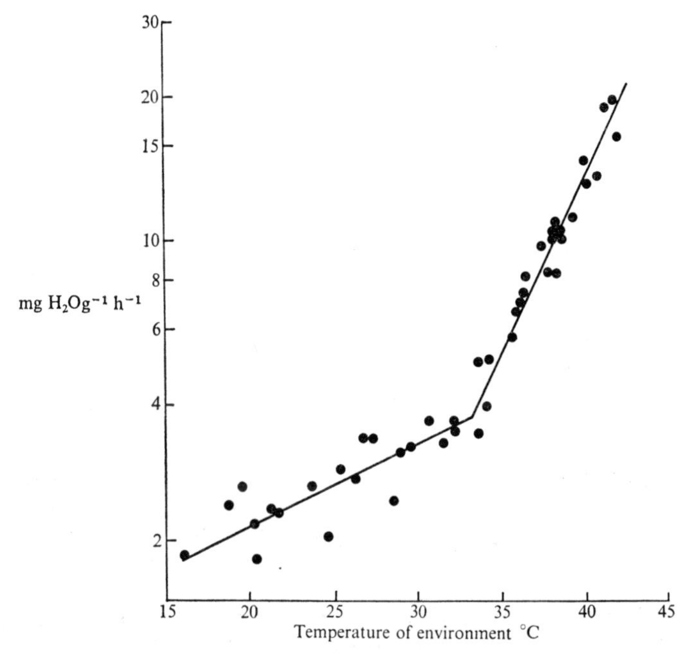

FIG. 9.6. The rate of loss of water in the cardinal at various environmental temperatures. Sixteen birds were used. Redrawn from Dawson, *Physiol. Zoöl.* **31** (1958).

and some can increase heat loss by tachypnoea, or by rapid movements of the throat (gular fluttering), but they do not sweat. The feathers are good insulators, and in cold weather are fluffed up so that the amount of air trapped by them is increased. Fig. 9.5 shows the oxygen consumption (which is proportional to heat production) of the cardinal (*Richmondena cardinalis*), a North American bird that lives in a wide range of environmental temperatures and has a wide range of thermal neutrality. Evaporative heat loss increases rapidly at ambient temperatures above 33 °C (Fig. 9.6), but at 41 °C it is still only half the metabolic production, so that the body temperature rises. While the upper critical temperature in birds is always about

33–38 °C, the lower critical temperature generally bears some relationship to the normal environment, being about 37 °C in the tropical widowbird *Steganura (Vidua) paradisaea* and below zero in the Canadian jay *Perisoreus canadensis*, and other boreal species. Many

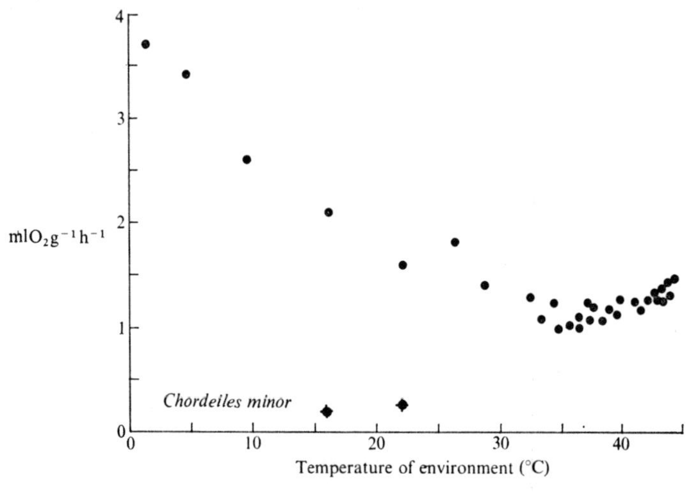

FIG. 9.7. The relation between oxygen consumption and ambient temperature in the common nighthawk of America (*Chordeiles minor*). Circles, six homothermic birds resting in the dark, stars, one torpid bird. From Lasiewski and Dawson, *Condor* **66** (1964).

small birds allow their body temperature to rise in hot weather, as the camel does. In some, such as the common nighthawk of North America (a nightjar, *Chordeiles minor*), there is no thermoneutral zone (Fig. 9.7). In this bird the evaporative water loss is constant at ambient temperatures up to about 35 °C, but rises rapidly above this, and gular fluttering begins at 42 °C. The evaporative heat loss is then greater than the metabolic production, so that the bird can take in heat from its surroundings without a significant rise in its body temperature. In sea birds the temperature of the legs and feet (which have very little muscle) may fall very low, so that they can stand on ice without losing much heat. In nidifugous birds, which leave the nest on hatching, regulation may be present at birth, but in nidicolous species, which remain in the nest and are at first helpless, it comes later (Table 8).

The control system seems to be similar to that of mammals. In fowls there is response especially to the temperature of the surface of the head, and there is also a receptor in the hypothalamus. Since changes in temperature of the hypothalamus of a turtle cause changes in blood pressure, it may be that this is the origin of the ability to regulate both in mammals and in birds.

TABLE 8. *Showing the forms of temperature regulation present at birth or hatching (from Needham)*

	Chemical	Physical
Guinea-pig	+	+
Rabbit	+	−
Mouse	−	−
Cat	+	−
Dog	−	−
Man	+	−
Chick	+	+
Pigeon	−	−

J. 9.2. TEMPERATURE CONTROL IN INVERTEBRATES

In the honey-bee in summer there is apparently some chemical regulation, similar to that of mammals, but it seems to be imperfect, and there is a report that the metabolic rate of a crab increases as the ambient temperature falls. Sphingid moths are unable to fly when their body temperature is below a certain value (20 to 30 °C, according to the species) and achieve this in cold weather by fanning with their wings until they have warmed up enough. In this they resemble bats, whose temperature falls in the daytime and has to be raised by shivering until it is high enough for them to fly.

G 9.3. BEHAVIOURAL ADAPTATIONS

Many animals behave in such a way as to reduce or avoid the effects of extreme environmental conditions. The simplest method, perhaps, is the avoiding reaction or phobotaxis of ciliates, in which the animal reverses and turns aside when it swims across an unfavourable line in a gradient. By this *Paramecium*, for example, collects in regions of moderate temperature or acidity, because it reacts when its random motion takes it out of such places but not when going in. Many invertebrates react by kineses, that is their rate of moving or their rate of turning depends on the magnitude of some of the physical factors of the environment. The woodlouse *Porcellio scaber*, for instance, moves slowly and is often stationary in damp air. Hence once it comes into such places, as under logs, it stays there for a long time, but if it does move away into the open the dryness of the air makes it move fast until by accident it once again comes into a damp region and slows down.

A similar type of behaviour, but probably under more complicated control, is shown by many desert-living insects and reptiles, which escape the high surface-temperatures of 50 °C or more by burrowing, and when above ground move rapidly to shelter. If they are tethered in the sun in these places they die within a few minutes. In many of the social Hymenoptera there is some communal regulation of the temperature of the nest by such methods as opening and closing the nest entrances and fanning with the wings. In many mammals, especially marsupials, but also rodents, cats, and to some extent cattle, cooling is increased by the production of copious saliva, which is spread over the body by licking.

MV 9.31. Hibernation

The term hibernation is best confined to the winter sleep, with complete or partial loss of temperature regulation, of animals that are normally homoiothermic. Apparent examples are found in most of the orders of mammals, but few have been critically examined and the best known are species of insectivore and small rodents. Bears become sleepy, but their temperature falls only to about 30 °C. Most hibernators appear to have a relatively low normal body temperature of about 36 °C, and to have greater variation in this than do other mammals.

In complete hibernation, such as that of the hedgehog or dormouse, the animal is in a deep sleep. The body temperature follows that of the atmosphere. The respiratory rate and pulse rate are low, the oxygen consumption is much less than normal. Like that of a poikilotherm, it is linear with temperature, but in the dormouse may be little more than 1 per cent of the basal metabolic rate. There are associated changes in metabolism. It is probable that uninterrupted periods of sleep are not more than a few weeks, and during the waking intervals some species, but not others, feed. Most of the small amount of energy required during hibernation comes from stored fat.

The simplest explanation of hibernation would be that it is induced by the falling temperatures of autumn, but this is not adequate. Specimens of the African insectivore *Centetes ecaudatus* at the London Zoological Gardens hibernated at the same time each year, irrespective of the environmental conditions. The golden-mantled ground-squirrel, *Citellus lateralis*, continued to hibernate at the proper time of year even when it was confined in the laboratory at a constant temperature of 2 °C and under 12 h daily illumination, with unlimited food, water, and bedding.

Experiments on photoperiod have mostly given negative results, but short days possibly induce hibernation in the edible dormouse, *Glis glis*, which adjusted its hibernation to reversed seasons after 10 months. In bats the accumulation of food reserves seems to be a predisposing cause. The onset is gradual, the temperature in *Citellus* falling more and more each night until hibernation starts. Arousal is more sudden. It is generally brought about by high environmental temperature (15 °C in American marmots, *Marmota*) and shivering rapidly brings the body temperature up to normal, but if the ambient temperature falls too low (to freezing point in the marmot) the animals may wake also. In other animals a low ambient temperature may produce some degree of regulation, so that the body remains warmer than the surroundings but the animal does not wake. The hormone balance of the animal is important, and during hibernation the anterior pituitary, thyroid, and adrenal cortex are reduced in size or activity or both, while the islets of Langerhans are increased. Experimentally, injection of insulin induces a state resembling, though not identical with, hibernation, while adrenaline and thyroxine may wake hibernants. In other species adrenalectomy prevents hibernation.

Aestivation, or summer sleep, appears to be physiologically similar to hibernation, and is sometimes indulged in by animals such as dormice and marmots that normally hibernate.

For a long time hibernation in birds, fervently believed in by eighteenth-century naturalists, was strongly denied by modern scientists. It is now known that many species of humming bird regularly become torpid and poikilothermic, and reduce their metabolism at night, and wake again before daybreak. This behaviour is perhaps connected with a very high resting metabolism during the day so that it acts as a system for saving food A few other species of bird, mainly American, are similar (Fig. 9.7), and at least one, the poor-will, *Phalaenoptilus nuttalii*, of Colorado, hibernates through the winter. This bird is known to the Indians by a name that means 'the sleeping one'. Swallows may become torpid in cold weather, but recovery is only possible if complete torpidity is not reached.

J 9.32. Diapause

Diapause is an arrest of development in cold-blooded animals, much greater and more prolonged than any that could be explained as a simple physical effect of low temperature or other adverse conditions. It includes the state of inactivity of the cysts of Protozoa and the winter eggs of many pond animals, but has been closely studied only in arthropods, and especially in insects. Since it often occurs in immature stages, it may appear as an arrest of development. In arthropods the stimulus for diapause is nearly always the length of day and night, but only rarely, as in the red locust, which diapauses if it emerges from the pupa in decreasing days, is the changing day-length important. Temperature and food supply may have subsidiary effects in the onset of diapause, but the former is the important agent in its cessation. A period at a low temperature is necessary before development can be resumed, and there is an optimum temperature, on either side of which diapause is prolonged. During diapause the oxygen consumption falls, but can usually be detected. The respiratory quotient is about 0·7, suggesting that fat is the substrate, and the occasional very low values that have been reported are probably due to the retention of carbon dioxide. Much of the respiration that is left is not sensitive to cyanide. The hormonal control of diapause is described in section J 6.19.

A further withdrawal from adverse conditions is shown by those invertebrates such as rotifers, tardigrades, and some insects, as well as many eggs, which reduce their metabolism effectively to zero. Such latent life is called cryptobiosis or anabiosis. It occurs chiefly as a response to drying. Perhaps the most striking example is the larva of a Nigerian chironomid (Diptera); when the pools in which it lives dry up, it shrivels to an apparently lifeless corpse containing sometimes less than 1 per cent of water, in which no gas exchange can be detected. When water returns it swells and swims again within an hour, and the process can be repeated.

G 9.33. Migration

One way of avoiding adverse conditions, especially low temperature and the reduction in food supply to which it often gives rise, is by migration. This may be defined as a periodic to-and-fro movement of a population between two areas, and is known in many fish and birds, in some mammals, and in a few insects and other invertebrates, notably butterflies. Three aspects of it are of special interest to physiologists—the system that controls its onset in the individual, the means by which the long journeys are sustained, and the sensory means by which the correct direction is determined.

All migration is probably connected with breeding, in that this takes place at one end only of the journey. The nesting of birds and the spawning of salmon and eels at the end of their migrations are well known, and breeding in a special area takes place similarly in seals, whales, and some herbivorous mammals, and in many marine fish such as herring and plaice. There is no doubt that the gonads begin to become active about the time when migration begins, and those of birds decrease in size at the end of the breeding season in time for the return migration. There can be little doubt that the physiology of migration is part of the general gonadal-pituitary cycle discussed in the last chapter, and that the initiation of migration depends partly on the phase reached in the cycle, which has been determined by the accumulated influence of the season in bringing the endocrine system into the appropriate state, and partly on the trigger-effect of particular conditions such as change in temperature and high barometric pressure. The state of the cycle makes migration possible, while the proximate environmental factors cause it to begin. Little experimental work has been done on the trigger-stimuli, but it has been shown that excess or deficiency of daylight will hasten or delay the migratory restlessness of caged robins and other birds, and it is known that migration of many birds occurs predominantly under anticyclones. Temperature is important, and the waves of advance of many species over Europe in the spring follow particular isotherms very closely. In salmon and other fish migration is accompanied by increased activity of the thyroid, and there is some slight evidence that the same is true of birds.

Birds that make long journeys over the sea cannot feed, and over the land there may also be difficulties. They lay down large quantities of fat before they migrate; in small birds this may amount to more than 50 per cent of total body weight, which may be compared with about 15 per cent for small rodents before hibernation. Calculation suggests that this is just about enough for the longest non-stop journey that the birds carry out.

Much attention has been paid in recent years to the navigational basis of migration. In salmon, the movement appears to be basically a rheotaxis, modified by the recognition of the chemical properties of particular rivers. In the sea the fish tend, at the appropriate time, to keep in water of low salinity, which will be found near the mouths of rivers, and especially in water of similar composition to that in which they were hatched. A sudden lowering of the salinity, caused by a freshet, makes them enter the river, and thenceforward the up-stream migration is a visual or tactile rheotaxis. Descent is largely passive, and must happen to any river-living animal if rheotactic responses are lost.

In the writings on the navigation of birds there has been much confusion between homing and migration. While migration may involve homing it does not necessarily do so; in some species, notably the cuckoos, the young emigrate alone and after their parents, so that they can neither have memory of where they are going nor passively follow the others; the young cuckoos have never even known their parents. While memory undoubtedly plays a part in the final few miles of migration—where, for instance, a swallow returns, as it often does, to the same barn that it occupied in the year before—it is extremely unlikely that the necessary cues for the rest of the journey could be learnt. This may be, for the swallow, 4000 miles or more, and although the average speed of northward travel may be only about 25 miles a day, many lengths of this (and many other migratory journeys) are carried out at high speeds. If, with the amount of stimulation that they can get in conditions like this, birds can form conditioned reflexes, they are performing very much better than any experimental animals known to us, and there is no evidence, in the experiments on learning in the laboratory, that this is so. Since most homing depends on previous learning of the territory to be covered it is probably quite different in nature from migration. The few cases of successful return of birds which have been displaced large distances over unfamiliar country are not more than are to be expected by chance wandering, followed by visual homing when the bird comes into a region that it knows.

There is little doubt that for every migratory population there is a standard direction at each season, for not only can this sometimes be seen, but experiments in the displacement of migrating birds have shown the usual direction to be generally maintained. Birds are often distracted from their standard direction by guiding lines such as coasts, river valleys, and mountains. How the bird determines the standard direction remains a mystery. Its persistence when the bird is displaced hundreds of miles (if the few experiments on storks, starlings, and crows may be generalized) shows that it cannot be determined by local clues, such as temperature gradients or winds, but that it must be dependent on something global or celestial in scale. Most of the hypotheses proposed, such as those dependent on terrestrial magnetism or on forces produced by the rotation of the earth, have been shown to be physically impossible or extremely improbable.

Most ornithologists believe that birds can navigate by the sun or the stars, for which they would need at least a chronometer and an artificial horizon, but the evidence on which this belief is based is statistically inadequate and different experimenters have drawn the same conclusion from contradictory observations. Such celestial navigation does not meet the difficulty that some migration continues when the sun is obscured by clouds and when the bird has crossed the equator. To account for the last point, either there should be a period of confusion, in which the bird has no bearings, or the bird must instantaneously reverse its reaction when it passes under a sun which is vertical at noon. The first does not seem to have been recorded, and the second seems unlikely.

G 9.34. Daily Rhythm

One of the most striking things in the activity of animals is that some species are active only at night, and others only during the day. It is easy to suggest that this rhythm depends on the light intensity, and the common observation that most birds are active during daylight and sleep in the dark, and begin roosting during an eclipse, seems to confirm this. But the matter is not so simple. In high latitudes the active day is extended, but in the Arctic summer, though some birds may be singing at any time during the 24 hours (as indeed they may be during an English June), there is a distinct lull in activity for a few hours around midnight, even though the sun is still shining. When animals are subjected to days of different lengths they may adjust very quickly, as do birds, or they may persist in approximately the old rhythm for months, as do many invertebrates.

In spite of a great deal of experiment it cannot be said that any clear physiological picture can yet be painted. It is not even always possible to see the advantage of the rhythm in the many species that are not dependent on vision. Nymphs of mayflies, such as *Ecdyo-*

nurus torrentis (which lives in fast-flowing streams) and others, are slightly more active during the night than during the day, and follow a 24-hour rhythm that does not develop if they are bred in continuous light, but is established by one experience of a normal day; it will then persist in continuous light for 5 months. A similar rhythm in cockroaches is mediated partly by a hormone, which is secreted in an underlying cycle, and partly by the nervous system. In many species the rhythm is not exactly based on a 24-hour cycle, and it tends to drift.

Index

Group names of animals which will be found in systematic order in the appropriate sections are not indexed. Generic names are fully indexed. **Heavy type** indicates an important reference. The first page only of a section is given.

abomasum 53
absorption in gut 22, 36, **55**
in kidney 140
in Malpighian tubules 152
acceleration, sensitivity to 211
accommodation 225, 231
acetic acid 54, 129, 134
acetyl acyl carrier protein 126
acetylcholine 34, 100, 187, 189, 243, 251, 257, 261, 267, 269, 277
acetylcoenzyme A 32, 118, 126
achroodextrin 52
Acidia, excretion 152
acidity, *see* hydrogen-ion concentration
acromegaly 164
ACTH, *see* adrenocorticotropic hormone
actin 251, 265
Actinophrys 44, 280, 285, 288
Actinosphaerium 44
action specific energy 189
'activation' 16
active transport **21,** 56, 65
actomyosin 251, 258
acyl carrier protein 126
acyl coenzyme A 126
adaptation 195
adenase 110
adenine 110, 112
adenohypophysis **163,** 294, 300
adenosine diphosphate (ADP) 107
adenosine monophosphate (AMP) 107
adenosine triphosphate (ATP) 17, 106, 109, 116, 124, 126, 158, 243, 247, 271
adenylic acid 107
adenyltriphosphate, *see* adenosine triphosphate
ADP, *see* adenosine diphosphate
adrenal glands 121, 143, **168,** 294
adrenaline 99, 100, 121, 158, **168,** 174, 243, 253, 269, 277, 327
adrenergic nerves 257
adrenocorticotropic hormone (ACTH) 121, **165**
Aeolosomatidae 282
aestivation 327
afferent fibre, nerve or neurone 177
Agriolimax 203
air-bladder 222
alanine 25
albedo 277
aldosterone 143, 170
alginic acid 64
alimentary canal in breathing 74, 78
allantoic acid 129, 151
allantoicase 130
allantoin 111, 129, 151
allantoinase 129
allantois, excretion by 145
Allolobophora 92, 208
all-or-nothing rule 182, 258
α-ketoglutaric acid 108, 118
alveolus 74
Ambystoma 76, 161
Ameiurus 204, 205, 222, 276
Amia 77
aminoacids 1, 25, 49, 107, 113
aminobutyric acid 199
aminopeptidase 50, 62, 63
aminotransferase 107
amitosis 280
ammocoete 43
Ammodytes 89
ammonia, excretion 130, 141, 146, 147
metabolism 26, 50, 107, 111, 130
ammoniotelic animals 130
Amoeba, behaviour 224
excretion 8, 146
locomotion 6, 244, **245**
nutrition 65
reproduction 280
amoebocytes 70, *see* phagocytes
amoeboid movement **245**
AMP, *see* adenosine monophosphate
ampullae of Lorenzini 202, 236
amylase **52,** 60

amylopectin 52
amylose 52
anabiosis 328
anabolism 107
Anadara 91
anaemia 34
anaerobiosis 104, 125, 130, 134
analgesia 201
anal gill 73
Ancylostoma 46, 66
androgamone 286
androgen 300, 302
androsterone 298, 301
Anemonia, 147
aneurin **31**
angiotensin 143
Anguilla 313
anisogametes 285
Anisops 91
Anisoptera 78
Anodon 133, 153, 312
anoestrus 291
Anolis 277
anosmatic animals 206
antennae 208
antennary gland 150, 313
antibodies 57
anticoagulin 46
antidiuretic hormone 142
antifertilizin 286
anti-hormones 158
antiperistalsis 59, 78
Aphelocheirus 81
Aphidae 288
Aphrodite 148
Aplysia 153
apnoea 76
apposition image 235
Apterygota 74
aquatic insects, respiration 80
aqueous humour 225
arachidonic acid 28
Arca 91
Arcella 281
Areneidae 100
Arenicola 64, 67, 90, 92, 150, 172
arginase 17
arginine phosphate 265
  vasopressin 166
Arion 303
Armadillidium 151
armadillo 290
arterial-venous difference 94
artiodactyls, nutrition 30, 121
Ascaris, 66, 91, 104, 133, 134, 148
ascidians, budding 283
ascorbic acid 29, 33, **34**, 38
Asellus 151
asexual reproduction **280**
assisted transport 21, 55, 57, 61
Astacus, 45, 79, 105, 312, 313
Asterias 44
athrocytosis 140
ATP, *see* adenosine triphosphate
auditory ossicles 218
Auerbach's plexus 59
Aurelia 208, 281
auriculo-ventricular node 98
autacoid 157
autogamy 289
automatic volume control 218
autonomic nervous system 257; *see* parasympathetic, sympathetic
avoiding reaction 325
axial filament 244
axolotl 74
axopodia 244

Bacteria 50, 54, 55, 67
Baetis 78
balance, organs of **211**
Balantiophorus 104
Balanus 264
basal body or granule 239
basal metabolism 27, 160
basal recess 218
basilar membrane 218
bees, feeding 46
  parthenogenesis 289
bees-wax 61, 64
behavioural adaptations **325**
beri-beri 31
β-oxidation of fats **125**, 134
bicarbonate 93, 317
bile 49, 55, 60
bile salts 49, 61
bilirubin 49
biliverdin 49
bioluminescence **269**
biotin **33**, 38
birth 297
birth rate 307
bitterling 175
bittersweet 205
bladder 144
blind spot 230

blood **83**
  carrying power 84
  corpuscles 85, 91
  oxygen tensions (Table) 87
  pigments (Table) 84
  sugar 121
blood-sucking animals 46
blowflies 45, 46, 67
BMR 28
Bohr effect 88, 90, 92
Bojanus, organ of 153
Bombyx 174
bone 162
Botryllus 283
Bowman's capsule 137
brain, 190
branchial heart 96
branching factor 116
Branchiostoma 48, 145, 159
breathing **74**
breeding season 291, 299, 302
bromine 160
brood-patches 301
brown 231
brown body 150
Brunner's glands 48
brush-border 10, 56, 57, 144, 151, 153
B-substance 277
buccal glands 48
budding 281, **283**
Buenoa 91
buffers 315
Bufo 43, 61, 163, 313
bugs, feeding 46
  parthenogenesis 288
bundle of His 98
butterflies, migration 329
butyric acid 54, 121
butyryl acyl carrier protein 127

caecum 54
Caenorhabditis 132
calciferol 36, 38
calcitonin 159, 162
calcium excretion 136, 144
  metabolism 39, 162
calcosphaerites 152
Calliactis 185, 190, 191
Calliphora 62, 63, 239
campaniform sensilla 199
Cancer 65, 151
cane sugar 52
Carassius 130
Carausius 279
carbamino-compound 94
carbamyl phosphate 109
carbohydrase 62, 63
carbohydrate 5
  as food 27
  digestion **52**
  metabolism **114,** 128
carbon dioxide, excretion of 136
  formation 118
  in metabolism 126, 130
  transport of 93
carbon monoxide 89, 92
carbonic acid 93
carbonic anhydrase 93
carboxylation 33
carboxypeptidase 51, 62, 63
Carcinus, breathing 79
  osmotic regulation 315
  vision 234
cardiac muscle 96, **258**
carnitine, 39
Carnivora 128
carnivores 45
carotene 35
carotid body 75, 99
carrying power of blood 84
cartilage 125
casein 26, **50**
caseinogen 50
Castor 141
cat, falling 211
  regulation 317
catalysts 15
catechol hormones 168, 321
catfish 204, 205, 222, 276
Cecidomyidae 289
celery-fly 152
cells 7
cellulase 64, 65, 66, 67
cellulose 53
Centetes 327
central nervous system 177, **189**
cephalopods, chromatophores 279
Cestoda 46
Ceteorhinus 43
Chaetogaster 283
Chaetopterus 269
Chaos chaos 6
Chelodina 277
chemical sense 203
chemoreceptors 99, 143, **203**
chemotaxis 287

chewing 58
chimpanzee 54
Chironomus 43, 74, 91, 105
chitin, digestion of 55, 64
in excretion 137
chitinase 55, 61, 64, 65, 66, 67
Chlamydomonas 243
Chloeon 78
chloragogen cells 132, 133, 149
chloride 52
chlorocruorin **84,** 91
chlorolabe 231
cholecalciferol **36**, 37
cholecystokinin 60
cholesterol 38, 49
choline 29, **33,** 37
cholinergic nerves 257
cholinesterase 187
in invertebrates 278
Choloepus 54
Chordeiles 324
chordotonal organs 199, 223
choroid 225
chromaffin tissue 168
chromatophores 166, 169, 171, 237, **273**
chromium 39
chromosome 12
chyle 55
chyme 49
chymotrypsin 51
chymotrypsinogen 51
cilia 237, **238**
in nutrition 40
ciliary muscle 225
ciliates, in gut 54
Cimex 203
Ciona 43
circulatory system **82**
Citellus 327
citric acid cycle **118**, 126, 134, 286
citrulline 109
Clavellina 283
clonus 249
clot 36
clothes-moth 63
cnidae 237, **271**
cnidoblast 271
cnidocil 272
cobalt 39
co-carboxylase 31
cochlea 218, 221
co-decarboxylase 32
codon 114
coelomoduct 137, 146, 153
Coenobita 74
co-enzyme A 32, 118, 126
I 31
II 32
Q 36, 40
co-enzymes 16
coition 290
cold-blooded animals 317
collagen 2, 7
colloids of blood 139
colony 284
colour change **273**
colour vision 229, 231, 232, 233, 234
Colpidium 146
columella auris 221
common chemical sense 204, **207**
compound eye **234**
conditioned reflex 59, 189, 195
conduction of nervous impulse **178**
cones 226
contractile vacuole 146, 312
control of breathing 75, 76, 77, 79
of circulatory system 96
of digestion 58
of metabolism 120, 131
of excretion 143
Convoluta 281
coordination **155**
copper 39
coprophagy 30
copulation 290, 297, 304
Corethra 67
Corixa 81, 223
cornea 201, 225
corpora allata 173
corpus cardiacum 174
corpus luteum 293
Corti, organ of 219
corticosteroids **170,** 294
corticosterone 170
corticotropin 165, 170
cortin 121
cortisol 170
coxal glands 153
Crago 278
Crangon 278
crayfish, respiration 9
creatine 145
phosphate 253, 265, 268
creatinine 145
Crepidula 70, 305

cretinism 160, 161
crickets 222
crista 211
crop 61, 67
Crotalinae 202
Crustacea, chromatophores 172
  hormones 172
cryptobiosis 328
cryptomitosis 280
crystalline style 69
cuckoos 330
Culex 43
cupula 213
cyanide 124
cyanocobalamin **34**
cyanolabe 231
Cynipidae 288
Cypridina 269, 271
cysteine 3
cystine linkage 3
cytochrome 2, 13, 123, 129, 133
cytochrome oxidase 123
cytochrome reductase 36
cytosine 112

daily rhythm 331
Dalmatian dog 140
Danielli and Davson 9
Daphnia, and light 224
  life cycle 288
  respiration 91, 98, 100, 103
dark-adaptation 230
Dasypeltis 45
deamination **107**, 129, 132
death rate 307
debranching factor 116
defecation 59
deficiency diseases 30
dehydrocholesterol 36
dehydrogenase 13
de-inhibition 17
deoxyribonucleic acid (DNA) 12, 112
deoxyribose 112
deoxyribose nucleic acid 33, 112
dermatitis 32
Dermestidae 63
desert-living animals 326
desmosome 10
detritus feeders 46
dextrins 52
diabetes mellitus 120
diabetogenic hormone 165
diacyl linkage 2
diamino-carbinol linkage 2
diapause 174, **328**
diaphragm 74
Didinium 38, 146
Difflugia 246
diffusion 19
digestion 24, **47**
  control of **58**
  extracellular 48, 67, 70
  in invertebrates **64**
  in mammals **48**
  in other vertebrates **60**
  in Protozoa 65
  intracellular 48, 65, 69
  of carbohydrates **52**
  of fat **55**
  of nucleoproteins 50
  of proteins **49**
Dinoflagellata 269
dioestrus **291**
dipeptidase 51, 62, 63
dipeptide 2, 51
diphosphopyridine nucleotide (DPN) 31
disaccharase 53, 63
disaccharides 53
discus proligerus 290
dissociation curves of oxyhaemoglobin 86
disulphide linkage 3
diving 76, 80, 214
Dixippus 279
Doliolum 283
Dolium 70
Donnan equilibrium 21
DPN 31
Drosophila 26, 265
drowning 82
duck 102
ductless glands **156**
dulcamarin 205
duocrinin 60
dynein 244
Dytiscus 44, 80, 151

ear 211, **218**
Ecdyonurus 331
ecdysial gland 173
ecdysiotropin 173
ecdysis 172
ecdysone (PGH) 172, 173
Echinus 286

echolocation 221
ectohormones **175**
ectoparasites 46
effectors **237**
efferent fibre, nerve or neurone 177
Eiseniella 73
elastase 51
elastin 7, 51
electric catfish 267
electric eel 236, 267
electric organs 236, 237, **267**
electric ray 267
electrical sense 236
electron micrographs 8
electron transfer 123
electron-transporting flavoprotein (ETF) 126
Electrophorus 267, 268
electroplaques or electroplates 267
elephants 52, 60
endergonic reactions 14
endo-amylase 52
endocrine glands 157
endolymph 211, 213
endomixis 288
endoparasites 46
endoplasmic reticulum 12, **13**
endostyle 159
energetics 13
energy 13, 71, 106, 121
  requirements of man 28
  value of foodstuffs 128
enterocrinin 60
enterokinase 51, 62
enzymes **15**
Ephemera 78
Ephemeroptera 78
Ephestia 28
epinephrin, *see* adrenaline
equilenin 293
equiline 293
erepsin 50
ergatula 43, 70
ergosterol 36
Eriocheir 151, 314
Erythrinus 81
erythrocytes 85, 93
erythrodextrin 52
erythrolabe 231
esterase 17, 64
Euglena, flagellum 239
  locomotion 240
  nutrition 65
Euglypha 244
Eunice 303
Euplotes 280
euryhaline animals 313
Eustachian canal 218
excretion **135**
  and habitat 130, 141
  in invertebrates 145
exergonic reactions 14
exophthalmia 160
external gills 76, 77, 78
exteroceptor 197
extracellular digestion 48, 67, 70
eye **225**, 233
eye-spot 224
eye-stalk hormone 172

faceted eyes 234
facilitation in nervous system 186
  in muscle 262
  in nerve net 191
FAD, *see* flavine adenine dinucleotide
faeces 58
Fasciola 66, 133, 148
fat **4,** 134
  as food 28
  as fuel 132, 134
  conversion to carbohydrate 129
  digestion 55
  excretion 136
  formation from carbohydrate 129
  metabolism 120, **125,** 131
  oxidation 125, 132
fat-body of insects 152
fatigue 186, 255
fatty acids 4, 54, 57, 121, 125, 127, 134
feathers 161
feeding 24, **40**
Felidae 136
fenestra ovalis 218
fenestra rotunda 219
fertilization **284**
  membrane 287
fertilizin 286
fibrinogen 4
filopodia 244
filter-feeding 43
firefly 269
fire-worm 304
fission 280
flagella 237, **238**
flame cells 145

flavine adenine dinucleotide (FAD) 31, 110, 118, 123, 133
flavin mononucleotide (FMN) 31
flavoprotein 13, 31, 123, 126
flavour 204
Fletcher and Hopkins 124
fluid feeders 46
FMN, *see* flavin mononucleotide
folic acid **32,** 38
folinic acid 33
follicle-stimulating hormone (FSH) 165, **294**
food 24
  absorption **55**
  and sexual development 296
  movement **58**
  requirements of man 26, 28
foodstuffs **24**
Foraminifera 244
foreplay 301
Forficula 68
Fourier images 235
fovea centralis 229, 232
fragmentation 281
free energy 14
fructose 27, 114, 121, 286
  diphosphate 116
FSH, *see* follicle-stimulating hormone
fumaric acid 128
Fundulus 276

galactogen 130, 133
galactose 27, 114, 133
gall-bladder 49, 60
gall-wasps 288
Galleria 64
gametes 284
Gammarus 79
ganglia 177
gastric glands 48
gastric juice 48
gastrin 59, 159
gastrocnemius of frog 247
Gastrophilus 105
gelatin 26
gemmation 283
generator potential 196, 201, 220, 225, 235
GH, *see* growth hormone
giant nerve fibres 184
gigantism 164
gills **76**
  anal 73
  excretion by 145
  external 76, 77, 78
  of lamellibranchs **41**
  regulation by 311, 313
gizzard 45
glands 237, 266
Glaucoma 27, 44, 146
Glis 327
globin 83
globulin 57
glomerulus 137
Glossina 62
glow-worm 44, 269
glucagon 121, 131, 158
glucocorticoids 170
glucose 27, 114, 121
glucose-6-phosphate 115
glucosidase 53
glutamic acid 32
glycerol 5, 128
glycerose 120
glycocyamine 265
glycogen, break-down 116, 121
  digestion 52
  formation 116
  storage 133
glycolysis **116,** 130, 133, 243, 253, 265, 286
glyoxylic acid 130
goitre 160
gonadotropins 165, 296, 300
gonads 290
gorilla 55
Graafian follicle 290
Grandry corpuscle 201
gravity 211
green gland 150
growth 161, 173
growth and differentiation hormone 173
growth hormone (GH) 121, **164,** 294
guanase 110
guanidine diphosphate (GDP) 118
guanidylethyl methylseryl phosphate 265
guanidylethylseryl phosphate 265
guanine 110, 112, 132, 153
guinea-pig 55
gular fluttering 323
Gunda 148, 310
Gymnarchus niloticus 236, 268
Gymnotus electricus 267, 268
gynogamone 286

gynogenesis 287
Gyrinus 223

haem 123
haematin 83
haemerythrin **84,** 91
haemocyanin **84,** 91
haemoglobin 2, 4, 39, 49, 82, **83,** 90, 103, 106
Haemonchus 104
halteres 217
hearing 217, **218**
heart 96, 258
hectocotylus 304
helicotrema 218
Helix, digestion 63, 70
  excretion 132, 153
  nutrition 38
  reserves 133
hemicellulose 53
herbivores 30, 53, 62
Herbst corpuscle 201
hermaphroditism 285, **305**
heteronereis 172, 303
hexahydroxycyclohexane 33
hexose 27
hibernation **326**
high-energy compounds 15, 106
His, bundle of 98
histamine 201, 207
hologametes 284
Holothuria 78, 260
Holotrichida 54
Homarus 313
homing 330
homoiosmotic animals 308, **310**
homoiothermic animals 308, **319**
honey-guide 61
hormones **156,** 269, 313, 327
  and chromatophores **276,** 279
  in digestion 59
  in invertebrates **171,** 279, 303, 313, 332
  in metabolism 114, 120
  in reproduction 294, 303
horned toad, *see Phrynosoma*
horse 54, 55, 57
humming birds 327
Hyalophora 174
hyaluronidase 287
Hydra, coordination 191
  feeding 44
  movement 245
  nematocysts 272
  reproduction 281, 283
  sensitivity 224
hydrion, *see* hydrogen-ion concentration
hydrochloric acid 48
hydrogen acceptors 31, 36, 108, **116,** 123
hydrogen-ion concentration **315**
  of gut contents 49, 52, 61
  of vacuoles 65
hydrolase 17
hydroxytryptamine 188, 261
Hymenolepis 38
hyperglycaemia 131, 133
hypoglycaemia 131
hypophysis 163
hypotaurocyamine phosphate 265
hypothalamus 164, 296, 300, 322, 325
hypoxanthine 110, 130

ICSH, *see* interstitial cell stimulating hormone
iminoacids 1, 108
immune reactions 27
inclusions 12
incus 218
independent effector 237
Indicator 61
inflammation 31
infraorbital gland 48
infra-red radiation 202
inhibition 186
innate releasing mechanism 189
inorganic substances as food 39
inosine 110
inosinic acid 110
inositol 33
insulin 4, **120,** 131, 159, 327
intermedin 165, 276
internal secretion 157
internuncial fibre, nerve or neurone 177
interoceptor 197
interrenal gland 168
intersexes 306
interstitial cell stimulating hormone (ICSH) 165
intestinal juice 49, 60
intestinal mucosa, hormones of 159
intestine, excretion by 144
  movements of 58
intracellular digestion **65,** 69

intrafusal muscle fibre 255
intrinsic factor 34
inulase 64
inulin 52
iodine 40, 160
iodopsin 233
ionic regulation **314**
iris 255
iron 39, 144
islets of Langerhans 120
isogametes 284
isomerase 17
isometric contraction of muscle 248
isotonic contraction of muscle 248
itch 207
Ixodes 203

Jacobson's organ 206
Jaculus 141
juvenile hormone 173
juxtaglomerular apparatus 143

kangaroo 54
katabolism 107
Keber's organ 153
keratin 2, 3, 7, 63
keratinization 35
keto-acid 107
ketogenic hormone 164
kidney **137,** 167
  regulation by 167
kidney of accumulation 136
kinaesthetic sense 197
kineses 325
kinetosome 239
kinocilium 215, 216, 223
kiwi 206

labial glands 67
labyrinth **211**
lactase 53
lactation 161, 165
lacteal 55
lactic acid 124
lactoflavin **31**
lactogenic hormone 294
lactose 53
lagena 214, 218
lagomorphs 54
Lampetra 276
Lampyris 44, 269
Langerhans, islets of 120
Lasaea 70
latent period of muscle 248
  of nerve 185
lateral-line organs 202, **215**
Latimeria 129, 312
Law of Mass Action 85
Leander 310
Lebistes 275
lecithin 49
leeches 46
lens 225
Leodice 303
Lepidoptera 26, 46, 68, 78
Lepidosiren 77
leucine 56
Leucophaea 174
Leuresthes 103
LH, *see* luteinizing hormone
lice 46
Lieberkuhn's follicles 49
ligase 17
light, and colour change 275
  and sexual cycle **295, 299,** 303
  production **269**
  receptors **224**
Ligia 151, 279
lignin 53
Limax 153
Limnaea 79, 302, 313
Limulus 92, 208
Lingula 91
linoleic acid 28
linolenic acid 28
lipase 55, 57, 61, 62, 64
lipid 45
lipohumours 158
lipoic acid 118
lipoprotein 74
liquor folliculi 290
Littorina 82, 233, 234
liver, biochemistry 49
  of Crustacea 151
liver-fluke 46
loading tension 87
lobopodia 244
Loligo 70, 183
loop of Henle 138
Lorenzini, ampullae of 202, 236
LTH, *see* luteotropic hormone
luciferase 270
luciferin 270
Lumbricus 73, 100, 149
  feeding 46

Lumbricus (*cont.*)
haemoglobin 90, 92
sense organs 208
luminescent organs 237, **269**
lumirhodopsin 231
lunar sexual rhythm 296, 303, 304
lung-books 79
lungs **74,** 79
luteinizing hormone (LH) 165, **294,** 298
luteosterone 294
luteotropic hormone (LTH) 165, 294
luteotropin 294
lyase 17
Lymantria 67
lymphatic system 96

macrophagous feeders 43
macrosmatic animals 206
macula 211
utriculi 214
magnesium 39, 51, 113, 118, 144, 153, 314
Maia 62
maize 26
malleus 218
malonyl acyl carrier protein 126
malonyl coenzyme A 126
Malopterurus 267
Malpighian body 137
Malpighian tubules 151
maltase 53
maltose 52
maltotriose 52
mammary glands 294
mammotropin 294
manganese 39, 51
mannose 27, 114
Marmota 327
mass-action law 85
mastication 58
Mastigamoeba 244
maxillary gland 150
mayflies 26; *see* Ephemeroptera
mechanical coordination 155
mechanoreceptor 197
medullin 143
megagamete 285
meiosis 285
Meissner's plexus 60
melanin 274
melanocyte 274
melanophore 163, 274
melanophore-stimulating hormone (MSH) 277
melatonin 171, 278
membrana granulosa 290
membranes **9,** 18, 19, 310
menstruation 297
Merino sheep 321
merogamete 284
metabolism **106**
and adrenals 169
and thyroid 160
metachronal rhythm 241
metamorphosis of Amphibia 161
of insects 173
metanephridium 145
methionine 26, 33
metoestrus 292
Metridium 224, 243, 260, 264, 281
Miastor 289
microgamete 284
microphagous feeders 40
microphonics 220
microsmatic animals 206
microvilli 10
micturition 144
migration **328**
milk 50, 295
mineralocorticoids 170
mitochondria **12,** 17, 35, 36, 118, 151 161, 266
Mixodiaptomus 224
modiolus 218
Mollienesia 289
molybdenum 39
Monocystis 281
monoglyceride 55
moon, sex and 296, 303, 304
mosquito 46
motor fibre, nerve or neurone 177
motor unit 178, 250
moulting 161, 172, 173
movement 18
of food 58
mucin 48
mucopolysaccharide 5, 10, 27, 34
mucus 43, 48
mud feeding 46
Müllerian duct 306
Müller's cells 229
Müller's law 193
Multicilia 238
Murex 62, 70

muscle 237, **247**
  cardiac 96, 258
  invertebrate 260
  involuntary, smooth, plain, or unstriped 255
  voluntary, striped, or skeletal 247
  — — — chemistry of 251
muscle haemoglobin 88
muscle-spindle 198, 254
muscle sugar 33
Mustelus 276
mycetocyte 38
myofibril 247
myoglobin 4, 88, 102
myonemes 237
myosin 251, 265
Mytilus 239, 243, 261
  feeding 40
Myxicola 184
myxoedema 160

NAD 31
NADP 32
Naididae 282
Naigleria 244
Nais 78
naphthaquinone, *see* vitamin K
nasal gland 170, 313
natriferin 167
navigation 330
Necturus 61, 142
nematocysts 237, **271**
Nematodirus 104
neoteny 289
neotenin 173
nephridiopore 145
nephridiostome 145
nephridium, and osmotic regulation 149, 310
  morphology 145
  of annelids 149
  of platyhelminths 148
nephrocytes 152
nephromixium 146
Nephthys 90, 171
Nereis, coordination 190
  feeding 43
  haemoglobin 92
  hormones 171
  regulation 310, 314
  reproduction 282, 304
nerve, parasympathetic 257, 277
  sympathetic 121, 143
nerve-fibres 10, 177
nerve-muscle junction 177, **250**
nerve net 177, **190**
nervous impulse **178**
  velocity of 179, 184, 190
nervous system **176,** 183, 189
neurohypophysis 163, **166**
neuroid transmission 154
neuromast 216
neurophysin 167
Neuropterus 288
neurosecretory cells 171
nicotinamide 31
  adenine dinucleotide (NAD) 31, 116, 118, 123, 133
  — — phosphate (NADP) 32, 118, 120, 123
nicotinic acid **31,** 37
night blindness 35
niobium 40
Nippostrongylus 104
nitric oxide 89
nitrogenous excretion 136, **137**
  metabolism 107
Noctiluca 270
node of Ranvier 181, 201
non-shivering thermogenesis 319, 321, 322, 323
noradrenaline 168, 188, 189, 257
Notonecta 80, 151, 217, 233
nucleic acid, 34, 50, **111,** 120
nucleoproteins 50, 109
nucleosidase 50
nucleotidase 50
nucleotide 50
nucleus 11
Nucula 70
nutrition 24
nystagmus 211

ocelli 233
Octopus 70, 153, 233, 303
Odonata 78
Odontosyllis 304
oestradiol 293
oestrins 293
oestriol 293
oestrogens 293, 301, 302, 306
oestrone 293
oestrous cycle **291**
oestrus 291
olfactory sense 205, 210

omasum 53
ω-oxidation 128
ommatidia 234
Oncorhynchus 43, 278
Oniscus 151
Opalina 282
orbital gland 313
organ of Bojanus 153
  of Corti 219
organs of balance 211
ornithine cycle **109,** 129, 132
osmoreceptors 143, 207
osmotic pressure 19, 310
  regulation 148, 149, 150, 168
osteomalacia 37
Ostraea 42
ovary 290
ovoviviparity 307
ovulation 290, 301, 302
ovum 285
oxaloacetic acid 118
oxidase 17
oxidative decarboxylation 118
oxidoreductase 17
oxygen
  debt 102, 125, 130, 134, 243
  store 92
  transport **82**
oxyhaemoglobin 88
  dissociation curve 86, 88
oxytocin 166, 297
oyster, feeding 42

pacemaker 97, 191, 264
Pacinian corpuscle 198, 200
paedogenesis 289
pain **199**
Palaemonetes 278, 311
Palolo worm 282, 303
pancreas, control of 60
  external secretion **49,** 60
  internal secretion 120, 131, 159
pancreatic juice 49
pancreozymin 60
pantothenic acid **32,** 37
papilla 211
para-aminobenzoic acid 32
paracasein 50
Paramecium, anaerobiosis 104
  behaviour 325
  contractile vacuole 147
  digestion 65
  excretion 146, 147
  feeding 40
  haemoglobin 91
  locomotion 240
  reproduction 285, 288, 305
  sensitivity 217
paramyosin 265
parapineal 170, 276
Parasilurus 205
parasites, anaerobiosis in 103
  nutrition 46, **66**
parasympathetic nerves 257, 277
parathormone 158, **162**
parathyroid 162
parotid gland 48
pars intercerebralis 173
parthenogenesis 287
Patella 70
Pecten 260
pellagra 32
Pelomyxa 133
penguin 102
penis 304
pentose 27, 120
pepsin 49
pepsinogen 49, 59
peptide linkage 51
peptones 49
pericardial gland 173
pericardial organ 173
perilymph 218
Peripatus 100
Periplaneta 133
Perisoreus 324
perissodactyls 60
peristalsis 59
peritracheal gland 173
peritrophic membrane 67
Petromyzon 43, 89
pH, *see* hydrogen-ion concentration
phagocytes 244
  in digestion 245
  in excretion 150, 152, 245
Phalaenoptilus 327
Phalangidae 100
pharynx 159, 162
phenylalanine 51
Pheretima 63, 149
pheromones 175
phobotaxis 325
Phoca 102
Pholas 269, 271
Phormia 209
phosphagen 253, 265

phosphate
  and absorption 57
  and carbon dioxide 93
  excretion 136
  high-energy 15
  metabolism 162
phospholipids 5, 33, 34, 286
phospho-nicotinamide-adenine dinucleotide 31
phosphorus 40, 136
phosphorylation 36
Photinus 269
photoreceptors 224
Photurus 269
Phoxinus 222
Phrynosoma 276, 277
pigeons' milk 301
pigs 52, 57
pineal gland **170,** 276, 278
pinna 218
pinocytosis **22,** 57
pitch 220
pitocin 166
pitressin 166
pituitary gland **163**
  anterior part 121, 161, **163**
  distal part 164
  intermediate lobe 276
  intermediate part 164, 166
  nervous part 164
  pars tuberalis 166
  posterior part 143, 164, 276
placenta 292, 294
Planaria 302
Planorbis 79, 91
plasmagel 246
plasmalemma 7
plasmasol 246
Plasmodium 281
plasmotomy 282
plastogamy 282
plastron respiration 81
Platynereis 303
Platysamia 174
Pleurobrachia 265
Poecilia 289
poikilosmotic animals 308, 310
poikilothermic animals 308, 317
polarized light 234, 236
polypeptides 2, 4
Polypterus 77
polysome 113
Polystomella 147, 246
Polytoma 284
Porcellio 151, 326
porphyrin 83
porphyropsin 232
portal system 163
postoestrus 292
posture 198, 254
potassium 179, 286
pregnancy 292
priapulids 91
Primates 52
procarboxypeptidase 51
proelastase 51
pro-enzymes 17, 51
progesterone 294, 301, 302
progestin 294, 301
prolactin 165, **294,** 301
proline **1,** 133
pro-oestrus 291
propionic acid 54, 55, 121, 134
proprioceptors 75, 77, **197,** 208, 210, 254
prorennin 50
prostaglandin 143, 171
prosthetic group 4, 16
protandry 305
protease 50, 61, 62
protein 1
  as food 25
  conversion to carbohydrate 129
  digestion **49**
  formation **111,** 129
  metabolism **107,** 120
  oxidation 128
proteinase 49, 62, 207
proteose 49
prothoracic gland 173
prothoracotropin 173
prothrombin 36
protogyny 305
protohaem 83
protonephridium 145
protoplasm 1
Protopterus 129
provitamin 29
psalterium 53
Psammomys 141
pseudobranch 163
pseudopodia 237, **244**
pseudopregnancy 292
pseudorumination 54
Pseudoscorpionidae 100
pterin 32, 137
pteroylglutamic acid 32

Ptinus 289
ptyalin 52
pupation 173
purines 33, 110, 129, 132, 153
pylorus 59
pyridoxal **32,** 38
  phosphate 108
pyridoxamine 32
pyridoxin 32
pyrimidine 112
pyruvic acid 108, 116, 124

queen-substance 176
Quelea 300

rabbit 54
radula 45, 70
Raman spectrum 207
Rana 43, 61, 161, 163, 233, 302
rattlesnakes 202
receptor 192
  potential 196
rectal gland 312
rectum 58
reflex arc 177, 185
reflexes, postural 198
  righting 211
  stretch 198, 254
refractory period of muscle 249
  of cardiac muscle 259
  of nerve 178
  of nerve net 192
regulation **308**
reingestion 54
Reissner's membrane 218
relaxin 297
renal portal system 142
renin 143
rennin 50
reproduction **280**
resonance 220
respiration **71,** 81
respiratory organs **73**
respiratory pigments **83**
respiratory quotient **72**
respiratory trees 78
reticulum 53
retina 226
retinal 35, 38, 231, 232, 234
retinene 35, 231
retinol 35, 38, 231, 232, 234
retinula 234
retrolingual gland 48
Rhabditis 148
rhabdome 234
rhabdomere 234
rheotaxis 330
rhinoceros 55
rhizoplast 239
rhizopodia 244
Rhodnius 151, 173, 203
rhodopsin 35, 231, 232, 234
riboflavin **31,** 37
  adenine dinucleotide (FAD) 31
  5-phosphate (FMN) 31
ribonucleic acid (RNA) 112
ribose 27
ribosome 12, **13,** 113
ribs 75
Richmondena 323
rickets 37
rodents 54
rods 226
rotifers 145, 288
R.Q. 72
rumen 53
ruminants 26, 53, 57, 121

Sabella 78, 92
Sacculina 46
sacculus 214, 218
salamanders 74
saliva 48, 59, 60, 326
salivary glands 48
Salmo 312, 313
sarcolemma 247
sarcoplasm 247
scala media 219
scala tympani 218
scala vestibuli 218
Scaphiopus 302
Schistocerca 133, 310
Schistosoma 104, 134, 148
Schwann cell 10
sclera 225
scorpion 79
scotopsin 231
scrotum 321
scurvy 35
Scyliorhinus 276
scyphistoma 281, 283
Scyphomedusae 83
sea-birds 313
seals 102
secretin 50, 62, 157
secretion 266

segmentation movements 59
selenium 40
self-fertilization 305
semen 286, 304
semicircular canals 214
sensation 193
sense organs **192**
sensory fibre, nerve or neurone 177
Sepia 63, 70
serotonin 188, 243
Serpula 91
setae 43
Setonix brachyurus 54
sex 285, 305
  and luminescence 270, 304
  hormones 294, 300, 303, 306
  reversal 306, 307
sexual dimorphism 285
sexual display 301
sexual reproduction 284
sexual rhythm **291**
shivering 319, 322
shrew 54
shunt, arteriovenous 102, 103
  pentose 120, 133
Sialis 151
sickle-cell anaemia 2
sight 225
silky feathers 161
Simulium 74, 78
simultaneous contrast 230
sinu-auricular node 97
sinus-gland 172, 303, 307
sipunculids 91
Sisyphus 217
size of animals and basal metabolism 28
  and oxygen supply 73
skeletons in excretion 137, 147, 151
skin 73, 125, 145, 199
slipper limpet 70, 305
sloth 54
small intestine 48, 50, 53, 55
smell **203**
social hierarchy 299, 301
sodium 179
sodium glycocholate 49
sodium pump 180
sodium taurocholate 49
solenocyte 145
Solifugae 100
somatotropic hormone 164
somatotropin (STH) 164
sound receptors **218**
sperm 285
spermatogenesis 161
spermatozoa 285
Spilopsyllus 303
spiracles of dogfish 76
  of insects 100
Spirographis 78
Spirogyra 44
Spirorbis 91
Spirostomum 146, 147
Spirotrichida 54
spores 281
sporulation 281
sprue 33
Squalus 312
stapes 218
starch 27, 52
statocysts 217
Steganura 324
stenohaline animals 310
stereocilia 206, 215, 216, 227, 272
sterility 36
steroids 170, 293, 294, 298
STH, *see* somatotropic hormone
stigma 224
stigmata 100
stimulation of nerve 194
stinging cells 271
stomach 50, 52, 61
  of artiodactyls 53
storage 133
stress 170
stretch receptors 98
  reflex 198, 254
stroboscope 240
strontium 147
strychnine 186
sublingual gland 48
submaxillary gland 48
succinic acid 119
succinic dehydrogenase 133
succus entericus 48
sucrose 53
sulphate 136
sulphide 145
sulphur, excretion of 136, 145
  in nutrition 40
sulphuric acid 70
summation of stimuli in nerve 179
superposition image 235
supraoesophageal ganglion 172
suprarenals **168**

suspensory ligament 225
swallowing 58
sweat 145, 321
sweat glands 320, 321
Sylvilagus 298
symbiosis 26, **46,** 53
sympathetic system 121, 143, 257, 277, 321
sympathin 169
synapse 177, **185,** 192
syngamy 284
synthetase 17

tachypnoea 320, 321, 323
tactile sense organs 199
tadpole feeding 43
Taenia 104, 133, 134, 147
tantalum 40
tapir 55
Tarsius 226
taste **203**
taste buds 204
taurocyamine 265
tectorial membrane 219
teeth 45
temperature, and biological reactions 15
  and chromatophores 275
  and cilia 243
  and nerve 179
  and reproduction 296, 300
  regulation **317**
  sense organs **199**
tendon 198
Tenebrio 174
Teredo 64
terminal bar 10
testis 297, 299
testosterone 298, 306
Testudo 61
tetanus 249, 262
Tetrahymena 27, 244
thermal neutrality 320, 323
Theromyzon 172
thiamine **31,** 37
  pyrophosphate **31,** 118
thiouracil 161
thixotropy 6, 13
thoracic gland 173
threshold 195
thymine 112
thymus 163
thyroglobulin 160
thyroid gland 130, **159**
thyroid stimulating hormone (TSH) 165
thyrotropin 165
thyroxine **160,** 327
ticks 46
Tilapia 43
Tineola 63
tissue respiration 71
tocopherol 30, **36,** 38, 40
tone or tonus 255
Torpedo 267, 268
touch receptors **199**
TPN 32
tracheae 79, **100**
tracheal cells 101
tracheal gills 78
tracheole 100
transaminase **17,** 32, 107
transamination **107,** 132
transferase 17
transport by circulatory system **82**
  by tracheae **100**
  of carbon dioxide **93**
  of oxygen **82**
trehalase 53
trehalose 53, 133
tricarboxylic acid cycle **118,** 128, 130, 133, 243
Trichinella 105, 134, 147
trichocysts 237
Trichogramma 210
Trichomonas 243, 244
Trichoptera 78
triglycerides 5, 128
triiodothyronine 160
trimethylamine oxide 145, 153, 312
triose 120
triose phosphate 116, 128
tripeptides 2
triphosphopyridine nucleotide 32
tristearin 72
Trypanosoma 39, 65
trypsin 50
trypsinogen 2, 50
tryptophan 25, 26, 32
TSH, *see* thyroid stimulating hormone
tube-building polychaetes 78
tube-feet 155, 265
Tubifex 92
tubule of kidney 137, 141, 144
tympanal organ 223
tympanum 218
tyrosine 26

ubiquinone 36
ultimobranchial body 159, 162
ultra-violet light, 224
undulating membrane 239, 244
unloading tension 87
unmasking 17
uracil 112
urates 111, 152
urea 26, 50, 312
  excretion 140, 145
  formation 108, 130, 151
Urechis 43, 78
ureotelic metabolism 130
ureter 138, 144
uric acid, excretion 110, 137, 151, 153, 309
uricase 111, 129
urico-oxidase 111
uricotelic metabolism 130
uridine triphosphate (UTP) 116
urine 140, 144
urohaemal organ 168
urohypophysis 168
uroid 245
uronic acids **64**
urophysis 168
urticators 237
utriculus 212

vagina 292
valine 57
Vampyrella 44
vanadium 40
vasopressin 158, 166
vegetative reproduction 280
Velella 191
veratrin 240
Vermetus 43
vertebrate kidney **137**
vertebrate labyrinth **211, 218**
vestibule 218
Vidua 324
villi 55
viscosity 5
vision **225**
visual purple 231
vitamin A 30, **35**, 38, 161, 231, 232
  B complex **30**, 38
  B_1, *see* thiamine
  B_2, *see* riboflavin, nicotinamide
  B_6 32
  B_{12} 34
  C 34
  D 36
  E 30, **36**
  F, *see* arachidonic, linoleic, linolenic acids
  H 33
  K 36
  M 32
vitamins **29**
  in invertebrates 37
  in other vertebrates 37
vitreous humour 226
viviparity 307
Volvox 240
vomeronasal organ 206
Vorticella 281

water, absorption 58
  conservation 309
  detection 205, 209
  excretion 135, 167
  in nutrition 39
  osmotic regulation **310**
water-vascular system **83**
wax 5
Weber's law 195
Weber's ossicles 222
whale 43, 103
Wolffian duct 306
W-substance 166, 277

xanthine 110
xanthine oxidase 110
Xenopsylla 80
Xenopus 89, 130, 142, 163, 233, 276, 301
X-rays 224

yeasts 26, 38, 67
yellow cells 150
yolk 6
Young's theory 231

zinc 39, 51
zygote 284

PRINTED IN GREAT BRITAIN
AT THE UNIVERSITY PRESS, OXFORD
BY VIVIAN RIDLER
PRINTER TO THE UNIVERSITY